The Palgrave Macmillan Animal Ethics Series

Series Editors
Andrew Linzey
Oxford Centre for Animal Ethics
Oxford, UK

Clair Linzey
Oxford Centre for Animal Ethics
Oxford, UK

In recent years, there has been a growing interest in the ethics of our treatment of animals. Philosophers have led the way, and now a range of other scholars have followed from historians to social scientists. From being a marginal issue, animals have become an emerging issue in ethics and in multidisciplinary inquiry. This series will explore the challenges that Animal Ethics poses, both conceptually and practically, to traditional understandings of human-animal relations. Specifically, the Series will:

- provide a range of key introductory and advanced texts that map out ethical positions on animals
- publish pioneering work written by new, as well as accomplished, scholars;
- produce texts from a variety of disciplines that are multidisciplinary in character or have multidisciplinary relevance.

More information about this series at
http://www.palgrave.com/gp/series/14421

Felice Cimatti • Carlo Salzani
Editors

Animality in Contemporary Italian Philosophy

palgrave
macmillan

Editors
Felice Cimatti
University of Calabria
Arcavacata di Rende, Italy

Carlo Salzani
Messerli Research Institute
Vienna, Austria

The Palgrave Macmillan Animal Ethics Series
ISBN 978-3-030-47506-2   ISBN 978-3-030-47507-9   (eBook)
https://doi.org/10.1007/978-3-030-47507-9

This Palgrave Macmillan imprint is published by the registered company Springer Nature Switzerland AG.
The registered company address is: Gewerbestrasse 11, 6330 Cham, Switzerland

CONTENTS

Notes on Contributors

Luisella Battaglia is Professor of Ethics and Bioethics at the University of Genoa and the University Suor Orsola Benincasa of Naples. In Genoa she founded the Istituto Italiano di Bioetica, of which she is scientific director, and from 1999 is a member of the Comitato Nazionale per la Bioetica. She is a co-founder and since 2017 director of the Festival di Bioetica. Her publications include *Etica e diritti degli animali* (1997); *Alle origini dell'etica ambientale. Uomo, natura, animali in Voltaire, Michelet, Thoreau, Gandhi* (2002); *Bioetica senza dogmi* (premio Le Due Culture 2010); *Un'etica per il mondo vivente. Questioni di bioetica medica, ambientale, animale* (2011); *Poterenegato. Approcci di genere al tema delle diseguaglianze* (2014); and *Uomo, Natura, Animali per una bioetica della complessità* (2016).

Laura Bazzicalupo was Full Professor of Political Philosophy at the University of Salerno and has retired in 2018. She works on the crossing of *aisthesis* and politics. Her main topics of investigation are biopolitics, the economy of the governance, the productivity of power, and the crisis of democracy. She is on the editorial board of several political philosophical journals and is editor-in-chief of *Soft Power: Euro-American Journal of Historical and Theoretical Studies of Politics and Law*. Her recent publications include *Il governo delle vite. Biopolitica ed economia* (2006); *Superbia* (2008, latest edition 2018; translated into Spanish and Bulgarian 2015); *Biopolitica, una mappa concettuale* (2010; translated into Spanish 2016 and Portuguese 2017); *Eroi della libertà* (2011); *Politica. Rappresentazioni e tecniche di governo* (2013); and *Dispositivi e soggettivazioni* (2014). Quite recently she has published in English the essays "Economy as Logic of Government" (*Paragraph*, 2016) and "The Scene of Politics in an Atonal World: Hegemony, Contagion, Spectrality" (*Politica Comùn* 9, 2016).

Niccolò Bertuzzi is a Post-doctoral Research Fellow at the Scuola Normale Superiore and a member of COSMOS (Centre on Social Movement Studies). His main research interests are political sociology and social movement studies. He recently published the monograph *I movimenti animalisti in Italia. Strategie, politiche e pratiche di attivismo* (2018). He also investigated other social movements, protests and forms of participation, and in particular the one against Expo 2015. His articles appeared in some international journals such as *Modern Italy* ("The Contemporary Italian Animal Advocacy," 2018), *Social Movement Studies* ("No Expo Network: A Failed Mobilization in a Post-political Frame," 2017), *Interface* ("No Expo Network: Multiple Subjectivities, Online Communication Strategies, and the World Outside," 2017), *Relations* ("Veganism: Lifestyle or Political Movement? Looking for Relations Beyond Antispeciesism," 2017), and *Revista Crítica de Ciências Sociais* ("Urban Regimes and the Right to the City: An Analysis of No Expo Network and Its Protest Frames," 2017).

Leonardo Caffo is Adjunct Professor of Ontology at the Polytechnic University of Turin and of Philosophy of Contemporary Art at Nuova Accademia di Belle Arti in Milan. His primary research is focused on animal philosophy, in the sense of a possible philosophy outside the human atmosphere. In this framework he has worked and is working on simplicity, the relation between form of life and space for life (philosophy of architecture), ontology and individuals versus ecology and relations, a new concept of posthuman and antispeciesism, and philosophy as a practice of life. His latest books include *An Art for the Other* (2015), *Only for Them* (2016), *La vita di ognigiorno* (Einaudi, 2016), *Fragile Umanità* (Einaudi, 2017), *Vegan* (Einaudi, 2018), and *Il cane e il filosofo* (Mondadori, 2020).

Felice Cimatti is Full Professor of Philosophy of Language and Mind at the University of Calabria and also teaches at the Istituto Freudiano in Rome. In 2012 he received the Premio Musatti from the Società Psicoanalitica Italiana. He is one of the radio hosts of the radio programs *Fahrenheit* and *Uomini e profeti* on Rai Radio 3 and of the TV program *Zettel* (*Fare filosofia* and *Debate*) for Rai (Radiotelevisione Italiana) Scuola. His research interests, moving from the semiological study of the languages of non-human animals, mainly concern the complicated relationships between language and human mind/body. Recently he concentrated himself on the concept of "animality," focusing in particular on what a *human* animality could be. On such a topic, he wrote *Filosofia dell'animalità* (2013) and several other texts. His many other publications include *Il volto e la parola* (2008), *La vita che verrà. Biopolitica per "Homo sapiens"* (2011), *Il taglio. Linguaggio e pulsione di morte* (2015), *Sguardi animali* (2018), *A Biosemiotic Ontology. The Philosophy of Giorgio Prodi* (2018), *Cose. Per una filosofia del reale* (2108), *La vita estrinseca. Dopo il linguaggio* (2018), and *Philosophy of Animality: Unbecoming Human* (in press). He also co-edited the volumes *Filosofia della psicoanalisi. Un'introduzione in ventuno passi* (with Silvia Vizzardelli, 2012); *Corpo, linguaggio e psicoanalisi* (with Alberto Luchetti, 2013); *A come animale. Per un bestiario dei sentimenti* (with Leonardo Caffo, 2015); and *Abbecedario del reale* (with Alex Pagliardini, 2019). Felice is present on Academia.edu.

Massimo Filippi is Full Professor of Neurology at Vita-Salute San Raffaele University, Milan, Italy; director of the Residency School in Neurology and president of the Bachelor's Degree in Physiotherapy at the same university; chair of the Neurology Unit; chair of the Neurophysiology Unit; director of the MS Center; and director of the Neuroimaging Research Unit (NRU), Department of Neurology, Institute of Experimental Neurology, San Raffaele Scientific Institute, Milan. He is member of various national and international scientific societies and boards where he covered or is covering institutional roles. He is author of over 1000 papers and editor of more than 20 books; he is editor-in-chief of the *Journal of Neurology* and member of the editorial boards of many international scientific journals. He is very often requested as speaker and/or chairman in national and international neurological congresses. In 2001, Prof. Filippi was awarded the "Rita Levi-Montalcini Prize" for his outstanding contributions to the study of MS. Massimo Filippi is also a thinker and militant antispecist and author or coauthor of several essays on the animal question including *Ai confini dell'umano. Gli animali e la morte* (ombre corte 2010); *I margini dei diritti animali* (Ortica 2011); *Natura infranta* (Ortica 2013); *Crimini in tempo di pace. La questione animale e l'ideologia del dominio* (Elèuthera 2013); *Penne e pellicole. Gli animali, la letteratura e il cinema* (Mimesis 2014); *Sento dunque sogno* (Ortica 2016); *Altre specie di politica* (Mimesis 2016); *L'invenzione della specie. Sovvertire la norma, divenire mostri* (ombre corte 2016); *Questioni di specie* (Elèuthera 2017); and *Genocidi animali* (Mimesis 2018). He co-edited *Corpi che non contano. Judith Butler e gli-animali* (Mimesis 2015), the monographic issue of *Aut Aut* (n. 380, December 2018) entitled *Mostri e altri animali*, and Jean-Luc Nancy's *La sofferenza è animale* (Mimesis 2019). He also translated several works of Charles Patterson, Chris De Rose, Tom Regan, Jim Mason, Ralph R. Acampora, Matthew Calarco, and Rasmus R. Simonsen. Massimo Filippi is a founder of the *Oltre la specie association* and scientific director of *Liberazioni. Rivista di critica antispecista* and collaborates regularly with *Il Corriere della Sera, il manifesto*, and *alfabeta2*. In 2020 his new book with Enrico Monacelli will appear: *Divenire invertebrato. Dalla Grande Scimmia all'antispecismo viscido* (ombre corte).

Federica Giardini teaches Political Philosophy at the University Roma Tre (Rome). She is the director of the Master's program in "Gender Studies and Policies" and has co-founded the Master's program in "Environmental Humanities." As the general coordinator of the IAPh Italia Research Center, she is supervising the EcoPol/Political Economics-Ecology Program. She has been working on the relational body confronting feminist difference thought, Husserlian phenomenology and Lacanian psychoanalysis (*Relazioni. Differenza sessuale e fenomenologia*, 2004); on feminist genealogies; on commons; and on social reproduction. Lately her research has been focusing on "cosmo-politics," the transitional space blurring the boundaries between nature and politics (*Cosmopolitiche. Ripensare la politica a partire dal kosmos*, 2013; *I nomi della crisi. Antropologia e politica*, 2017).

Giovanni Leghissa is Associated Professor of Epistemology of the Humanities in the Department of Philosophy at the University of Turin. He graduated in Philosophy from the University of Trieste, and from the same university, he holds a PhD in Philosophy. He was visiting professor at the Institut für Philosophie of the University of Vienna and at the Hochschule für Gestaltung in Karlsruhe, Germany. His work focuses on phenomenology, continental philosophy, and psychoanalysis; postcolonial, gender, and cultural studies; comparative philosophy; posthumanism; epistemology of economics; and theory of organizations. At present his main focus concerns a critique of the neoliberal model of rationality, as it has been developed both by the School of Chicago and by the theory of organizations. He has authored five books: *L'evidenza impossibile. Saggio sull'immaginazione in Husserl* (1999); *Il dio mortale. Ipotesi sulla religi-*

osità moderna (2004); *Il gioco dell'identità. Differenza, alterità, rappresentazione* (2005); *Incorporare l'antico. Filologia classica e invenzione della modernità* (2007); *Neoliberalismo. Un'introduzione critica* (2012); *Postumani per scelta. Verso un'ecosofia dei collettivi* (2015); and with Giandomenica Becchio, *The Origins of Neoliberalism* (2017). He has co-edited with Enrico Manera *Filosofie del mito nel Novecento* (2015). He edited six collective volumes and special issues of journals, as well as the Italian translation of works by Husserl, Derrida, Blumenberg, Hall, de Certeau, Overbeck, and Tempels. He is member of the editorial board of the journal *Aut Aut* and director of the online journal *Philosophy Kitchen. Rivista di filosofia contemporanea.*

Giorgio Losi is a graduate student and assistant instructor at Indiana University Bloomington. His interests lie in critical animal studies, Italian literature, and cinema. Since 2013 Giorgio has been an activist with the Animal Liberation Group "Oltre la Specie," and he is part of the editorial board of *Liberazioni – Rivista di critica antispecista.* In 2016 he got a master's degree in Classics, with a thesis on Aristotle's biological and political works. In 2018 he taught a class entitled "Crossing Animal Borders: Animal Issues in the Italian and International Debate" at IU Bloomington.

Roberto Marchesini is the director of SIUA (School of Human-Animal Interaction) and of the Centro Studi Filosofia Postumanista (Center for the Study of Posthumanist Philosophy), both based in Bologna, Italy. He has been a prominent voice in the development of zooanthropology and posthumanism in Italy and teaches human-animal interactions as dialogues between minded interlocutors in courses around the country. He has written or co-written more than 30 books and 100 scientific essays. Among his main publications for the English-speaking audience are *Over the Human. Post-humanism and the Concept of Animal Epiphany* (2017); *The Philosophical Ethology of Roberto Marchesini* (collected essays edited by Jeffrey Bussolini, Brett Buchanan, and Matthew Chrulew, 2017); and *Beyond Humanism* (2018). He runs an ethology blog on the major Italian newspaper *Il Corriere della Sera,* and he is also member of the following scientific and editorial boards: Minding Animals International, the World Phenomenology Institute, and the book series Numanities (Springer).

Alma Massaro received a PhD in Philosophy at the University of Genoa with a dissertation devoted to the study of Christian animal ethics in eighteenth-century England. She now teaches History and Philosophy in different high schools in Genoa. Her research interests range from animal ethics to food ethics and from moral philosophy to veterinary ethics. Her publications include two single-authored books, *I diritti degli animali. Una riflessione cristiana* (2018) and *Alle origini dei diritti degli animali. Il dibattito sull'etica animale nella cultura inglese del XVIII secolo* (forthcoming), the edited book *L'anima del cibo. Percorsi fra emozioni e coscienza* (2014) and the edited issue *Animal Mundi. Le grandi religioni e gli animali* of the journal *Animal Studies* (issue 13, 2015), and a number of articles and book chapters. She translated into Italian the books *Why We Love Dogs, Eat Pigs, and Wear Cows* by Melanie Joy (2012) and *Concern for Animals* by Deborah Jones (2013). She is also president of the Centro Studi Cristiani Vegetariani (CSCV).

Marco Maurizi collaborates with the chair of Theoretical Philosophy at the University of Tor Vergata in Rome. From 2007 to 2014, he was teaching assistant at the University of Bergamo (Theoretical Philosophy, Epistemology of Social Sciences, Social Philosophy) and at the University of Tor Vergata (Institutions of Philosophy, Philosophy of Religions, Philosophical Hermeneutics). In 2014 he won a scholarship from the University of Bergamo to research on "The question of animality in Adorno, Derrida and Heidegger." His research interests lie in the field of philosophy of history from the perspective of critical theory (Adorno, Horkheimer, Marcuse), Marxism (Marx, Lukács, Žižek), and dialectical philosophy (Cusanus and Hegel), with a focus on the nature/culture problem and its relevance for the definition of the status of non-human animals in the contemporary bioethical discourse. He was a member of the editorial board of the journals *Liberazioni* and *Animal Studies*. His publications include *Adorno e il tempo del non-identico* (2004); *Al di là della Natura: glia nimali, il capitale e la libertà* (2012); *Cos'è l'antispecismo politico* (2012); *The Dialectical Animal: Nature and Philosophy of History in Adorno, Horkheimer and Marcuse* (2012); *Chimere e passaggi. Cinque attraversamenti del pensiero di Adorno* (2015); *Altra specie di politica* (with Michael Hardt and Massimo Filippi, 2016); and *Quanto lucente la tua inesistenza. L'Ottobre, il Sessantotto e il socialismo che viene* (2018).

Diego Rossello holds an MA and PhD in Political Science from Northwestern University (specialization in political theory) and is an associate professor at the Department of Philosophy, Adolfo Ibáñez University, in Santiago, Chile. Former editor of *Revista de Ciencia Política* (2012–2016) and current co-editor of *Economía y Política*, his work focuses on the intersection between political theory and the critical humanities, with emphasis on early modern political thought and contemporary critical political theory. He has been a Fellow of the Paris Program in Critical Theory (Northwestern) and of the Law and Humanities Junior Scholars Workshop (Columbia). His work has appeared in journals such as *Ideas y Valores, Society & Animals, Philosophy Today, New Literary History, Contemporary Political Theory*, and *Political Theory*, among others. He is finishing his book project, entitled *Political Theory at the Limits of the Human: Sovereignty, Animality, Rights*.

Matías Saidel holds a PhD in Theoretical and Political Philosophy (2011) from the Italian Institute of Human Sciences, with a thesis on the ontological and impolitical perspectives on the common developed by Nancy, Agamben, and Esposito. He works as a researcher at the Argentine National Scientific and Technical Research Council (CONICET) and the Catholic University of Santa Fe (Argentina) and as Tenured Professor of Political Philosophy at the Faculty of Social Work of the National University of Entre Ríos, Argentina. He has also taught postgraduate seminars on the Common and on Neoliberal Capitalism. He has been visiting PhD student at Cornell University (2010) and postdoctoral researcher at Complutense University (Madrid, Spain, 2018). He was also a Fellow of the Summer School in Global Studies and Critical Theory (Bologna, Italy, 2017). In the last few years he has done research on the common and biopolitics in contemporary thought. His work has appeared in journals such as *Revista de Estudios Sociales, Res Publica. Revista de Historia de las Ideas Políticas, Isegoría, Temas y Debates, Las torres de Lucca, Fragmentos de Filosofía, Ecopolítica, Eikasia. Revista de Filosofía, Soft Power*, and *TRANS/FORM/AÇÃO*, among others. He was co-editor of *Roberto Esposito: dall'impolitico all'impersonale. Conversazioni filosofiche* and has also published book chapters on Agamben, Nancy, Esposito, and neoliberal capitalism.

Carlo Salzani is *Gastwissenschaftler* at the Messerli Research Institute of the University of Vienna, Austria. He has widely published, both in Italian and English, on Benjamin, Musil, Kafka, and Agamben—and also on the animal question. His research interests presently focus on animal ethics, posthumanism, and biopolitics. His publications include the single authored books *Constellations of Reading: Walter Benjamin in Figures of Commonality* (2009); *Crisi e possibilità: Robert Musil e il tramonto dell'Occidente* (2010); and *Introduzione a Giorgio Agamben* (2013) and the edited collections *Philosophy and Kafka* (2013, with Brendan Moran), *Towards the Critique of Violence: Walter Benjamin and Giorgio Agamben* (2015, with Brendan Moran), *Agamben's Philosophical Lineage* (2017, with Adam Kotsko), and *Saramago's Philosophical Heritage* (2018, with Kristof Vanhoutte).

Valentina Sonzogni is an art and architecture historian. She obtained her PhD at the Universität für Angewandte Kunst in Vienna and is presently director of the Archivio Piero Dorazio in Milan. She has worked in several institutions such as the Museum of Contemporary Art at the Castello di Rivoli, Rivoli-Turin; the Kiesler Foundation, Vienna; and the Guggenheim Foundation, New York. Her articles have appeared in various journals and books, and she has held conferences at international universities. She is the co-founder and co-director, with Leonardo Caffo, of the animal studies journal *Animot. L'altra filosofia*. With Caffo she has published the book *An Art for the Other: The Animal in Philosophy and Art* (2012).

LIST OF FIGURES

CHAPTER 1

Introduction: The Italian Animal—A Heterodox Tradition

Felice Cimatti and Carlo Salzani

1   LITTLE HISTORY OF A BELATEDNESS

A few years ago, John Simons asserted the superiority of the Anglo-American approach to the animal question, claiming that "most wealthy western societies outside the Anglo-American nexus have not developed similar consciousness" and singling out Spain and Italy as examples of attitudes to animals "long conditioned by the Roman Catholic Church following the extremely animal-hostile theology of Thomas Aquinas" (2002: 11). Simons' thesis is accompanied by stereotypical statements bordering the ridiculous, such as "it is, I believe, true that no woman in France has ever won a case for sexual harassment at work," or "it is clear that health consciousness is far more a matter of public debate in the Anglo-American sphere than it is more generally. The issue of tobacco smoking is the best example here, but a concern with dietary matters also

F. Cimatti
University of Calabria, Arcavacata di Rende, Italy

C. Salzani (✉)
Messerli Research Institute, Vienna, Austria

© The Author(s) 2020
F. Cimatti, C. Salzani (eds.), *Animality in Contemporary Italian Philosophy*, The Palgrave Macmillan Animal Ethics Series,
https://doi.org/10.1007/978-3-030-47507-9_1

stands out" (2002: 5, 11). However, and despite the justified remonstrations of Damiano Benvegnù (2016: 42), from a purely historical point of view, this thesis is not entirely false: on the one hand, the precedence and primacy of British animal protection movements and associations in modern history is indisputable; on the other, these movements also established a sort of philosophical "orthodoxy," which has marked for a long time the history of animal advocacy—even in the sexist, chain-smoking countries of Southern Europe. Simons' thesis, moreover, reflects a long-standing bias, and in order to dispute it, one needs more than outraged and righteous protests.[1]

A cursory look at the history of animal protection in Italy in a sense even confirms Simons' prejudices. Giulia Guazzaloca has thoroughly researched this history and repeats Simons' argument that a deeply rooted Catholic, anthropocentric, and creationist tradition played against the animal protection cause—together with the persistence of a predominantly peasant society, economic backwardness, widespread illiteracy, and the proud defense of local traditions (2018: 45–46 and passim).[2] All of this reflected abroad into the image of a country essentially disrespectful of animal welfare, a bias which has evidently persisted to these days. Moreover, Guazzaloca repeatedly insists that "it was very often thanks to British noblemen and noblewomen that an animal-friendly sensibility was brought to Italy" and that Italian animal protection societies "for a long time benefited from the financial and organizational support of foreigners" (2018: 17–18). The foremost example is the foundation of the Società torinese per la protezione degli animali (Turinese Animal Welfare Society, later to become the Ente Nazionale Protezione Animali—ENPA), whence customarily the history of animal protection in Italy is said to begin: the Society was created on April 1, 1871, by the physician Timoteo Riboli at the instigation of national hero Giuseppe Garibaldi and of Lady Anna Winter, Countess of Sutherland, thanks to whom it established a fruitful relationship with the Royal Society for the Prevention of Cruelty to Animals, and in 1897 even received the honorary patronage of Queen Victoria. And this was not an isolated case: the analogous Roman Society was founded in 1874 by Terenzio Mamiani and Lady Paget, wife of the British ambassador and vice-president of the London Vegetarian Society; the Neapolitan Society was created in 1891 at the initiative of Elizabeth Mackworth-Praed; and even much later, in 1952, the Società vegetariana italiana (Italian Vegetarian Society) was founded in Perugia by Aldo Capitini and the British citizen Emma Thomas (Guazzaloca 2018: 18–20,

110). The essential and enduring involvement of British gentry had the effect that animal protection was long perceived in Italy as a foreign phenomenon, supported mainly by bourgeois and liberal elites—and to some extent this holds even today (Guazzaloca 2018: 61).

Guazzaloca argues however that, despite this little delay (the Royal Society for the Prevention of Cruelty to Animals was founded in Great Britain in 1824 and the Société Protectrice des Animaux in France in 1845), the Italian animal protection movements basically followed a development similar to that of their Anglo-American counterparts. If the animal cause was highjacked by official propaganda during the Fascist era (not, however, to the extent of the German case), the economic boom of the postwar years brought both the country and animal advocacy on a par with the other "civilized" countries. On a philosophical level, animal advocacy remained obviously a fringe phenomenon, but also presented some emblematic figures: for example, Piero Martinetti (1872–1943), an anti-Catholic and antifascist, better remembered for being one of the very few academics who, in 1931, refused to swear an oath of allegiance to the Fascist Party. The same did (or rather didn't) also Aldo Capitini (1899–1968), like Martinetti anti-Catholic and an advocate of vegetarianism and a major figure in Italy's postwar nonviolent movement. As Luisella Battaglia argues in her contribution to this volume, though perhaps marginal figures, Capitini (and Martinetti) laid the groundwork for what will later become the Italian philosophical reflection in animal ethics.

The emancipative unrest of the late 1960s and 1970s resulted, in Italy just as in all other Western societies, in the flourishing of many "liberation" movements, among which also appeared a galaxy of animal protection groups (mostly directed against vivisection). If Peter Singer's *Animal Liberation* (1975) was translated into Italian only in the late 1980s,[3] the antivivisection pamphlet *Imperatrice Nuda* (*Naked Empress*, initially translated into English as *Slaughter of the Innocent*) by the Italian-Swiss activist Hans Ruesch was published in 1976 with great national and international impact, and in 1982 the architect and cofounder of the LAV (Anti-Vivisection League), Alberto Pontillo, even coined a new term, "animalismo" (today the most used in animal advocacy discourses), to identify a new, rational rather than merely compassionate and emotional way of relating to the animal question (cf. Guazzaloca 2018: 124).[4]

The delay in the translation of Singer's founding text seems therefore to reflect again the general (albeit slight) delay of Italian thought in absorbing, and conforming to, the Anglo-American "orthodoxy." The texts of the new wave of animal advocacy that hit Western societies since the 1970s were nonetheless well disseminated and debated also in Italy. The first anthology collecting essays by Singer, Ruesch, Tom Regan, and others was published in 1985, edited by philosopher of law Silvana Castiglione, and pivotal to this end was of course the work of Paola Cavalieri, who edited and introduced the Italian translation of *Animal Liberation* and was then to collaborate with Singer in *The Great Ape Project* (1993) (and therefore is perhaps the most internationally known figure of Italian animal advocacy). From 1988 to 1998, Cavalieri also founded and edited an influential journal, *Etica & Animali*, which introduced to the Italian audience the conceptuality of the (mostly analytical) Anglo-American tradition. Since the 1990s and the 2000s, with the so-called animal turn in the humanities and social sciences and the exponential growth of publications and discussions on the subject, Italy fully absorbed the terms of a debate by now become international and interdisciplinary, and too many would be here the names of people, movements, or schools of thought to be listed. A few journals deserve however to be mentioned, which, like Cavalieri's *Etica & Animali* in the 1990s, greatly helped in disseminating animal philosophy in Italian academia and society at large in the 2000s and 2010s: *Animal Studies. Rivista italiana di antispecismo* (*Animal Studies: Italian Antispeciesist Journal*, 2012–present), *Animot. L'altra filosofia* (*Animot: The Other Philosophy*, 2014–present), and especially *Liberazioni. Rivista di critica antispecista* (*Liberations: Journal of Antispeciesist Critique*, 2010–present). The group editing the latter journal, animated by Massimo Filippi, is particularly noteworthy and representative of the current Italian peculiarity insofar as, unlike the mostly analytical *Etica & Animali*, pursues a mostly continental path which overcomes the boundaries of traditional, "orthodox" animal philosophy, even summoning renowned scholars such as Judith Butler (Filippi and Reggio 2015), Michael Hardt et al. (2016), Alessandro Dal Lago et al. (2018), or Jean-Luc Nancy (2019) to a fruitful dialogue with the animal question.[5]

However, the question of Italy's belatedness is not exhausted by its coming up to date with the terms of the Anglo-American debate. Rather— and this is the very reason and rationale of this volume—the temporal gap seems to hide a more essential, philosophical difference, a "heterodoxy" which does not simply challenge, but also problematizes and enriches the terms of an orthodoxy that, as such, all too often takes its limits for granted and universal.

2   PHILOSOPHY AND THE OUTSIDE

But is this delay really a matter of heterodoxy or rather, more simply, one of derivation and adaptation? What would be the alleged heterodox tradition and conceptuality that Italy proposes as alternative to Anglo-American orthodoxy? The question of the identity of Italian philosophy (and not merely with reference to the animal question) has been raised at different times and has lately become quite actual; we also need to briefly address it here in order to sustain our proposal. Already in the early 1990s, Deleuze and Guattari (1994: 102–3) argued that in early modernity Italy missed the beginning of an "incomparable philosophy" and ended up, like Spain (!), without a proper "milieu" for it (interestingly, just like Simons, they also blame the subjection to the Catholic Church). More recently, Antonio Negri (2009: 13) pontificated that "there hasn't been any philosophy in Italy in the twentieth century" in the sense of a "critical activity which allows one to grasp one's time and orientate oneself within it, creating a common destiny and witnessing the world for this purpose" (yet he identifies an "Italian difference" in the marriage of theory and praxis).[6] In the end, it cannot be denied that Italy never produced (until very recently) philosophical "schools" or "traditions" able to impose their conceptuality onto the international debate, and its philosophy always remained as fragmented and deinstitutionalized as its territory, producing great but isolated "comets"—which the institutions often liked to burn (Deleuze and Guattari 1994: 103).

What interests us here—and warrants a whole volume on animality specifically in *Italian* philosophy—is however what Roberto Esposito, among others, has made of this institutional weakness in his theorization of the so-called Italian Thought. Esposito turns the terms of the question upside down and argues that precisely this lack of consistency and of a center, this centrifugal penchant for the other, is what constitutes the most original feature of Italian philosophy—and not merely in the past 20 years or so, when the international fame of a few Italian thinkers (Agamben, Negri, Esposito himself) has led to the creation of a new, unlikely label: "Italian Theory." The "transcendental fold" in which modern philosophy enveloped itself from Descartes to Kant and onward, with its emphasis on the constitution of subjectivity, the problems of consciousness, and the theory of knowledge, never fully absorbed Italian philosophy, which has rather always been characterized by a propensity to escape from this self-referential loop and to be contaminated by something outside itself

(Esposito 2012: 10). This outside as the "nonphilosophical" is for Esposito "life": biological, animal life, which, unlike in the dominant traditions, in Italian thought was never opposed to a "form" extrinsic to it. And this is why there has never been a specific "philosophy of life" in Italy: "because the entirety of Italian thought is traversed and determined by it" (Esposito 2012: 31). This is also why it is in Italy, and not in France where it was first named and defined, that biopolitics has been made one of the world's major themes in the philosophy of the new century: only a thought intrinsically stretched outside itself toward life could fully develop Foucault's intuition, coined in the 1970s but remained dormant for more than 20 years (Esposito 2012: 3–4).[7]

The contemporary prominence of Italian thought means that what Michael Hardt (1996) once defined as "laboratory Italy" has ceased to be an exception and has become the rule (this was indeed Hardt's thesis already more than 20 years ago). If, as Felice Cimatti thoroughly explains in his contribution to this volume, Italian philosophy has never been Cartesian—and never Kantian either: its reason, as Remo Bodei (2009) argues, has always been "impure"[8] —then it is the international debate that today has taken up the once "untimely" traits of Italian Theory. Esposito (2018: 11) speaks in this respect of a "biological turn" in contemporary philosophy, resulting from the wearing away of all those schools of thought originated in the "linguistic turn" of the mid-twentieth century (deconstruction, hermeneutics, but also postmodernism and, in Italy, the "Weak Thought"). This shift in focus from the sphere of language to the horizon of life spells out what Kari Weil (2012: 11–16) has called a "counterlinguistic turn," which brings to the fore the materiality of biological and animal life that precedes and exceeds language, and that has characterized Italian thought from the very beginning.[9]

However, as Esposito also acknowledges, philosophy *always* presents an essential and constitutive relation with its own outside and intrinsically needs to break—or traverse—the mirror in which the subject admires its own reflection, lest it becomes a neutralizing textual science. There obviously exists a whole alternative tradition in Western philosophy (Esposito explores its Italian side), to which this brief sketch cannot certainly do justice and that breaks the self-referential mirror, opening up to life. Deleuze—who, Esposito argues, together with Foucault partially anticipated the "biological turn" (cf. Esposito 2018: 11)—went as far as naming this nonphilosophical outside, not only as *human* animality but also as *the animals that are not human*. He wrote together with Guattari:

We think and write for animals themselves. We become animal so that the animal also becomes something else. The agony of a rat or the slaughter of a calf remains present in thought not through pity but as the zone of exchange between man and animal in which something of one passes into the other. This is the constitutive relationship of philosophy with nonphilosophy. Becoming is always double, and it is this double becoming that constitutes the people to come and the new earth. (Deleuze and Guattari 1994: 109)

A similar insight can be also found in Derrida (2008: 29), who, confronted in his bathroom by the gaze of his cat, exclaims: "The animal looks at us, and we are naked before it. Thinking perhaps begins there" (to soon forget however the cat and concentrate, rather predictably, on his own nudity). These philosophers do not belong to the field of "animal advocacy" or to its "orthodox" philosophical compartment, but precisely for this reason their remarks are important: they open up the analysis of the animal question to a wider consideration, breaking the narrow limits of orthodoxy. And this is also what this volume aims at.

Thinking begins with, arises from, and is in a constant and essential relation with animality. Italian contemporary philosophy, more than other traditions, has fleshed out, in different and often discordant fashions, this underlying need of philosophy to come to terms with its outside in the form of animality itself, and the goal of the present volume is to propose an overview of some of the ways in which this has been done.

3   Res Naturalis

However, before exploring these issues, we need to anticipate the theses of Cimatti's chapter and clarify what constitutes the anti-Cartesianism of Italian philosophy. What does it mean not to be a Cartesian? It means that consciousness and language are not used as criteria for distinguishing between what is a subject and what is a simple thing. In fact, what one can call the original Cartesian gesture consists in the sovereign act of deciding what a subject is—with all the ethical and juridical consequences that such a metaphysical status implies—and what it is not, that is, what is instead reduced to a simple and worthless thing, a humble portion of *res extensa*. Descartes uses language and consciousness (which for him are basically one and same thing, since the famous and autocratic "*Cogito, ergo sum*" is a linguistic act, a performative) not to register a preexisting difference between a non-human animal and a member of the species *Homo sapiens*, instead, the Cartesian gesture *is what establishes this very difference*.

From this point of view, the notorious Cartesian dualism is not a philosophical (or scientific) assertion; rather, it resembles much more a legal judgment. In this sense, most criticisms of this dualism completely miss the point, because they try to show that non-human animals too have in fact (some more or less complex forms of) language and consciousness (from simple sentience to self-consciousness). These observations do not pay attention to the fact that what Descartes aims at is not to ascertain what the actual cognitive capacities of non-human animals are; what Descartes wants is instead to trace a neat separation between human beings and the rest of the natural world. His dualism is not based on scientific reasons, but on religion: Descartes *cannot but* be a dualist, because he wants to save the metaphysical status of human beings from the threatening rise of science in his times, since he realized that the new Galilean science would soon equate human beings to any other animate being. The *Cogito*, therefore, is not an empirical assertion: it is a definition of what a human being is. From this point of view, Descartes cannot but place non-human animals in a subordinate position in respect to *Homo sapiens*.

One of the unexpected consequences of this dualism is that, when one wants to criticize it, one must first accept it. Take the case of someone who advances the praiseworthy argument that at least some non-human animals show the behavioral signs of being conscious. Chimpanzees, for example, are frequently described as animals which are able to "recognize" themselves in a mirror (though the literature about this issue is controversial). The idea is that, since they have such a capacity, they are conscious—and therefore Descartes is wrong: at least some non-human animals partake in *res cogitans*. In this operation, however, instead of completely deconstructing the Cartesian dualism, one fully accepts it, simply moving the boundaries between *res cogitans* and *res extensa*; now *Pan troglodytes* falls into the scope of the first, that's all. However, the point is not to redefine the boundaries between who should be included into the scope of *res cogitans* and what pertains only to res extensa; the philosophical question is rather to make this dualism collapse. For example, what is at stake is not whether the communicative dance of bees is truly language or not; the point is to make inoperative the privileged role language and consciousness always played within Western metaphysics.

This kind of dualism does not properly apply to the Italian tradition: here such a neat boundary between human and non-human, consciousness and unconsciousness, *res cogitans* and *res extensa* never existed. In the Italian tradition, the question of animality is posed differently in respect to the way this problem is tackled, for example, by the analytical tradition. While for the latter the question of animality mainly pertains to ethics, in the Italian tradition, the problem of animality pertains to the deconstruction of the metaphysical dualism of the Western philosophical tradition. Take the exemplar case of Francis of Assisi: when he preached to the birds, Francis implicitly assumed that the difference between their life and his own life was not metaphysical; bird life is simply different from human life. This is not a metaphysical statement, there is no dualistic stance. According to Francis, all forms of life are worth consideration. Whereas the Cartesian gesture—a gesture which still marks Western metaphysics—produces a radical dualism, what one can call the Franciscan gesture, by contrast, opens to the diversity of forms of lives (cf. Agamben 2015). Just like the great diversity that marks the physical Italian landscape (cf. Cuniberto 2018), the key character of the Italian cultural tradition is the centrality—at the same time biological, political, and linguistic—of the notion of "difference." Difference does not produce hierarchy, as in the Cartesian dualism. One could imagine that instead of the dualism of the two *res, extensa* and *cogitans,* the only *res* that one can find in this tradition is what we could call *res naturalis,* where one cannot distinguish between what is human and what is non-human, what is subjective and what is objective, what is alive and what is "simply" material.[10]

A similar stance can be identified in the main figures of the Italian tradition, from Saint Francis to Giacomo Leopardi and up to Pier Paolo Pasolini. A last example: in *Petrolio*, Pasolini's last, unfinished novel (1997), the today's much-debated question of the Anthropocene is somewhat anticipated when the author on the one hand describes the process of destruction of the planet operated by the human species and on the other seems to envisage a posthuman horizon where the metaphysical dualism of what life is and what is not alive finally vanishes. The Pasolini example is very interesting, because Pasolini was, properly speaking, *not a philosopher*. According to Esposito (2012), this is a peculiar trait of the Italian tradition, which is not composed by "professional" philosophers. Whereas philosophers like Descartes aim at establishing neat distinctions, poets aim on the contrary at a natural blending, where all human categories get confused. In this perspective, the question of animality is much more than a "simple" ethical question; to take seriously the question posed by animality means to change the basic categories of our understanding of the world and of ourselves. From this point of view, the Italian tradition can offer new and fresh perspectives on animality, at the same time philosophical and political: philosophical, because it focuses on, and deploys, new authors and themes, which are more or less unknown to the actual philosophical debate, in particular in the Anglo-Saxon literature; political, because it offers new ways of taking into account the question of animality. Focusing on animality does not simply mean to extend the boundaries of the "moral circle"; quite the contrary: it means to call into question the privileged position that Western metaphysic assigns to consciousness and subjectivity. The question posed by animality is not that of transforming non-human animals into quasi-subjects; it is instead that of transforming human subjects into quasi-animals, or still better, it is the question of how to deactivate such metaphysical distinction between the animal and the human.

4   On the Structure of This Volume

A very last issue needs to be briefly tackled: what is truly "animality"? In a recent book, Roberto Marchesini (2017: 7) remarks that, though apparently a familiar concept always taken for granted, animality is actually an extremely confused construction, a maze of clichés, preconceptions, biases, and vested interests, and the task of the coming philosophy is precisely that of emancipating animality from the straitjacket imposed on it by the Western cultural tradition. In the same vein, Felice Cimatti (2013: viii–ix) defines animality as the sum of what we do not recognize as human, the photographic negative of the definition of the human—in fact, "the animal" as such does not even exist, and "animality" is an all too vague and imprecise taxonomic category. In this short introduction, we cannot certainly take up the task of providing some clarity on this issue, but we must nonetheless warn the reader that the concept this volume purports to explore in contemporary Italian philosophy is ambiguous and elusive and is used in philosophical discourses in ways that are neither univocal nor evident. Accordingly, the present volume will be dishomogeneous and inconsistent in its uses of the term, and consequently also in its overall tone: neither animality nor Italian philosophy presents a uniform and compact profile, hence the result will be heterogeneous.

*   *   *

The volume is divided into three parts.[11] The first part, *Animality in the Italian Tradition*, is more historico-descriptive and includes three chapters elucidating some historico-philosophical aspects of the Italian tradition. The first chapter, "Animality and Immanence in Italian Thought" by Felice Cimatti, constitutes a sort of meta-introduction to the rationale of the volume and argues that the Italian tradition has been anti-Cartesian even before Descartes: its anti-Cartesianism is an anti-dualism of principle that involves a completely different perspective on the animal world. Whereas all Cartesians (including traditional anti-Cartesians) start off with a contraposition, the Italian tradition rests upon the observation of the *diversity* of the various animal forms of life. Whereas dualism is hierarchical, the principle of diversity is "naturally" egalitarian. The chapter traces the outlines of this tradition, briefly considering figures such as Francis of Assisi, Dante Alighieri, Hieronymus Rorarius, Bernardino Telesio, Giordano Bruno, Giambattista Vico, Giacomo Leopardi, and Giorgio Prodi. In the second chapter, "Aldo Capitini, Animal Ethics, and Nonviolence: The Expanding Circle," Luisella Battaglia focuses, from a different perspective, on a major figure of twentieth-century animal advocacy in Italy, Aldo Capitini, who represented a qualitative change in the Italian movement: he submitted to thorough critical examination the very categories of humanity and animality, in order both to verify their consistency and theoretical adequacy with respect to current scientific parameters and to derive normative implications from such reflections. In this way Capitini opened the way to animal bioethics, that particular field of applied ethics that deals with the moral, social, and juridical aspects of human relations with other species and therefore addresses the complex problem of animal rights, animal husbandry, medical experimentation with animals, and genetic engineering. Finally, the third chapter, "What Is Italian Antispeciesism? An Overview of Recent Tendencies in Animal Advocacy" by Giorgio Losi and Niccolò Bertuzzi, offers an overview of the different agents currently operating in Italy as part of the Animal Liberation movement and analyzes the impact on this social reality by books and essays published or translated in Italian in the last 20 years in the field of animal ethics.

*   *   *

The second part, *Animality in Perspective*, endeavors to present an overview of how (if ever) animality is considered in several branches and orientations of contemporary Italian philosophy, from biopolitics to posthumanism, feminism, Marxism, and theology. This part does not aim at completeness, but at giving an overall idea of what is happening in contemporary Italian philosophy in regard to animality. In "Beyond Human and Animal: Giorgio Agamben and Life as Potential," Carlo Salzani reconstructs Giorgio Agamben's relation to the animal question throughout his whole carrier, culminating in the publication of *The Open* in 2002. Salzani argues that the animal question is pivotal to the comprehension of the whole philosophical project of Agamben insofar as the rift dividing human and animal represents and constitutes the main structure of Western metaphysics, which always presupposes an unknowable and unnameable substrate (here the animal) supporting a knowable and nameable "substance" (the human). This presuppositional structure always leads to the subjection and dominion of one part over the other and to the deadly production of "bare life," a life stripped of its qualities and consigned to death. The overcoming of this structure is what Agamben calls "form of life," a life of "potential" that would deactivate the caesura dividing human and animal and open both to a new understanding and a new relationship.

The examination of the biopolitical take on animality continues in the following chapter, "Deconstructing the *Dispositif* of the Person: Animality and the Politics of Life in the Philosophy of Roberto Esposito" by Matías Saidel and Diego Rossello, who, like Salzani with Agamben, propose an overview of Esposito's philosophy in respect to animality. Rather than having a frontal approach toward animality or reflecting on the animal as such, Esposito introduces this problematic as an element that traditional metaphysics and political philosophy have not been able to properly reflect upon. Esposito's philosophy suggests that if we want to leave behind an immunitarian politics over life that has produced submission, suffering, and senseless death of human beings, animality should not be thought anymore as what can be destroyed with impunity, nor as that part of humanity that must be rejected and controlled by our higher spiritual and mental abilities. On the contrary, he shows—reading together Italian "Living Thought" and French Post-structuralism—that it is only by establishing a new political and philosophical framing of existence in which corporeality and spirituality, animality and humanness, *zoé, techné,* and *bios* cannot be severed, that a politics of life can be thought and experienced.

In "Animality Between Italian Theory and Posthumanism," Giovanni Leghissa considers instead the question of animality in the debate on posthumanism, focusing in particular on the intersections between the living and the machine. What is really at stake when we attempt to define the theoretical gains coming from a posthuman philosophy, he argues, is the abandonment of the difference between the living and the machine, and not between the animal and the human. Technological devices, even the most sophisticated, are not able, of course, to reproduce themselves, while self-reproduction is the main property of living matter. Both the living and some complex artifacts are, however, autopoietic systems that share a very important property: they manage their own performances by reacting to environmental stimuli. But, although crucial and necessary, this point is not enough if one really aims at defining the relation between the living and the machine within the paradigm of posthuman philosophy. What counts more is to understand to which extent humans, living entities, inanimate things, and artifacts produced by living entities (both human and belonging to other living species) are entangled.

In the following chapter, "For the Critique of Political Anthropocentrism: Italian Marxism and the Animal Question," Marco Maurizi focuses on the reception of the concept of "animality" in the Italian Marxist tradition. Maurizi argues that, while the "orthodox" agenda of traditional Marxism seems to show very little interest in the subject of animality, this theme becomes more and more important in post-Marxist thinkers. He first focuses on those authors who challenged the anthropocentrism of orthodox Marxism, from Timpanaro's Materialism to the introduction of the debate on Animal Rights by Bartolommeo and Veca in the 1980s, and then develops the theoretical problems at stake in such reception, showing the distance between the Italian scene and the Anglo-Saxon debate. Finally, the chapter shows the original path taken by Italian Marxism in the last decades, especially in the works of Post-Operaists and scholars of the Frankfurt School, where the rigid contradiction between animality and humanism seems to be superseded by a dynamic approach which does not give up the central assumptions of Marx's critique of capitalism.

Federica Giardini, in "Experiencing Oneself in One's Constitutive Relation: Unfolding Italian Sexual Difference," moves the focus to feminism and animality and develops the premises of the feminist difference paradigm, as it was defined in Italy in the 1980s. Taking advantage of the three different readings by Luisa Muraro, Adriana Cavarero, and Rosi Braidotti of Clarice Lispector's novel, *The Passion According to G.H.*—on the encounter between a woman and a cockroach—the feminist difference paradigm becomes the opportunity to outline an ontological and epistemological shift toward the dynamics and topology of a *constitutive relation*. The symbolical isotopy of both woman and animal—unlike the constructivist approach of authors such as Simone de Beauvoir and Judith

Butler—becomes the opportunity to reconsider the political realm in an innovative way. In the perspective of the constitutive relation, "animal" appears then not as a topic but rather as a relation by which human can position and conceive itself. Animal is a "political operator," defining the historical meaning and functioning of "human," in a topological space, where the inside and the outside, proximity and distance, similarity and alterity/strangeness are not fixed properties but dynamic outcomes.

Finally, the last chapter of this part, "Paolo De Benedetti: For an Animal Theology" by Alma Massaro, analyzes the original theological view of animals by theologian Paolo De Benedetti. Bypassing Western anthropocentric bias, de Benedetti embraced a reading of the Holy Scriptures and of the Jewish commentaries on the Bible capable of showing animals as direct recipient of God's love, where animal suffering becomes the biggest conundrum theology has to face. To speak about animal theology is not a vain ambition of theology or a mere sentimental act; it is rather an important tool that could help human beings in finding a more "just" relationship between the *finite world* that they live in and that *infinite one* that they do not know. Humans, animals, and plants are called to the same destiny. By shading light on the articulated existential sense the whole creation has inside the Jew-Christian tradition, the work of de Benedetti has deeply marked the Italian debate on animal and environmental ethics.

*   *   *

The third part, *Fragments of a Contemporary Debate*, is more speculative and includes more theoretical chapters, which propose some "fragments" of an extremely diverse and articulated debate. Massimo Filippi's "'*Il faut bien tuer*,' or The Calculation of the Abattoir," begins this part analyzing the precise algorithm which builds, step by step, this strange architecture we call the Self. This algorithm, Filippi argues, is not a painless one, even though it pretends to be the most natural thing in the world: its operations are constructed upon and justified by a noncriminal sacrificial norm which puts the other (non-human others, sexualized and racialized others, and many other butcherable bodies) to death, devouring their flesh and their presence and flushing them out, as if they were never there. The Self is created devouring the other and denying this massacre. Filippi goes through the various steps which compose this algorithm, analyzing some of the most crucial figures which deconstructed this mechanism—from Jacques Derrida and Giles Deleuze to Elias Canetti, Francis Bacon, and

Slavoj Žižek—trying to account for the murderous construction of the most serious fable of all time: our own subjectivity.

In "Philosophical Ethology and Animal Subjectivities," Roberto Marchesini endeavors instead to explain animal subjectivity without considering consciousness. Marchesini structures his proposal in two passages: first, the idea that inside animality there exists a principle of projection toward the world, a tensionality linked to the fact that being animal means being desiring, projected toward an "elsewhere." The second main aspect is to consider innate and learned dotations no longer as automatisms, but as a map, which the subject uses to realize a package of possible functions. The map allows to manage also any novelty in circumstances, just like a city map. If automatism was what moved the animal puppet, in this theory the animal becomes the owner of its innate and learned cognitive tools.

Laura Bazzicalupo then analyzes animality as a political paradigm in "From Renaissance Ferinity to the Biopolitics of the Animal-Man: Animality as Political Battlefield in the Anthropocene." The specific political outlook, focused here on the Italian tradition, highlights the pivotal importance that animal figures—mute yet powerful—have had in the devices of power for a long time, from Machiavelli to Vico up to positivism and sociobiology: both as a surplus exceeding the language on which humanist anthropology is based and as a mark of a power installed on the objectification of the body and its governability. Moving from this double ambivalent marker, animality becomes more and more a figure of immanence, in the trail of a Deleuzian reading of Foucault, which captures the material space of the multitude's biopower.

In the following chapter, "The Animal Is Present: Non-Human Animal Bodies in Recent Italian Art," Valentina Sonzogni surveys the use of non-human animals in Italian art from the post-World War II period until today, building on the analysis of some case studies. Starting from the performance by Jannis Kounellis *Senza titolo* (Untitled, 1969) at the Galleria l'Attico in Rome involving some live horses, Sonzogni analyzes works containing taxidermized animals (e.g., Maurizio Cattelan), which have often been banned and attacked by animal rights activists, and finally exposes the case of Gino De Dominicis, who exhibited at the 1972 Venice Biennale the work *Seconda soluzione di immortalità (L'universo è immobile)* (Second Solution of Immortality: The Universe Is Immobile), in which the artist had a person with Down syndrome involved.

The volume closes with the chapter "Animality Now" by Leonardo Caffo, who, in a literary and evocative style, reflects on the fate of human and non-human animals in a human-dominated world and emphasizes the importance and urgency of thinking about animality today. Thinking about animality means thinking about the humanity we want to become: what we have been, what we are, and, of course, what we could be. A humanity that lives by crushing the corpses of history, like Paul Klee's *Angelus Novus*, is a false humanity: it is bestiality that tries to be different from animality—which is precisely what we should go back to in order to save the world and ourselves from the unbearable lightness of the animal slaughterhouse.

NOTES

1. Historian Harriet Ritvo (1987: 126ff), however, has shown long ago that this long-standing bias is baseless and a cultural and nationalist prejudice: "as early as the 1830s, despite the circumambient evidence to the contrary, the English humane movement had begun to claim kindness to animals as a native trait and to associate cruelty to animals with foreigners, especially those from southern, Catholic countries" (127).
2. Other historical works on modern animal protection in Italy include Tonutti (2007), Maori (2016), and the first chapter of Bertuzzi (2018).
3. The first translation appeared in 1987 published by the Antivivisection League (LAV), followed by the second one in 1991 with a wide circulation by the major publisher Mondadori.
4. It is impossible to translate *animalismo* and *animalista* into English without recurring to inaccurate periphrases such as "animal right or animal protection activism/activist." The false friend "animalism" in English refers instead to animal qualities or behaviors, particularly emphasizing their physicality or instinctuality in contraposition to (human) spiritual, moral, or intellectual qualities.
5. Filippi's own work is representative—though not exclusively—of a certain "Italian way" to contemporary (continental) antispeciesism and animal philosophy. Cf. Filippi (2010, 2011, 2016, 2017), Filippi and Trasatti (2010, 2013).
6. Negri too remarks the weakness of Italian philosophy in the face of popes (and/as bosses and dictators)!
7. This thesis has since been also adopted by others: cf., for example, Campbell and Size (2013: 4).
8. "Never Cartesian" and "never Kantian" could amount to: Italian philosophy has never been *modern*, to quote Bruno Latour.
9. Cf. also Gentili (2012: 7ff).
10. For a new analysis of this point, cf. *Cimatti* (2018).

11. We would like to thank Dave Mesing, who proofread a number of chapters.

Works Cited

Agamben, Giorgio. 2015. *The Use of Bodies*. Trans. Adam Kotsko. Stanford: Stanford University Press.

Benvegnù, Damiano. 2016. The Tortured Animals of Modernity: Animal Studies and Italian Literature. In *Creatural Fictions: Human-Animal Relationships in Twentieth- and Twenty-First-Century Literature*, ed. David Herman, 41–63. Basingstoke: Palgrave Macmillan.

Bertuzzi, Niccolò. 2018. *I movimenti animalisti in Italia. Strategie, politiche e pratiche di attivismo*. Milan: Meltemi.

Bodei, Remo. 2009. Goodbye to Community: Exile and Separation. Trans. Sylvia Hakopian. *Diacritics* 39(4): 178–184.

Campbell, Timothy, and Adam Size. 2013. Biopolitics: An Encounter. In *Biopolitics: A Reader*, ed. Timothy Campbell and Adam Size, 1–40. Durham: Duke University Press.

Castiglione, Silvana, ed. 1985. *I diritti degli animali. Prospettive bioetiche e giuridiche*. Bologna: Il Mulino.

Cimatti, Felice. 2013. *Filosofia dell'animalità*. Rome-Bari: Laterza.

———. 2018. *Cose: Per una filosofia del reale*. Turin: Bollati Boringhieri.

Cuniberto, Flavio Pietro. 2018. *Paesaggi del Regno*. Vicenza: Neri Pozza.

Deleuze, Gilles, and Félix Guattari. 1994. *What Is Philosophy?* Trans. Hugh Tomlinson and Graham Burchel. New York: Columbia University Press.

Derrida, Jacques. 2008. *The Animal that Therefore I Am*. Trans. David Wills. New York: Fordham University Press.

Esposito, Roberto. 2012. *Living Thought: The Origins and Actuality of Italian Philosophy*. Trans. Zakiya Hanafi. Stanford: Stanford University Press.

———. 2018. *A Philosophy for Europe: From the Outside*. Trans. Zakiya Hanafi. Cambridge: Polity Press.

Filippi, Massimo. 2010. *Ai confini dell'umano. Gli animali e la morte*. Verona: Ombre corte.

———. 2011. *I margini dei diritti animali*. Aprilia: Ortica.

———. 2016. *L'invenzione della specie. Sovvertire la norma, divenire mostri*. Verona: Ombre corte.

———. 2017. *Questioni di specie*. Milan: Elèuthera.

Filippi, Massimo, and Marco Reggio, eds. 2015. *Corpi che non contano. Judith Butler e gli animali*. With an Interview with Judith Butler. Milan: Mimesis.

Filippi, Massimo, and Filippo Trasatti, eds. 2010. *Nell'albergo di Adamo. Gli animali, la questione animale e la filosofia*. Milan: Mimesis.

———. 2013. *Crimini in tempo di pace. La questione animale e l'ideologia del dominio*. Milan: Elèuthera.

Filippi, Massimo, Michael Hardt, and Marco Maurizi. 2016. *Altre specie di politica*. Milan: Mimesis.

Gentili, Dario. 2012. *Italian Theory. Dall'operaismo alla biopolitica*. Bologna: Il Mulino.

Guazzaloca, Giulia. 2018. *Primo: non maltrattare. Storia della protezione degli animali in Italia*. Bari-Rome: Laterza.

Hardt, Michael. 1996. Laboratory Italy. In *Introduction to Radical Thought in Italy: A Potential Politics*, ed. Paolo Virno and Michael Hardt, 1–12. Minneapolis: University of Minnesota Press.

Dal Lago, Alessandro, Massimo Filippi, and Antonio Volpe. 2018. *Genocidi animali*. Milan: Mimesis.

Maori, Andrea. 2016. *La protezione degli animali in Italia. Storia dell'Enpa e dei movimenti zoofili e animalisti dalla metà dell'Ottocento alle soglie del Duemila*. Rome: Enpa.

Marchesini, Roberto. 2017. *Emancipazione dell'animalità*. Sesto San Giovanni: Mimesis.

Nancy, Jean-Luc. 2019. *La Sofferenza è animale*. Ed. Massimo Filippi and Antonio Volpe. Milan: Mimesis.

Negri, Antonio. 2009. The Italian Difference. Trans. Lorenzo Chiesa. In *The Italian Difference: Between Nihilism and Biopolitics*, eds. Lorenzo Chiesa and Alberto Toscano, 13–24. Melbourne: re.press.

Pasolini, Pier Paolo. 1997. *Petrolio. A Novel*. Trans. Anna Goldstein. New York: Pantheon Books.

Ritvo, Harriet. 1987. *The Animal Estate: The English and Other Creatures in the Victorian Age*. Cambridge, MA: Harvard University Press.

Simons, John. 2002. *Animal Rights and the Politics of Literary Representation*. Basingstoke: Palgrave Macmillan.

Tonutti, Sabrina. 2007. *Diritti animali: storia e antropologia di un movimento*. Udine: Forum.

Weil, Kari. 2012. *Thinking Animals: Why Animal Studies Now?* New York: Columbia University Press.

PART I

Animality in the Italian Tradition

CHAPTER 2

Animality and Immanence in Italian Thought

Felice Cimatti

1   ANTI-CARTESIANS BEFORE DESCARTES

Western and modern thought about the animal is indelibly marked by the radical Cartesian distinction, which establishes the human subject only because, in the same originary gesture, it deprives the human from every animal characteristic (the human is human *insofar* as it is conscious; the body—from the moment that it is an animal—is not properly human). The *ego cogitans* can exist only because there exists a living being that does not *cogitat*—because there is the animal as the radical other of the human. In this sense, more than the human, Descartes invents the animal as absolute alterity with respect to the human. The entire tradition that follows is based on this dualism, both those who more or less accept it and those who criticize it. Those who accept it view the human as a radically different living being, belonging to a substance (*res cogitans*) different from that of animals; those who criticize it, on the contrary, try to overturn this position, seeking to attribute to non-humans those characteristics (thought, language, consciousness, will) which Descartes had instead

Trans. Dave Mesing

F. Cimatti (✉)
University of Calabria, Arcavacata di Rende, Italy

© The Author(s) 2020
F. Cimatti, C. Salzani (eds.), *Animality in Contemporary Italian Philosophy*, The Palgrave Macmillan Animal Ethics Series,
https://doi.org/10.1007/978-3-030-47507-9_2

negated. The contemporary movement for animal rights, for example, is explicitly anti-Cartesian and indeed proposes to even consider non-human animals (or at least some among them) as subjectivities, if not as persons in the juridical sense. What this animalism does not see is that in this way it continues, paradoxically, to remain trapped in Descartes' dualistic apparatus. To maintain that an anchovy is in some way a "person" in fact means agreeing with Descartes when he attributes an exceptional position to consciousness with respect to the rest of the natural world. The issue at hand is not to transform the animal into a quasi-human, but on the contrary, to deactivate the dualistic Cartesian machine. In this sense, anti-Cartesian *animalism* always remains within the field of ethics and the law and therefore the *human* world.

On the contrary, the tradition of Italian Thought occupies a different position (and precisely for this reason remains mostly marginal), because as a tradition it has always been radically anti-dualistic. In this sense, the Italian tradition was anti-Cartesian *even before Descartes.* An anti-Cartesianism by principle, that is, a radical anti-dualism. This entails an entirely different perspective on the animal world: while all Cartesians (including the anti-Cartesians) start from a dualistic contraposition, the Italian tradition instead considers animality as the common background of all possible forms of life. For whatever epoch, Cartesian thought is a thought of division and hierarchy; the Italian tradition, starting from the writings of Francis of Assisi, is a thought of immanence and life. And in this way while the Cartesian views the animal always and only as an abstraction (Descartes' parrot, Heidegger's bee), that is, as an *animot,* in the Italian tradition, from the very beginning, there does not exist the Animal, but rather the extraordinary *diversity* of living forms.

Accordingly, even when human supremacy over the animal world is affirmed (e.g., in Pico della Mirandola's *Oration on the Dignity of Man*), this supremacy is a de facto observation rather than a *de jure* stipulation. Human beings live in a different way than brutes (i.e., living beings without articulate language); this is a fact, which should be no more surprising than the fact that human beings do not have gills or horns. If the Cartesian tradition is one of *dualism,* the Italian tradition is one of *diversity.* While dualism is hierarchical, the principle of diversity is "naturally" egalitarian. An equality of principle which thus does not fear emphasizing even the differences among diverse forms of life.

It is an anti-dualistic thought that on the one hand does not transform differences into hierarchies and on the other always maintains the links between pre-human origins and the historical present. As Roberto Esposito observes, one of the major "paradigms" of the Italian tradition is what defines the "historicization of the non-historical"; the wild and animal dimension is not located in an outmoded past that is to be overcome (as in the metaphysical caesura in Hobbes between "state of nature"—marked by the terrible condition of *homo homini lupus*—and "social contract"). On the contrary, the originary "non-historical element" of the human condition always remains even within human civilization, and indeed humanity cannot "fully resolve itself in the historical dimension, thus opening up a problematic, indecipherable area as its interior" (Esposito 2012: 26). Animality is not the contrary of rationality, an other that must be overcome and abandoned. This means that animality is in some way also rational, but also that rationality never ceases to be innervated by animality.

In this chapter, I reconstruct the major lines of this tradition,[1] starting from the figure of Francis of Assisi and what could be called his "mystical animalism." I then turn to the animal figures in the *Divine Comedy* and attempt to show that even if Dante remains a "hierarchical" thinker in many ways, his descriptions of particular animals actually reveal a thinker of diversity. An extremely interesting case is that of Hieronymus Rorarius, author of a treatise—*Quod animalia bruta ratione utantur melius homine* (1543–1544)—which was widely circulated in the seventeenth-century European debate on the soul of beasts. The major figures of Italian Renaissance naturalism, Bernardino Telesio (1509–1588), Giordano Bruno (1548–1600), and Tommaso Campanella (1568–1639), occupy a central position in this reconstruction. In the eighteenth century, the figure of Giambattista Vico stands out, with his "beasts" and "giants," which highlight the thousand threads that link reason to its animal and irrational origins. The other central figure in this reconstruction is the great nineteenth-century Italian poet and philosopher Giacomo Leopardi, who in the *Zibaldone* confronts the humanity-animality nexus several times, starting from a sensualist and radically anti-Cartesian perspective. The final figure I will discuss is another "marginal" figure (marginality is the hallmark of this tradition, which however also represents the reason for its current timeliness), the scientist-philosopher Giorgio Prodi (1928–1987), who proposes an extremely original reconstruction of the inseparable links between the inorganic and organic world (on a semiotic basis), from proteins to the human mind—a reconstruction based on the radical *continuity* of all forms of life.

2   ANIMAL BROTHER

In his *Vita prima S. Francisci* (1228–1229), Thomas of Celano recalls an episode about Francis, which will then be depicted in one of Giotto's frescoes (1295–1298) in the *Storie di san Francesco* series in the Upper Basilica of Assisi.[2] In Giotto's representation Francis—amidst a barren Apennine landscape—bends down to "speak" to the birds, while another friar observes with amazement if not irritation. Here is Celano's narration, which we should carefully read on its own terms, rather than what the hagiographic tradition of St. Francis has accustomed us to seeing:

> Francis walked along the Spoleto valley. Arriving near Bevagna, he saw many birds of every kind of species gathered together: doves, crows, and "*monachine.*"[3] The servant of God, Francis, a man full of ardent love who had great piety and tender love even for the inferior and irrational creatures [*inferiores et irrationabiles creaturas*], ran to them in haste, leaving his companions on the road. Drawing near to them, and seeing that they were waiting for him, he greeted them according to his custom. But noting with great amazement that they did not want to fly away, as they usually did, completely happy, he exhorted them to desire to hear the word of God. And among other things, he said to them: "My brother birds [*Frates mei volucres*], you must constantly and highly praise your Creator because he gave you feathers to dress you, wings to fly, and everything else that you need. God made you nobles among the other creatures and granted you to wander through the clean air: you do not sow and reap, yet He helps you and guides you, dispensing you from all worries." (Caroli 1990: 458)

Francis, a man, speaks to the animals. The situation Giotto depicts is as old as humanity,[4] which has always tried to "speak" with non-human animals. In this case, however, what is at stake is not communication so much as fraternity. Indeed, Francis addresses the birds on a basis of absolute equality—the birds are his "brothers." And precisely because they are brothers, it is not important that they are irrational and inferior creatures. A brother remains a brother even when, for example, he is a newborn baby, or if he is old and sick. It is in this sense that the birds are "inferior." Their inferiority does not question the fact that birds and humans are brothers, that is, creatures of the *same* world. What does Francis propose to the birds? Simply that they "praise" [*laudare*] the creator Lord. Here it is important to grasp the non-religious sense of his exhortation. *Laudare Dominum creatorem* means nothing else, fundamentally, than to love life as it is. Birds are creatures that live the life that they happen to live. Indeed, a life is holy when it desires nothing more than what that same life offers. A holy life is the life of birds, that is, a life that lacks nothing, in the sense that it is a life *sine sollicitudine*, without worries. Indeed, the worries of animals—for example, food, rest, and survival—are however worries that they are capable of facing. A bird "knows" what dangers it can encounter, and it tries to avoid them. Dangers as well as pleasures belong to its command. In this sense a bird's life, for Francis, is a blessed life, because it does not desire anything else than what is already at its disposal.

In general, for Francis animal life is a life of complete holiness in the sense of a completely *immanent* life. A life in immanence, or better yet a life of immanence, is a life beyond, a life without, time—precisely, a life of immanence is a life without transcendence (a holy life has no need of God, because it already coincides with God). Gilles Deleuze, in the final text he wrote before his death, "Immanence, a life...," describes this condition in words that would please Francis: "Absolute immanence is in itself: it is not in something, to something; it does not depend on an object or belong to a subject" (Deleuze 2001: 26). The life of birds, namely the simply animal life, is a life of immanence. More precisely, the life of birds represents, for a human being, the exemplary life exemplified in immanence. We can now see how the passage in Thomas of Celano's *Vita prima* continues:

> To these words, as [Francis] himself as well as the friars who were present told, the birds expressed their joy according to their own nature, with various signs – stretching their necks, spreading their wings, opening their beaks and looking at him. He then came and went freely among them, brushing their heads and bodies with his frock. Finally he blessed them with the sign of the cross and gave them license to resume flying. Then he, too, together with his companions continued their journey, full of joy and thanked the Lord, who is revered by all creatures with such devout acknowledgment. Since he was a simple man, not by nature but by divine grace [*simplex gratia, non natura*], he began to accuse himself of negligence for not preaching to the birds earlier, because they listened so devoutly to the word of God; and from that day on he began to invite all the birds, all the animals, all the reptiles and even inanimate creatures [*creaturas quae non sentiunt*] to praise and love the Creator, because every day, invoking the name of the Lord, he realized through personal experience that they were obedient. (Caroli 1990: 459)

The life of birds, like the life of all animals, but in general the life of the world, even of that so-called "inanimate" world, is a simply living life. For Francis life is not life in the way that a biologist or zoologist might think. "Life" is a term that gets beyond the distinction between biology and geology; it is life understood as the pure unfolding of the events of the world. The life of birds (as we will see when we turn to another great Italian poet-philosopher, Leopardi) is thus the prototype of the life that lives, of the life that is happy to live, of the life that demands nothing other than the same life that it is already living (cf. Cimatti 2014).

This explains Francis' passion for the animals. The point in question is never the abstract juridical claim of the equivalence of living beings. Francis' perspective, and herein lies his radical modernity, is radically anti-juridical (cf. Agamben 2014). Animals "matter" not because they are persons, that is, because they are like us, but on the contrary, we humans matter like them only because we are creatures of the same Lord, of one and the same world. Indeed, for Francis, animality immediately means world:

> How to describe his ineffable love for the creatures of God and with how much sweetness does he contemplate the wisdom, power, and goodness of the Creator in them? Precisely for this reason, when he admires the sun, moon, and stars of the firmament, his soul floods with joy. Oh simple piety and pious simplicity! He felt great affection even for the worms [*vermiculos*] [...]; therefore he took the trouble to remove them from the road, so that they would not be crushed by passers-by. And what to say about other inferior creatures, when we know that, during the winter, he was even worried about how to prepare honey and wine for the bees so that they would not die from the cold? (Caroli 1990: 474).

The Cartesian glance divides and classifies, *res cogitans* and *res extensa*, human and animal, language and instinct. Francis, on the contrary, places himself on the level of the life of the world, where these distinctions do not exist and have no meaning. On this level the sun is worth just as much as a worm. They are always creatures of the Lord. Deleuze will call this level, in which all anthropocentric and animalphobic distinctions collapse, the "plane of immanence," understood as "pure immanence," that is, as "A LIFE, and nothing else" (Deleuze 2001: 27). In Francis' non-theoretical language, this is expressed as a simple love for all creatures (animate and otherwise) in the world: "if he saw expanses of flowers, he stopped to preach to them and invited them to praise and love God, as beings endowed with reason [*ac si ratione vigerent*], and likewise the crops and vineyards, the stones and woods and beautiful countryside, the flowing waters and green gardens, the air and wind – with simplicity and purity of heart, he invited them to love and praise the Lord" (Caroli 1990: 474). In this sense the Italian tradition, which we begin here with Francis, is originarily anti-dualistic and thus anti-Cartesian. This is an inspiration that powerfully resounds in the *Canticle of the Creatures* (*Laudes Creaturarum*, ca. 1226), wherein from the title Francis puts all creatures together: "Praise be to You, my Lord, with all Your creatures, / especially Sir Brother Sun, / Who is the day and through whom You give us light. / And he is beautiful and radiant with great splendor, / and bears a likeness of you, Most High One" (Caroli 1990: 178). Francis uses the concept "creature" in order to deactivate all of the distinctions that our "natural" (see Bloom 2005) dualistic thought ceaselessly produces. In this sense the concept "creature" is not so much a religious concept (religion is perhaps the most powerful dualistic apparatus that there is), but rather what serves to remove any desire for transcendence; for this reason, "he called all creatures [*omnes creaturas*] with the name brother and sister [*fraterno nomine*]" (Caroli 1990: 475).

However, the relevance of this position would not be understood if we do not consider that for Francis, animality also represents the model of a perfect human life. For modern political thought, beginning with Hobbes, the animal is at the antipodes of politics; on the contrary, for Francis the holy life, that is, the life of Francis and his disciples, is properly an animal life, precisely like the life of the birds. This point is clear particularly in the *Regola non bollata* (1221), the first rule (not officially ratified by the papal authority). It is a completely paradoxical rule, because it ultimately maintains that those who lead their entire existence on the basis of Christ's example in fact do not need any rule. The first and fundamental point of this rule states: "the rule and the life of brothers is this, namely to live in obedience, chastity, and with nothing of their own, and to follow the doctrine and example of our Lord Jesus Christ [*regula et vita istorum fratrum haec est, scilicet vivere in obedientia, in castitate et sine proprio, et Domini nostri Jusu Christi doctrinam et vestigia sequi*]" (Caroli 1990: 99). All of those who want to live according to this rule are, first of all, brothers. Just like the birds and the sun are brothers. "Brother" means that there is no hierarchy in Franciscan life, that all is on the same level. The sixth rule explicitly prescribes: "no one should be called prior, but all should be simply called friars minor [*nullus vocetur prior, sed generaliter omnes vocentur fratres minores*]" (Caroli 1990: 105). What else does "friars minor" mean if not to lead a blessed life like the birds, or the worms, namely a perfect life of immanence? Again, Deleuze immediately comes to mind, when, together with Guattari, he speaks of a "minor literature," that is, precisely a writing that escapes every temptation of power and hierarchy: "the writer," like the friars minor, "ceases to be a man in order to become an ape or a beetle, or a dog, or mouse, a becoming-animal, a becoming-inhuman" (Deleuze and Guattari 1986: 7).

If we turn now to the first rule, it is clear that the absolutely decisive point, the one that in fact immediately triggers conflicts and disagreements within the Franciscan community, is the one that prescribes that the friar minor should not possess anything, that he lives *sine proprio*. For this same reason the friars minor must avoid taking possession of any place or contending with anyone, but above all must avoid any contact with money: "no brother, wherever he is, and wherever he goes, should in any way take with himself or receive from others or allow money to be received [*nullus fratrum, ubicumque sit et quocumque vadit, aliquo modo tollat nec recipat nec recipi faciat pecuniam aut denarios*]" (Caroli 1990: 106). Money, in fact, is the paradoxical material manifestation of transcendence and value: money is something that in itself is worth nothing, and yet at the same time it has an extremely large incorporeal value. Believing in money means, simply, not believing in the world. It is not by chance that non-human animals do not know what to do with money. There is no transcendence in animal life. For this same reason, because the life of the minor brother is an entirely worldly life, he must possess neither things nor money.

But who is it, once again, that lives without having anything, because they receive whatever they need [*quidquid necesse vobis*] from the Lord? It is thus shown how the reference to animality, for Francis, does not serve as an allegory of a virtue, as it does in the tradition of medieval bestiaries. The animal is instead a "conceptual personae," that is, a "power of concepts" which forces us to think of human life in a completely different way (Deleuze and Guattari 1994: 64–65). Not having anything of one's own means living like an animal, living like a bird, like a stone. It is also in this sense that the reference to chastity is understood: the stakes are not so much sexual lust as desire as an apparatus that produces transcendence and therefore detachment from the immanence of life. The minor brother has no desires, because, like the sun or the birds, *he is* already all that is the case. A holy life is indeed a life that lacks nothing. On the contrary, the brothers will have to attribute "to the Lord God [...] all goods" because "they are His" (Caroli 1990: 114). The animal model is also evident in Rule XI, which prescribes renouncing, if possible, language, precisely what marks the difference between humans and other animals: "all the brothers should beware of slandering anyone, and avoid the disputes of words, or rather try to be silent [*omnes fratres caveant sibi, ut non calumnientur neque contendant verbis immo studeant retinere silentium*]" (Caroli 1990: 109). Without property, without housing, without language. Like the animals, precisely. And again, "when the brothers go through the world they bring nothing for the journey [*Quando fratres vadunt per mundum, nihil portent per viam*], neither sack nor haversack, nor bread, nor money, nor staff" (Caroli 1990: 111). Just like the birds, which have in themselves, in their own bodies, all that they need, and the rest they find in the world. In the same way the brothers cannot "have either themselves, nor from others, any beast [*bestiam aliquam habeant*]" (Caroli 1990: 111). Francis' order is not inspired by the entirely modern, and Cartesian, idea that animals also have rights; on the contrary, since no one is a "person," that is, a transcendent entity, while everything is body and life, then no one can consider another living entity as personal property. Just like the animals, the friars also do not know shame and can eat "whenever the need arises [...] all the food [*omnibuscibis*] that men can eat" (Caroli 1990: 108). Exactly like the animals, who eat what they find when they need it. This is an important point, which denotes the differences between animality and animalism, which burdened by an unthought Cartesian prejudice, considers living beings superior to non-living beings and animals endowed with sensory capacities superior to those who are not. If everything lives, or instead, everything is life, then everything can eat, because "necessity has no law [*necessitas non habet legem*]" (Caroli 1990: 109). Animal life knows nothing but necessity, just like the life of the brothers who follow Francis' example in choosing to live according to the same necessity that marks the life of birds. Francis' life is literally the life of an outlaw. Finally, just as the birds were joyful in listening to Francis' words, so also the brothers must "beware of showing sadness [...] but rather be glad in the Lord and playfully cheerful [*non se ostendant tristes* [...] *sedostendant se guadentes in Domino et hilares convenienter gratiosos*]" (Caroli 1990: 106).

With Francis, the philosophical concept of "animality" (cf. Cimatti 2020) begins to take shape. This concept certainly concerns animals, although not exclusively or in the first place. Animality means, and in the Italian tradition will mean, the thought of an entirely immanent, worldly life, just like the life of non-human animals. This is an animality that has nothing to do with the law and morality. Francis, in order to define a human life inspired by animal life, that is, his own life, writes in the *Testamentum* (1226) that the same God "revealed to him" that he had "to live according to the form of the holy Gospel [*vivere secundum formam sancti Evangelii*]," namely a life in poverty, or more precisely a life without any property. Francis writes to Clare, the founder of the Order of Saint Clare, hoping that the "form of life [*forma vivendi*]" of this new group of sisters is inspired by the same principle that inspired her life: "I advise you to live always in this most holy life and in poverty [*in ista sanctissima vita et paupertate semper vivatis*]" (Caroli 1990: 132).

In Francis we find, in conclusion, all of the elements of the concept of animality proper to the Italian tradition: absolute immanence, animal life, absence of transcendence, and impersonality. The life of the Franciscan brothers is thus a "form-of-life," as Giorgio Agamben defines it, namely, "a life that can never be separated from its form, a life in which it is never possible to isolate and keep distinct something like a bare life" (Agamben 2015: 207). Actually, the Franciscan brothers are not animals like the birds, yet they choose to live like animals. But since they live like all the non-human creatures of the world live, that is, *sine proprio*, in this case the "form" (the rule) of their existence coincides with that same existence. "Form-of-life" deactivates the dualism between body and mind, as well as the dualism between animal and human. "Form-of-life" is a human life that does not cease to be animal. Cartesian dualism instead postulates a thinking subject separate from its own body. A dualism that is not overcome by maintaining that the body coincides with the mind (in this way only the body remains, but without the mind; the dualistic prejudice remains intact) but with a human life in the manner of the animals. This, and nothing else, is what *vivere secundum formam sancti Evangelii* means.

3   ANGELS, BEES, AND WORMS

In the medieval tradition, the animal is typically an allegory of a virtue or a vice. In this sense the animal is always a sign that stands for something else. Thus the animal is not really an animal, because its life—contrary to what we have seen in Francis—is always in the service of a transcendent value. For example, in the Φυσιολόγος (*Physiologus*), an anonymous text written around the second century CE, the beaver is described as an animal whose "sexual organs are effective in medical care." For this reason, when the beaver realizes it is being hunted, "it cuts off its sexual organs and throws them to the hunter." The same must be done, the text continues, by the believer: "render to the hunter what he deserves. The hunter is the devil, and in you there is lust, adultery, and greed. Extirpate these things from yourself and give them to the hunter devil, and these will leave you" (Zambon 2018: 39).

The metaphorical use of the animal is to some extent inevitable (actually this use never ceases; this very chapter, after all, is a sort of modern *Physiologus*). However, the use of the animal found in the Italian tradition often indicates a different direction: no longer from the animal to the human, but rather from the human to the animal. It is not so much a matter of a becoming-human of the animal, but rather a becoming-animal of the human. Francis' birds, for example, are examples of blessed life but also a political model for the establishment of the Franciscan rule. In this way, in *Paradiso*, Dante[5] observes the blessed (the "sacred ranks") and the angels who fly without ever stopping, like the bees of a flower in bloom, from God to the blessed and from the blessed to God:

> In form, then, as a rose, pure, brilliant, white,
> there stood before me now the sacred ranks
> that Christ, by His own blood, has made His bride.
> The other force that, flying, sees and sings
> the glory that so stirs their love of Him –
> the goodness, too, that makes them all they are –
> came down, as might a swarm of bees that first
> en-flower themselves, returning, afterwards,
> to where their efforts are made sweet to taste.
> They search the utmost depths of that great flower,
> with all its many petals. Then they rise
> once more to where their love will always dwell.[6]

Dante's image is at the same time mundane and mystical, with bees in a field and angels in the empyrean. We do not see bees that become other, but angels that become insects. In *Paradiso* the same birds that Francis addressed seem to return: a lark who after it wandered in the sky stops "content" with the "sweetness" that opens up to its gaze. This image allows Dante to represent the relationship between "everlasting joy" and all of creation, in which "each thing becomes what it truly is":

A lark, as first it mounts through airy space,
soars upward singing but is silent then,
flush with the sweetness of its highest reach.
So, too, it seemed, that image of the print
of everlasting joy, at whose desire
each thing becomes what truly each thing is.[7]

To become what one is, that is, to coincide with one's own nature. It is this condition that Deleuze calls "absolute immanence," that is, "the immanence of immanence": precisely to become what one is. Immanence, in this sense, means a radical incarnation. Animality is nothing but this complete adhesion to corporeal condition. This process is pushed to such a point that "the life of the individual," that is, of the person, "gives way to an impersonal and yet singular life" (Deleuze 2001: 27–28). If understood as absolute immanence, it then becomes possible to include in the field that we are outlining a text such as Pico della Mirandola's *Oration on the Dignity of Man* (1486–1487), which otherwise would seem to represent a radical refutation of the thesis maintained in this chapter.

At the moment the Architect must create the human being, he realizes, Pico writes, that "there was nothing among His archetypes [*non erat in archetipis*] from which He could mould a new progeny, nor was there anything in his storehouses that He might bestow upon His new son as an inheritance" (2012, 115). What does the fact that no archetype was available mean except that there was no abstract, transcendent principle on which to base the creation of the human? But if there is no model, it means that only the singular body remains, and that every body, as Dante says, "becomes what it is." Thus God says to Adam: "We have given you, Adam, no fixed seat or form of your own, no talent peculiar to you alone [*nec certam sedem, nec propriam faciem, nec munus ullum peculiare tibi dedimus*]. This we have done so that whatever seat, whatever form, whatever talent you may judge desirable, these same may you have and possess according to your desire and judgment" (2012: 117). To be devoid of a natural essence means, ultimately, to be nothing other than the body that one is, in the completely contingent use that can be made from time to time. For this reason the human being is "neither of heaven nor of earth, neither mortal nor immortal," precisely because the human is simply and absolutely the body that it is, with neither models nor nature (2012, 117). Against the common way in which this text is understood, here Pico is claiming that the only animal is the human animal, precisely because it has no nature of its own. The privilege of the human being is that of a being whose living is the most animal living that there is, in the sense of being that living being that is nothing but the body that it is, in a complete and absolute coincidence between body and life. This is an important point to emphasize: the "natural" life of the human being is not simply that of an animal, because otherwise it would not also be a *human* life; it is a matter of finding a human way of continuing to be animal: for this reason "the Father infused in man, at his birth, every sort of seed and all sprouts of every kind of life [*omnifaria semina et omnigenae vitae germina*]" (2012, 121). The human being is literally full of life—it is human precisely because it is full of non-human seeds and sprouts.

This co-presence between animality and humanity, which is actually a genuinely uninterrupted mutual contamination, is the specific characteristic of the way in which the theme of animality and nature is treated in Italian Renaissance naturalism, particularly in Bernardino Telesio (1509–1588), Giordano Bruno (1548–1600), and Tommaso Campanella (1568–1639).[8] The starting point of Telesio's naturalism in *De rerum natura iuxta propria principia* (1570) is the inexhaustible richness and diversity of forms of life. Nature is nothing other than this infinite production of vital diversities: "since therefore extremely different entities are produced and changed in conformity to a nature that adds and moves away, this cannot at all be considered unique, but rather multiple and as divided and distinct in many forms" (Telesio 2009: 17).[9] Nature means diversity. There is no mention of a hierarchical vision in this "multiplex" nature, because these diversities are all located on the same level, exactly what Deleuze calls the "plane of immanence." Nature is unique and diversified within itself. There is no space for any dualism in this vision of the world: "the nature that remains is entirely unique and identical [*remanent natura una esse omnino*], and neither rejects nor hates any natural actor, but is equally proper to all and common to all and willingly, with all equally united and all together and all received and preserved, becoming so to speak their dwelling and their home" (Telesio 2009: 19). On the basis of this radical naturalism, for Telesio, it is on the one hand out of the question that animals can have a "soul" (2009: 429). On the other hand, it is also excluded that there can be a soul without body: indeed, the soul "absolutely [cannot] subsist and exist if it does not pertain to a body" (2009: 429). Yet again, we find that Gilles Deleuze is close to this intransigent monism of Telesio's. In *Difference and Repetition*, Deleuze argues that difference is not difference with respect to something (e.g., a

transcendent model), but rather difference is "affirmation [...] is difference itself" (Deleuze 1994: 55). Deleuze describes the field of these differences with words that resonate with Telesio's: "difference [...]
presupposes a swarm of differences, a pluralism of free, wild, or untamed
differences; a properly differential and original space and time" (Deleuze
1994: 50). In this sense *Homo sapiens* and *Lumbricus terrestris* are differences in themselves, without hierarchies or gradations.[10]

Telesio's naturalistic monism is fully developed in the cosmic vision of
Giordano Bruno. In the dialogue between Dicsono and Teofilo in *Cause,
Principle, and Unity*, Bruno presents the radical thesis that the whole universe is an animal entity: "Dicsono: the spirit or the soul, or the universal
form of all things is in all things [...] Teofilo: that spirit is found in all
things which, even if they are not animals, are animate. If not according to
the perceptible presence of life and animation, then according to the principle, and certain primary act of life and animality" (Bruno 1998: 44,
translation modified). The world itself is animality, because it is animated
and therefore alive. The dualism between spirit and matter that in a few
years would forever mark Western philosophy and science is entirely
rejected. All things, for Bruno—those things that biology distinguishes
between as living and non-living—contain one and the same "animal and
vital symbolic principle" (1998: 44). Animality is not the other of the
human, the enemy to crush and overcome, but on the contrary, *the world
is animal*.[11] Accordingly in *On the Infinite Universe and Worlds*, Filoteo
explicitly says: "I say that this infinite and immense [world] is an animal
although it has no determinate figure, and meaning that refers to external
things: because it has all soul in itself, and everything animated understands, and it is all of that" (Bruno 2000: 373). This is a genuine "animal
principle" of the Italian philosophical tradition. Animality as model of
explanation of the world, and not as a zoological category, let alone an
ethical one. In Bruno it is clear how the philosophical concept of "animality" acquires a general value. First philosophy is animal "since the earth is
an animal" (2000: 79). The first obvious consequence of this approach is
that every hierarchy among the living falls, as explained by the donkey
Onorio in *The Cabala of Pegasus*:

> I came to consider how, in terms of my spiritual essence, I was no different
> in genus nor in species from all the other spirits who had transmigrated from
> their animal and corruptible bodies; and I saw how in the genus of corporeal
> matter Fate not only fails to differentiate the human body from the ass, and

the body of animals from the body of things thought to be without soul, but even in the genus of spiritual matter Fate treats the asinine soul no differently than the human, and the spirit that constitutes those so-called animals than what is found in all things. (Bruno 2002: 54)

It is difficult not to find in these words an echo of Francis' "sermons" to the birds and the sun, what we might call a radical egalitarian monism. If the earth is "an animal," then everything is in some way living, even if its living does not mean living in the way biology describes living phenomena. Indeed the "animal model" precedes the distinction between biology and geology, that is, it precedes the dualistic vision of the world which Descartes will crystallize in the distinction between *res cogitans* and *res extensa*: the soul "of the human is the same in specific and generic essence as that of flies, sea oysters, and plants, and of anything whatsoever that one finds animated or having a soul" (2002: 56). What is particularly interesting is the theme, which not by chance returns in Deleuze, of metamorphosis, that is, of the becoming animal of an animal, but also of the possible contrary movement, of the becoming animal of a human. The point in question is why this passage is possible: because there are no absolute differences between different forms of life and therefore a contamination is always possible. Life is this omnipresent reality of contamination. Bruno presents the case of the serpent:

I conclude that if it were possible, or in fact came about, that the head of a serpent were to be formed or modified into the shape of a human head, and the chest were expanded to the extent satisfactory for such species, if the tongue were enlarged, the shoulders amplified, if the arms and hands were to branch out, and the legs sprouted from the place where the tail terminates; it would understand, would appear, would breathe, would speak, would operate and would walk as none other than a man because it would be none other than man. Conversely, the man would be none other than a serpent if he were to retract his arms and legs, like a stump, and his bones were to all converge in the formation of a spine, and were to viperize and suppress all the forms of limbs and the qualities of those behaviors. Then he would have a more or less lively mind: in place of speaking, he would hiss; in place of walking, he would slither; in place of building a home, he would dig a hollow (and not a room, but a hole, would suit him). (2002: 57)

The same "animal model"[12] is found in Tommaso Campanella, who in *De sensu rerum et magia* (1609) strictly clarifies that "we must therefore affirm that the world is an entirely sentient animal and that it enjoys all of the parts of common life" (2007: 24). From this perspective, just as was the case in Francis, the simplest creature is worth as much as the most presumptuous. Reversing Pico della Mirandola (really only in appearance, because if the stakes are absolute immanence, then as we have seen *Homo sapiens* is to be admired precisely because it is the perfect animal), Campanella proposes an extraordinary analogy between worms and the animal bodies they live within and animals with respect to the world. Worms that we could consider the exemplary animal of Italian Thought, an entirely earthly animal, made of the earth, which continuously regenerates the earth:

> All of the animals in the world are like worms inside of an animal, in that they should be thought of as feeling like the worms of our stomach who do not think that we feel and have a soul which is superior to theirs, nor that they are animated by the common blessed soul of the world but each by its own, like the worms in us, who do not have our minds for their soul, but the very spirit. (2007: 235)

4   "Poetic Wisdom"

Within Italian Thought, what we have called the "animal model" is most precisely a conceptual operator, which at least from Francis onward deactivates any dualistic temptation. The fundamental idea is that the rational dimension (mind, calculation, language, ethics) is never separated radically from the corporeal-sensible dimension (affects, passions, desires). However, it is necessary to avoid equivocating on this point, since there are recurrent attempts to dismantle the metaphysical mind-body couple across many philosophical traditions. The "animal model" does not simply maintain that reason in fact coincides with the body, nor that reason is impregnated with corporeality. The "animal model" defines a field of forces that is neither corporeal nor rational, beyond the mind/body distinction. It is a conceptual operator that locates in a single point what simultaneously separates and unites the terms of the dualism (Agamben 1998). In this sense the "animal model" is something more than a reference to the world of genuine animals (such as the world of a squid or an elephant). Indeed, to maintain that the human is *also* an animal, or that an animal is *also* endowed with reason, does not in any way dismantle the dualism of animal and human. It is not a matter of attributing a bit of animality to human beings, or a bit of humanity to animals. Instead it is a question of disactivating this dualism. For this reason, within this tradition animality has simultaneously had to do with the constitution of the world as well as the way of thinking it: in this way, on the one hand animality is the living world before Adam gives names to it, and on the other it is a way of seeing the world that always tries to go back to the unitary world that precedes our conceptual distinctions. When Francis speaks to the birds, on the one hand he attests that between *Homo sapiens* and *Apus apus*, for example, there is no metaphysical difference, and on the other, he tries to establish a form of human life capable of being in the world like the swifts. But because animals are in the world in a direct way—without ever thinking of themselves as someone in the world, that is, as *subjects*—a human form of life inspired by animality is a way of living beyond the distinction between subject and object. Animality thus means "absolute immanence," an entirely and solely earthly life, like that of the birds, who in fact live in the sky without thinking of it as an allegory for Heaven, or like that of the worms, who are literally made of the earth. The animality of Italian Thought is therefore the peculiar way in which this tradition of thought has tried to imagine a human life that has definitively deactivated the dualisms of mind and body, time and space, spirit and matter.

The philosopher known to have thought about this condition with great consistency is Giambattista Vico, particularly with the concept of "poetic wisdom," which makes up the theoretical heart of the *New Science* (1744). This is a "wisdom," that is, a sensible knowledge, which is the basis of human culture in all its forms, from language to law. "Poetic wisdom" is precisely this pre-human background which renders the development of human civilizations possible. In this sense Vico dismantles every dualistic hypothesis from the beginning, since there is no moment in his reconstruction of human history that is completely separate from its non-historical, animal origins. In other words, the human never ceases to be non-human, or the human never ceases becoming human. Animality is not a past condition that has now been entirely overcome, but rather, the human is *human* precisely because it never ceases relating to its non-human background.

In the *New Science*, Vico presents a model of historical development which is divided into three stages: the age of gods, heroes, and men. Each of these phases is marked by a particular way of experience, thinking, and communicating. Here we will only discuss the case of the relation between mind and language,[13] because it is exemplary in showing how for Vico, human language, often identified as the characteristic that clearly distinguishes the human from other animals (cf. Hauser et al. 2014), is never properly only human, and thus arbitrary, but always an animal behavior at the same time. Each of these ages is characterized by a particular way of communicating: thus "the first [hieroglyphic] language [...]; the second, symbolic, by signs [...]; the third, epistolary, for men at a distance to communicate to each other the current needs of their lives" (1984: 126). The first language is gestural; the second is based on "metaphors, images, similitudes, or comparisons"; the third is "epistolary speech," that of conventional languages, which however never become entirely arbitrary, because "*metaphor* makes up the great body of the language among all nations" (1984: 129, 130, 132). Languages, in other words, never break off relations with their sensory and corporeal basis.

The first age, for Vico, is that of the "*mutes*," that is, animals still incapable of expressing themselves through an articulate language; these primordial beings, neither animals nor humans, "make themselves understood by gestures or objects that have *natural relations* with the ideas they wish to signify" (1984: 68). At the beginning of human language and history, there is a living being deprived of language, a "mute," who establishes relations with his peers by "communicating" with actions and corporeal postures. In such "language" the "content" that one wants to "express" is naturally linked to the way in which it is expressed, such as in the case of onomatopoeia. These first human beings were "beasts," who had a relation with the world only through the senses; they were in other words living beings that did not have a cognitive or mental relation with the world, but only a practico-sensible one: "*human nature*, so far as it is like that of *animals*, carries with it this property, that the *senses* are its sole way of *knowing things*." At the beginning and within reason, there is a "beast," that is, there is animality. A "beast" that incarnates a:

> Poetic Wisdom, which was the first wisdom of the gentile world [that] must have begun with a *metaphysic* not rational and abstract like that of the learned men now, but *felt* and *imagined* as that of *these first men* must have been, who, *without power of ratiocination*, were all *robust sense* and *vigorous imagination* [...]. This metaphysic was their *poetry*, a *faculty born with them* (for they were furnished by nature with these *senses* and *imaginations*); born of their *ignorance of causes*, for ignorance, the *mother of wonder*, made everything *wonderful* to men who were *ignorant of everything*. (1984: 104)

"Poetic Wisdom" represents the first modality, sensorial and practical, of experiencing the world. In this way, when a "beast" encounters something that it does not understand, it assimilates it to something it already knows, in particular to the actions of its own body. In this way the entire natural world is felt as a great living body, based precisely on the unconscious analogy with the body of the "beast." One of Vico's favorite examples is lightning. The "beasts" have no idea what lightning is, and therefore they compare it to the sign of anger of a superhuman body, just as they, when they are angry, give rise to fury. Here then is the idea of a divinity such as Jove:

> [A]t last the *sky* fearfully *rolled* with thunder and *flashed* with lightning, as could not but follow from the bursting upon the air for the first time of an impression so violent. [...] Thereupon *a few giants* [another name for the "beasts"], who must have been the *most robust*, and who were dispersed through the forests on the mountain heights where the strongest beasts have their dens, were *frightened* and *astonished* by the great *effect* whose *cause they did not know*, and raised their eyes and became aware of the *sky*. [...] [T]he nature of the human mind leads it to *attribute its own nature to the effect*, and because in that state their nature was that of men all robust bodily strength, who expressed their very violent passions by *shouting* and *grumbling*, they pictured the sky to themselves as a great animated body, which in that aspect they called Jove, the first god of the so-called *gentes maiores*, who by the whistling of his bolts and the noise of his thunder was attempting to tell them something. (1984: 105)

At the beginning, for Vico, there is no reason and no science, but rather there is the body with its passions—at the beginning, there is animality. The body of the "beasts" becomes in this way the analogical model for "understanding" the world, which in fact is first perceived as "a great animated Body." Jove is the first and fundamental step for beginning to "think," albeit in a bodily and concrete way, the world of experience. If we recall Descartes' Cogito *ergo sum*, the reversal cannot be more radical; for Descartes at the beginning there is a thought, while for Vico there is "wonder" and then an expressive waking. Because the minds of the first human beings "were not in the least abstract, refined, or spiritualized, because they were entirely *immersed in the senses, buffeted by the passions, buried in the body*" (1984: 106). In this sense "Poetic Wisdom" is the wisdom of the sensible and passionate body of the "beasts," who begin to experience the world and "understand" it by starting from themselves.

Beyond simply representing a primitive way of knowing the world and thus a cognitive modality surpassed by the development of the mind and human culture, "Poetic Wisdom" is rather the initial, and recurrent, moment of *every* human experience. Indeed, each time a human being finds itself confronted with something that arouses "wonder," at that moment it is a "beast." The "mutes" or "beasts" are in fact like "children of nascent mankind" (1984: 105). Every new specimen of the species *Homo sapiens* is first of all a child and therefore is simultaneously "mute" and "beast." *Homo sapiens* never ceases to also be a "beast"; in this sense even in the rational, evolved, and abstract world of contemporaneity "the world's childhood" and the originary "thinking [as] beasts" never ceases to be operative (1984: 64, 90). Indeed:

> As *gods, heroes,* and *men* began at the same time (for they were after all *men* who imagined the *gods* and believed their own *heroic nature* to be a mixture of divine and human natures), so these three languages began at the *same time* [...]. They began, however, with these *three* very great *differences:* that the *language of the gods* was almost entirely *mute*, only very slightly *articulate*; the *language of the heroes*, an equal mixture of *articulate* and *mute* [...]; the *language of men*, almost entirely *articulate* and only very slightly *mute*. (1984: 134)

This contemporaneity of the three phases of the historical-cognitive development of the human mind is perhaps the most important theoretical advance of the *New Science*. The originary, that is, animal-like rather than non-human, is not relegated to the past of humankind, as an overcome and forgotten stage of its development; on the contrary, "the origin, with all the energy and violence it bears with it, is [...] the hollowed out bedrock underlying all of human history" (Esposito 2012: 71).

The stakes of the question of animality, then, entirely revolve around this relation with life and the earth. The "beasts," for Vico, are in fact "sons of the Earth" (1984: 101). Since we never cease to be "beasts" (and "heroes" and "men"), the theoretical question that this tradition of thought poses is how to keep the links with originary "Poetic Wisdom" always open, or in other words, precisely how to keep open links with the earth. For Leopardi, on the other hand, it is clear that the technological and scientific development that began to show its unstoppable power in the eighteenth century has precisely as its principal effect the strict separation of current humanity from the "wholly corporeal imagination" (1984:

105) of the "beasts." In this sense, Leopardi's "philosophical" theme is how to deal with the progressive and inexorable distance that opens—to continue using Vico's categories—between the age of the "gods" and the age of "men." Between "Poetic Wisdom" and what will then be called abstract "instrumental reason." Once again the question that arises is how to imagine a human condition that does not cease, however, to be simultaneously animal-like, that is, earthly. The theme is always that of a human life of immanence. Hence in the *Zibaldone*, Leopardi describes this progressive and inexorable process of the "spiritualization" of the human animal, which is precisely the distancing from the earth that is the foundation of Vico's "Poetic Wisdom":

> As men become more and more civilized, and as the spiritual and inward side of humanity acquired proportionately and grew in consistency, efficacy, value, importance, extent, activity, influence, strength, and power, and capacity in relation to civilization, we began to first recognize and suppose a hidden and invisible part in man which primitives either did not suppose at all or only glancingly and without making much distinction from the eternal and perceptible side, and then to regard it as much as the external side, and then more than the external, and little by little so much more, that today – unless nature rebelled (and in the last analysis she can never be wholly extinguished or overcome) – no consideration would be given to anything in humanity or each individual other than the internal, and never by the word "man" would we understand anything other than his spirit. (2015: 1627)

Homo sapiens is precisely this continuous process of "spiritualization," that is, the techno-scientific process of the "artificialization" of human life. From this point of view, these analyses cannot but resonate with contemporary debates on "biopolitics" (cf. Esposito 2008), which is fundamentally nothing but the set of political and economic mechanisms which capture and tame human life. It is thus not by chance that in another passage of the *Zibaldone*, where he discusses the current situation of the human condition, which is always "more spirit than body," Leopardi immediately goes on to discuss the techniques for teaching animals: "Now who can say whether such abilities which increase the faculties of those animals, etc., were prescribed for what purpose by nature, either general or their own particular nature, etc., whether they are beneficial for their happiness, etc., and whether their respective species would be more perfect or less imperfect, if such abilities in them would be more common, or universal, etc." (2015: 1665). The animal that is trained and adapted to human needs is like the "primitive" human trained and adapted to the political-economic needs of civil society.

If for Vico the "beast" is a still-living presence within the cycles of human history, for Leopardi it is instead a condition now lost forever. Indeed the primitive, which in the *Zibaldone* is incarnated by the figure of the Californian Indian, lives beyond civilization and therefore "spiritualization." For this reason he can lead a happy existence, however hard the conditions of his life might be. A happy life precisely because his life has not yet been captured by the apparatus of civilization; he is happy because he has never stopped being a "son of the Earth." Here again the theme of animality returns as *immanence*, which fundamentally means fullness, absence of doubts, sense certainty. On the contrary, Leopardi observes, the development of knowledge fundamentally shows nothing more than that common sense is wrong and does not grasp the truth of the world: "every step forward of modern philosophy eradicates an error. [...] Therefore if man had not erred, he would have already been supremely wise [as indeed were Vico's poetic "beasts"], and reached that goal towards which modern philosophy proceeds with so much sweat and difficulty. But he who does not reason, does not err. Therefore he who does not reason, or [...] does not think, is supremely wise. Therefore men were supremely wise before the birth of wisdom, and of reasoning about things: and the child is supremely wise, and the savage of California, who does not know *thinking*" (Leopardi 2015: 1127–28). The "savage of California" and the "child" are two ways of being in the world without abandoning what Deleuze calls the "plane of immanence." But there is a third form of life that "does not reason" and therefore "does not err": the animal.[14] Leopardi clearly does not argue that animals have no mental capacity. This idea is not in question, but rather it is a kind of thought that does not think itself as thinking. The animal like the "child" or the "savage of California" incarnates forms of life, such as that of Francis, in which there are those who do not think the world, but rather directly live the world. These are non-dualistic forms of life, in which on the one hand there are those who think and on the other there is what is thought. What we have called the "animal model" in Italian Thought specifically has the function of imagining a life not marked by dualism, in this sense an animal life, precisely like that of the "beasts": "since man is perfect in nature, the more he moves away from it, the more his unhappiness increases." On the contrary, as long as the human remains within "the bounds of human nature," he is "capable of complete happiness" (2015: 519).

In this way, Leopardi constructs a conceptual nexus (which he takes up from Vico) in which children, primitive peoples, and animals take part. All three are forms of life of immanence. And among the animals, it is not surprising that this reconstruction begins with Francis and in particular the birds: "Quickness is none other than liveliness. Liveliness is pleasing [...] so quickness is, too. So that the pleasure man normally feels at the sight of birds [...] belongs to the most intimate inclinations and qualities of human nature, that is the inclination to life" (2015: 783). To these animals Leopardi dedicates a short treatise, entitled "In Praise of Birds," in his *Small Moral Works* (*Operette morali*, 1826). This work is not as much about animals, particularly birds, as it is what we have called the "animal model." Indeed, the birds exemplify the possibility of an "exterior life [*vita estrinseca*]," that is, precisely a life of immanence (1983: 168). As was the case for Francis, so also for Leopardi the "birds are naturally the most joyous creatures in the world." As we have already seen in the *Regola non bollata*, the brothers "show themselves joyful in the world and playfully cheerful," just as the birds, who "for the most part demonstrate themselves through very happy movement and looks." A happiness, clearly, that has nothing to do with the fact that their life is free of difficulties or dangers: they are not happy because their life is happy, but on the contrary, their life is happy because "by nature they denote a special ability and disposition to experience enjoyment and gaiety" (1983: 164). A happiness of a full life, a life that lacks nothing, as happened to the "beasts" insofar as they were "sons of the Earth." This is what their "exterior life" consists in, a life entirely in the open, beyond the dualistic distinction between interiority and exteriority, mind and body. It is in fact a life similar to the life of children, who, like birds, are endowed with a "very great use of imagination." An imagination, however, that does not distance them from life as it is lived, but on the contrary fills even the most simple and taken for granted actions with happiness: an imagination that is "rich and varied [...], light, changeable, and childlike; which is a bountiful wellspring of pleasing and joyous thoughts, of sweet errors, of varied delights and comforts; and the greatest and most fruitful gift that nature ever lavishes on living souls" (1983: 168). Animality is thus not the life of the animal, but rather life as a human who could have the capacity for the blessed passion of animal life, such as the flock in "Night Song of a Wandering Shepherd in Asia" (1996: 103, translation modified):

O peaceful flock of mine, O happy ones,
I feel you do not sense your misery!
How much I envy you!
Not just because you lead
an almost carefree life
immediately forgetting,
all needs, all harm, all terrifying fears,
but rather since you never suffer boredom.
When you lie down upon the shady grass,
you are at peace, contented;
and almost all the year
you spend like this, untroubled by that boredom.

For Leopardi, "boredom" is the main effect of the "spiritualization" of human life, whereas on the contrary, animality means a life that does not know boredom, not because its life is interesting, but because only those who desire to live a different life from the one they are living can be bored. But this is precisely the essence of every dualism, the separateness between life as lived and life as thought, between body and mind.

This tradition is still living in the present, as is shown by another Italian philosopher and scientist, Giorgio Prodi. Throughout the 1970s and 1980s, Prodi developed an extremely original although little-known biosemiotic theory, in a polemic with the linguistic turn of the era.[15] Prodi's starting point is a close critique of a fundamental semiotic distinction, the separation between "signification" and "communication," formalized by Umberto Eco in *A Theory of Semiotics* (1979: 8–9). In the former case, for Eco, there is a simple and immediate passage of information, whereas in the latter there is effective communication because of the mediation of an interpretant between sign and signified. This dualism leaves all of those "communicative" phenomena that do not involve the trace of mental mediation outside the field of semiotics. For example, it entirely leaves out the immense field of cellular and intracellular phenomena. Biology, from this point of view, remains beyond what Eco himself defines as the "semiotic threshold," which separates genuinely semiotic (and therefore properly human) phenomena from non-semiotic phenomena, such as "stimuli" and "signals." Prodi's critique begins precisely by questioning this dualistic premise, which separates semiotics (typically human) from the rest of the living world. For Prodi, such a position is entirely unacceptable, because it breaks the continuity of the natural world, placing human semiosis in a separate field. To return briefly to

Vico's categories, semiotics, insofar as it is intentional and arbitrary, would have nothing to do with "Poetic Wisdom."

Prodi's thesis, on the contrary, is "the biological genesis of semiotico-cognitive functions" (Prodi 1977: 13). The first consequence of this premise is a radical critique of the "mythology of the radical novelty of man" (1977: 19). It is therefore a matter of starting from the most elementary phenomena, such as the complementary relations between molecules, in order to arrive, through a process of increasing complication, to the most complex forms of semiosis. At no moment, however, does this chain break. The author of the *Divine Comedy* is linked by an uninterrupted biological process to the simplest forms of molecular contact. There could be no Dante without Campanella's worms. Semiosis never loses its contact with the world, with the earth from which the "beasts" come: "communicating means being in solidarity with the world, which is in itself (because we are phylogenetically derived from it) a semiotic world" (1977: 164). A world that is "semiotic" because it is nothing but relations and contacts; semiosis does not come from nothing, but rather from these material contacts. Prodi uses scientific categories, yet his approach is the same as Vico's, if not Francis': the human is an earthly creature; there exist no qualitative hierarchies between different forms of life: "communicating does not mean intervening into extrasemiotic circumstances, but rather settling into [*calarsi*] the world that itself is semiotic, insofar as it has generated us as readers; to lower ourselves factually and genetically, tracing the path of the formation of codes, and even before that the path of the formation of natural structures in their complexity" (1977: 164–5). In this sense the "animal model" is also active in Prodi; the stakes are overcoming dualism, and the only way to do this is to show that it has never existed—that just as the three ages of "gods, heroes, and humans" are always operative at the "same time"—so also in Prodi the originary does not cease to be vital even within the most complex and sophisticated forms of human communication: it is a matter of "pointing out that, with all the enormous complexity and at times inextricable facts of language on a human level, their interpretation must pass through the facts of elementary complementarity, because the very things of language have passed beyond, at a certain point in their development" (1977: 28). Human language never ceases to be an animal language. But this does not mean that there are no differences between the communicative dance of bees and the *Divine Comedy*; it means rather that the dualism between human and animal ceases, finally, to be operative. Animality means, fundamentally, nothing but this deactivation.

Animality goes back to the originary condition that precedes the establishment of this dualism. Even Agamben, finally, although perhaps he does not dedicate many reflections to the theme of animality,[16] locates a possible *human animality* precisely at the intersection between animal and human, between biology and culture, where both terms of this opposition disappear in indistinction. A life of this kind "is no longer human, because it has perfectly forgotten every rational element, every project for mastering its animal life; but [...] this life cannot be called animal either" (Agamben 2004: 90). Indeed, a life in this manner, like that of Francis and his brothers, is a human life, because Francis is a living being who unlike the birds is endowed with language; at the same time, it is no longer a uniquely human life, precisely because it has renounced the distinctive characteristic of human humanity, the dualism of mind and body: "To render inoperative the machine that governs our conception of man will therefore mean no longer to seek new – more effective or more authentic – articulations, but rather to show the central emptiness, the hiatus that – within man – separates man and animal, and to risk ourselves in this emptiness: the suspension of the suspension, Shabbat of both animal and man" (Agamben 2004: 92).

NOTES

1. Actually, such a work remains to be written, because if in one sense there is an abundance of minute research on the way in which individual authors talk about animals – research which is interesting, but too often lapses into a paternalistic and stereotypical image of the animal – what is lacking is a complete theoretical reconstruction of what we could call the "animal model" in Italian Thought. The thesis of this chapter, then, is that in this tradition of thought, a theoretical-philosophical notion of "animality" begins to develop. This is a notion that has much broader scope than a set of animals. Indeed, this marks a recurrent misunderstanding: there are many authors who speak of dogs and lizards, but they have no notion of "animality"; there are authors who never speak of animals in their writings and yet develop the notion of "animality." Therefore, the critical bibliography will be limited to the few pertinent contributions which take this approach.
2. Courtesy of Wikimedia Commons. For a general situating of the theme of the "animal" in the medieval world, see Frugoni (2018) and Zambon (2018).
3. The editors of the text suggest that this is perhaps a term specific to the dialect, "mulacchie," i.e., crows which have a brilliant black color.

4. See Sebeok and Rosenthal (1981), Russell (2010), and Kulick (2017).
5. On animals in the *Divine Comedy*, see Schildgen (2010).
6. Dante, Canto XXXI, 1–12. 935. Penguin, Kirkpatrick, 2013
7. Dante Canto XX, 73–78. Kirkpatrick, Penguin. 846
8. On animality in this tradition, see Bondi (2011).
9. The famous conclusion to Darwin's *The Origin of Species* cannot but come to mind here: "there is grandeur in this view of life, with its several powers, having been originally breathed into a few forms or into one; and that, whilst this planet has gone cycling on according to the fixed law of gravity, from so simple a beginning endless forms most beautiful and most wonderful have been, and are being, evolved" (Darwin 1859: 490).
10. It is not by chance that Darwin's final major work is dedicated to worms. See Darwin (1881).
11. The origin of this tradition is in Plato's *Timaeus*, and then again in Plotinus, who writes in the *Enneads*: "this universe is 'one living being encompassing all the living beings inside itself'" (Plotinus 2018: 452). Cf. Gregory 2006.
12. In this sense an otherwise extremely interesting book, such as Hieronymus Rorarius' *Quod animalia bruta ratione utantur melius homine* (ca. 1539?), which was widely circulated until well into the eighteenth century, actually falls within a sort of *ante litteram* "animalism," rather than the philosophical use of the concept of animality.
13. There is an expansive literature on this theme. A few fundamental treatments include Pagliaro (1961), Danesi (1993), and Trabant (1996).
14. On this theme, see Prete and Aloisi (2010).
15. For an overall reconstruction of Prodi's work, see Cimatti (2018).
16. On the relation between Agamben and animality, see Castanò (2018).

WORKS CITED

Agamben, Giorgio. 1998. *Homo Sacer: Sovereign Power and Bare Life*. Trans. Daniel Heller-Roazen. Stanford: Stanford University Press.

———. 2004. *The Open: Man and Animal*. Trans. Kevin Attell. Stanford: Stanford University Press.

———. 2014. *The Highest Poverty: Monastic Rules and Form-of-Life*. Trans. Adam Kotsko. Stanford: Stanford University Press.

———. 2015. *The Use of Bodies*. Trans. Adam Kotsko. Stanford: Stanford University Press.

Alighieri, Dante. 2013. *The Divine Comedy*, Robin Kirkpatrick. Ed. New York: Penguin Random House.

Bloom, Paul. 2005. *Descartes' Baby: How the Science of Child Development Explains What Makes Us Human*. New York: Basic Books.

Bondì, Roberto. 2011. Telesio, Campanella e gli animali. In *Passaggi. Pianta, animale, uomo*, ed. Berenice Cavarra and Vallori Rasini, 145–164. Milano: Mimesis.

Bruno, Giordano. 1998. *Cause, Principle and Unity: And Essays on Magic*, ed. R. Bruno, Cambridge: Cambridge University Press.

———. 2000. *Dialoghi filosofici italiani*. Milano: Mondadori.

———. 2002. *Cabala of Pegasus*. Trans. and Annotated by Sidney L. Sondergard and Madison U. Sowell. New Haven: Yale University Press.

Campanella, Tommaso. 2007. *Del senso delle cose e della magia*. Roma-Bari: Laterza.

Caroli, Ernesto, ed. 1990. *Fonti Francescane: Scritti e biografie di san Francesco d'Assisi. Cronache e altre testimonianze del primo secolo francescano*. Padova: Edizione Messaggero.

Castanò, Ermanno. 2018. *Agamben e l'animale. La politica dalla norma all'eccezione*. Roma: Novalogos.

Cimatti, Felice. 2020. *Unbecoming Human: Philosophy of Animality After Deleuze*. Edinburgh. Edinburgh University Press.

———. 2014. Linguaggio e immanenza. Kierkegaard e Deleuze sul *divenir-animale*. *aut aut* 263: 189–208.

———. 2018. *A Biosemiotic Ontology: The Philosophy of Giorgio Prodi*. Berlin: Springer.

Danesi, Marcel. 1993. *Vico: Metaphor and The Origin of Language*. Indianapolis: Indiana University Press.

Darwin, Charles. 1859. *On the Origin of Species by Means of Natural Selection*. London: Murray.

———. 1881. *The Formation of Vegetable Mould, Through the Action of Worms*. London: Murray.

Deleuze, Gilles. 1994. *Difference and Repetition*. Trans. Paul Patton. New York: Columbia University Press.

———. 2001. *Pure Immanence: Essays on a Life*. Trans. Anne Boyman. New York: Zone Books.

Deleuze, Gilles and FelixGuattari. 1986. *Kafka: Toward a Minor Literature*. Trans. Dana Polan. Minneapolis: University of Minnesota Press.

———. 1994. *What Is Philosophy?* Trans.Hugh Tomlinson and Graham Burchell III. New York: Columbia University Press.

Eco, Umberto. 1979. *A Theory of Semiotics*. Bloomington: Indiana University Press.

Esposito, Roberto. 2008. *Bíos: Biopolitics and Philosophy*. Trans. Timothy C. Campbell. Minneapolis: University of Minnesota Press.

———. 2012. *Living Thought: The Origins and Actuality of Italian Philosophy*. Trans. Zakiya Hanafi. Stanford: Stanford University Press.

Frugoni, Chiara. 2018. *Uomini e animali nel Medioevo: Storie fantastiche e feroci*. Bologna: Il Mulino.

Granatella, Mariagrazia. 2015. Imaginative Universals and Human Cognition in *The New Science* of Giambattista Vico. *Culture & Psychology* 21 (2): 185–206.

Gregory, Tullio. (2006), "Anima del mondo" Bruniana & Campanelliana 12.2, pp. 525–535.

Hauser, Marc, Charles Yang, Robert Berwick, Ian Tattersall, Michael Ryan, Jeffrey Watumull, Noam Chomsky, and Richard Lewontin. 2014. The Mystery of Language Evolution. *Frontiers in Psychology* 5: 1–12.

Kulick, Don. 2017. Human-Animal Communication. *Annual Review of Anthropology* 46 (1): 357–378.

Leopardi, Giacomo. 1983. *The Moral Essays: Operette Morali.* Trans. Patrick Creagh. New York: Columbia University Press.

———. 1996. *Canti.*. Selected and Introduced by Franco Fortini. Trans. Paul Lawton. Dublin: UCD Foundation for Italian Studies.

———. 2015. *Zibaldone.* Trans. Kathleen Baldwin, Richard Dixon, David Gibbons, Ann Goldstein, Gerard Slowey. Martin Thom, and Pamela Williams. Ed. Michael Caesar and Franco D'Intino. New York: Farrar, Straus and Giroux.

Pagliaro, Antonio. 1961. Lingua e poesia secondo G.B. Vico. In *Altri saggi di critica semantica.* Messina-Firenze: D'Anna.

Pico della Mirandola. 2012. *Oration on the Dignity of Man.* Ed. and Trans. Francesco Borghesi, Michael Papio, and Massimo Riva. Cambridge: Cambridge University Press.

Plotinus. 2018. *The Enneads.* Ed. Lloyd P. Gerson and Trans. George Boys-Stones, John M. Dillon, R.A.H. King, Andrew Smith, and James Wilberding. Cambridge: Cambridge University Press.

Prete, Antonio, and Alessandra Aloisi, eds. 2010. *Giacomo Leopardi, Il gallo silvestre e altri animali.* Lecce: Manni.

Prodi, Giordio. 1977. *Le basi materiali della significazione.* Milano: Bompiani.

Russell, Nerissa. 2010. Navigating the Human-Animal Boundary. *Reviews in Anthropology* 39: 3–24.

Schildgen, Brenda Deen. 2010. Animals, Poetry, Philosophy, and Dante's Commedia. *Modern Philosophy* 108 (1): 20–44.

Sebeok, Thomas, and Robert Rosenthal eds. 1981. The Clever Hans Phenomenon: Communication with Horses, Whales, Apes, and People. *Annals of the New York Academy of Sciences* 364.

Telesio, Bernadino. 2009. *La natura secondo i suoi principi.* Milano: Bompiani.

Trabant, Jurgen. 1996. *La scienza nuova dei segni antichi: La sematologia di Vico.* Bari-Roma: Laterza.

Vico, Giambattista. 1984. *The New Science.* Trans. Thomas Goddard Bergin and Max Harold Fisch. Princeton: Princeton University Press.

Zambon, Francesco, ed. 2018. *Bestiari antichi e tardomedievali.* Milano: Bompiani.

CHAPTER 3

Aldo Capitini, Animal Ethics, and Nonviolence: The Expanding Circle

Luisella Battaglia

Born in Perugia in 1899 into a family of modest social conditions, Aldo Capitini initially took up technical studies for economic reasons, later pursuing literary studies as an autodidact. In 1924 he won a scholarship at the Scuola Normale Superiore in Pisa, where he would later be appointed secretary in 1930. During this period, he chose to embrace vegetarianism for ethical reasons. In 1931 he was among the 14 university professors who refused to take the Oath of Fidelity to Fascism and for this reason was removed from teaching. His commitment to anti-fascism, for which he was twice imprisoned, continued until the fall of the regime. After joining the Action Party, where, together with Guido Calogero, he represented the liberal-socialist wing, in 1944 he tried to carry out an experiment of direct democracy and the decentralization of power, founding the first "Social Orientation Centre" (Centro di Orientamento Sociale, COS) in Perugia, a political space open to the free participation of citizens.

After World War II, Capitini became rector of the University for Foreigners of Perugia, a position that he was forced to abandon because

L. Battaglia (✉)
University of Genoa, Genoa, Italy

© The Author(s) 2020
F. Cimatti, C. Salzani (eds.), *Animality in Contemporary Italian Philosophy*, The Palgrave Macmillan Animal Ethics Series,
https://doi.org/10.1007/978-3-030-47507-9_3

of the strong pressure of the local Catholic Church. During this time, he promoted a series of activities for the recognition of conscientious objection; in 1950, he organized the first Italian conference on the subject in Rome. In 1952 Capitini added the "Religious Orientation Centre" (Centro di Orientamento Religioso, COR) to the Social Orientation Centre, in order to promote the knowledge of universal religions. He became chair of Pedagogy at the University of Cagliari in 1956 and then in 1965 at the University of Perugia. In 1961 he organized the "March for Peace and Brotherhood Among Peoples" from Perugia to Assisi, a nonviolent procession that is still carried out every 2 years by pacifist associations. In the final years of his life, he founded and directed the periodical *Il potere di tutti* (*The Power of Everyone*), inspired by the principles of *omnicrazia* (omnicracy), the widespread and delocalized management of power opposed to the centralism of the traditional political parties, as well as the monthly *Azione nonviolenta* (*Nonviolent Action*), a part of the "Movimento nonviolento per la Pace" (Nonviolent Movement for Peace). He died in Perugia in 1968.[1]

1   OPEN RELIGION

A heterodox thinker within Italian philosophy, Capitini lives as a foreigner in his time, but writes for future times. As we will see, this is particularly true for his complex vision of animal ethics in which his vocation of "nonviolent revolutionary" is expressed and which has earned him the definition of "Italian Gandhi." For Capitini, nonviolence is not an ethics of intention, aimed at safeguarding the intangible purity of principles, but rather, as in Gandhi, an ethics of responsibility, attentive to the consequences of acting and therefore aware of the inevitable gradualness in the pursuit of aims. The reference to the context and situation in which our moral decisions are placed certainly invites us to a realistic consideration of our possibilities ("doing better than we can"), but above all it urges our "creativity," that drive to "do more" that should nourish our commitment to an "open" ethics. As we read in "Religione aperta" (Open Religion):

> Until now, the animal field has been regarded as a free field, where one could carry out massacres; nonviolence ushers in the plan of an agreement with the animal field that can go very far. [...] Towards non-human beings, nonviolence also has great value, precisely as an extension of love and collaboration. (Capitini 1998: 556)[2]

"Open religion" is the attitude of openness toward a "you"; it is a "calling you" (*dire tu*)—an expression that refers to an ethics of recognition that identifies in the other a "you," a person and not a thing—in every human and non-human being: the neighbor. To a religiosity traditionally conceived as acceptance, as patient resignation, Capitini opposes a vision that has at its core the radical rejection of everything that is morally wrong, and therefore the indignation at the scandal of suffering, the commitment to correct and modify situations in which "others suffer while I do not suffer, or suffer more than I suffer" (Capitini 1990: 37). An "open religion" must not accept a reality in which some individuals lose out. This is an explicit invitation for everyone to take one's own responsibility, not to tolerate the pain of others as more bearable than our own pain, not to accept reality as it is now, wherein force, power, and arrogance prevail everywhere: "this kind of reality does not deserve to last" (Capitini 1990: 38).

The conflict between what is "natural" and what is "right" reveals the deep utopian tension, the permanent protest of the Capitinian discourse. If human beings have the opportunity to transcend nature, rather than be predators among others, they must recognize themselves as the only beings able to put themselves on a different plane with regard to the "struggle for life" and therefore as able to replace the natural law with a moral law: ultimately, this means restoring one's full humanity.

The duty to respect the lives of non-human beings is also accompanied by the ability to imaginatively identify with them. This ability allows us to take an interest in their needs and their destiny, arousing a feeling of commonality and solidarity similar to what we have for humans. The pain of animals is mute and innocent. These are the traits—the most immediately evident and at the same time the most disturbing—which question (or should question) philosophy and theology. Why does animal pain exist? What are its reasons? What are its justifications? If we assume as possible interpretative categories for the relationship between humans and animals (at least as far as the history of Western philosophy is concerned) two antagonistic models, *domination* (inspired by a vision that, since animals lack reason and immortal soul, consigns them to the sphere of means at the service of man) and *fraternity* (inspired by a vision that, instead of isolating man as being apart from other creatures, affirms the kinship between all living beings, beyond the frontiers of species), then within the first model, dominant in our culture, we find three main answers:

1. *Denial.* Animals do not "feel"; they are machines or automatons lacking any sensitivity. Such is the well-known Cartesian mechanistic thesis, taken up in particular by Malebranche to refute the opinion "in conformity with prejudices" that they have the capacity to suffer. Any empirical evidence of animal pain is therefore considered fallacious on the basis of the argument that God would never allow the suffering of innocent creatures.

2. *Minimization.* Animals do "feel," but have no rational affinity with us. As unreasonable creatures subordinate to man, they cannot be understood as neighbors. Therefore, only indirect duties exist toward them. Consequently, what follows is the so-called thesis of cruelty, formulated both by Thomas Aquinas and by Immanuel Kant, according to which man must show goodness of heart toward animals, because whoever is cruel to them becomes equally insensitive toward men.

3. *Valorization.* Animals "feel" like us, but their innocent suffering takes on a symbolic value, finding its "justification" in the sign of redemption from the sin of Adam. This is the dominant thesis in the Christian tradition of the sacrificial lamb, symbol of Christ.

Animals would therefore be creatures of a world that does not concern us, beings condemned to the darkness of silence because their pain is not in full light, and does not cause scandal because it does not offer itself to our gaze? Or perhaps it is our gaze that turns away? But do we really see their pain?:

> When I see an animal that writhes in pain and dies, I can stretch out and try to be close to it [...]. It is not a question of the simple empathy one can feel when seeing an animal suffering – there is something more. There is a closeness to it, coming from the innermost self, that contributes to a feeling of co-presence. (Capitini 1950: 188)[3]

When we truly love someone, it is not possible to ignore their suffering, and if love embraces all creation religiously, the commitment must be to eliminate suffering wherever it occurs. In "Open Religion" Capitini writes:

> Above all, it will be a question of a constant and enlightened work to alleviate every pain wherever we find it, to wrest living beings from suffering [...]. Pain must be faced universally, turning to those in suffering, seeking ways to

alleviate it, being inspired by the purpose of not inflicting it and with the conviction that pain does not need to exist. (Capitini 1998: 531–32)[4]

On reflection, we see in reality only what we are willing to see, while we often do not even see what is evident. It is culture that puts greater or lesser stress on pain, sometimes making it even invisible and unspeakable, removing it or denying it. Is it a coincidence that the first animal rights theorist, Henry Salt, was also a defender of the rights of slaves, women, and children, of all the oppressed, beyond differences of race, sex, and species? The experience of pain is common and shared: only if recognized in its different "faces" can it be fought.

This, then, is the reason for insisting on the importance of each individual, of the individual *you*, all seen as subjects and not as objects, in clear opposition to the traditional theological vision which asserts that, with regard to animals, God takes care of the species but not of the individual beings and destines all non-human species to human utility. The individuality of every singular existence is therefore what must be safeguarded: "I do not care about the dog species (as it is described in the zoology books)," Capitini objects in "Vita religiosa" (Religious Life), "but rather I care much, much more about this dog identified here, alive and also dead, as it is remembered by others or by myself" (Capitini 1998: 97).[5] But what matters most is that, beyond any difference between different beings, it is always possible to see a similarity. As he states in "La realtà di tutti" (Everyone's Reality): "The dissimilarity of others does not overcome the similarity that I discover, similarity even to the body of every being, animal, plant, thing" (Capitini 1998: 97).[6]

2   Vegetarianism and Nonviolence

It is against this background that Capitini's choice of vegetarianism—at that time an object of curiosity and infinite ironies, including in academic circles—should be situated. As in Gandhi's case, for Capitini this choice represents the consequent outcome of the philosophy of nonviolence. The rejection of the meat diet corresponds to the global rejection of violence. It is a "saying no" to a multiplication of violent acts, a recognition of the value of the existence of every being, and at the same time a conviction that violence against humans should not be used. We can identify three essential elements in Capitini's vegetarianism: an ethical conception, a theory of human nature, and a vision of the social world—which together constitute a highly articulated philosophical vision:

1. The ethical conception refers to an "open morality" in the Bergsonian sense, capable of overcoming all species selfishness. Vegetarianism is the most radical application of the golden rule, because it involves the decisive enlargement of the notion of neighbor with the inclusion of non-human animals as living beings capable of experiencing the pain of violence. In *Elementi di un'esperienza religiosa* (*Elements of a Religious Experience*, 1937), we read: "What takes place in vegetarianism (that is, in the choice of not eating the meat of slaughtered animals, but rather the products of the earth and those derived from animals without killing them) is essentially the recognition of the value of the existence of those animal beings against which one decides not to wage murder and, by extension, a stronger persuasion that violence against human beings should not be used is also accomplished" (Capitini 1990: 53).[7] The rule of respecting animal life brings with it a greater attention to human life and makes indifference and cruelty toward human beings even more difficult. From this perspective, vegetarianism means to say no to a multiplication of violent acts and to initiate oneself into a different education.

2. The theory of human nature emphasizes the aspects of "continuity" and similarity between humans and non-humans in order to define the relationships between different species in terms of collaboration rather than antagonism. In Capitini we find the recovery of a classical tradition which recognized itself in the model of *kinship* or *fraternity*, in opposition to the anthropocentric model, and that ideally brings together Plutarch and the Neoplatonists, Auguste Comte and Jules Michelet. This tradition insists on the possibility of a communication between species, on the role of animals as "auxiliaries of man," and appeals to human responsibility in the enhancement and "upgrading" of the animal world.[8] The Kantian division between persons (belonging to the realm of ends) and animals (belonging to the realm of means) is critically revised and understood in an evolutionary sense: moral and social progress consists precisely in "expanding the circle" of people, including those who once were considered mere "means"—slaves, women, and now animals. With regard to the issue of distinguishing between "ends" and "means," Capitini explicitly refers to the Kantian distinction to affirm that it is

not at all a fixed and eternal division, because progress lies precisely in expanding the circle of what is an end (and Capitini argued for this way before Peter Singer [1981]!) Slaves—for example—were once simply a means, but now, after the abolition of slavery, they have become an end: "This is what happened with legal slavery [...], this is what happened with anthropophagy, and so it can also happen with carnivorism" (Capitini 1959: 4).[9]

3. The vision of the social world opens up a palingenetic perspective, coinciding with the end of human violence upon humans and animals. Vegetarianism is understood in a political sense as an extension of solidarity beyond the barriers of species and in relation with the choice of a method of political and social action. Hence the idea of a just society not only for human beings but for all living beings:

> With the meat diet we exploit a large group of living beings, and we exploit them in the most radical way, because our actions towards them are all aimed at using them, ending their lives. Now, while we strive to remove social exploitation between man and man, rising up a degree towards mutual respect, it is already something to rise up a degree for many living beings and use their products (it can be a form of collaboration, while waiting for further and less restrictive forms) without destroying their existence. A socialist feels compelled to be a vegetarian: are not animals also a subaltern and oppressed class? ("Lettera di religione n.18," in Capitini 1969: 258)[10]

On this basis, how can we define Capitini's animal ethics? What elements characterize it? What is its framework? Referring to his general ethical-philosophical vision, animal ethics is configured as a coherent application of the religious principle of nonviolence, to be understood positively—as it has been pointed out—as love, dedication, absolute closeness to all creatures. To the order of nature founded on the law of the strongest, Capitini opposes the persuasion of nonviolence which refuses to accept such an order as unsurpassable and brings love and respect for life as much as possible into the depths of nature itself.

Religious experience is, with a strong utopian tension, the struggle to the limit, the dissatisfaction with the insufficiency of the reality in which we live, of society for its unfreedom and injustice, which is expressed religiously in the tension toward its end and in view of a new man, a new reality, a new society. Capitini deeply senses the drama of limitations experienced by every being who suffers in its "incarnation-crucifixion" in the world: this applies both to human and animal suffering. His project is therefore inspired by the liberation of animals, but at the same time also strives—and this is its peculiarity—to their elevation through the work of human beings, following an inspiration already present in the work of Jules Michelet. The latter's idea of the "beautiful universal city" (2018: 8) appears in many respects close to the Capitinian utopia of a "liberated reality that includes everyone" (1959: 3):

> Were we to see them this way, animals would be enclosed within an inferior destiny, surpassed by us and unable to reach our level and to be free with us. But in contrast to this *closed* way of considering them, there is an *open* way, which considers every being as the beginning of an opening and of further possibilities [...]. Hence a different way of leading the struggle: either animals are like a leftover from the past, or they have joined us in a destiny of liberation. In the latter case, we take initiatives of hope, of approach, of language, of harmony, with the conviction that, even if at the present moment they are not exactly aware of the horizon towards which they can go, eventually they will be. (1959: 4)[11]

For Capitini, it is a matter of making animals move as much as possible "into the circle of our life," promoting a collaboration by which they bring their work into the common life: "We ask them to cooperate with us and we will study what they can do best, so that even for them life will be an orderly fulfillment of a job, as it is for us" (1990: 37).[12] In this way—which can be considered a true and proper cosmic ethics—one does not see the other as a means, an instrument, a thing, but as an end:

> Projecting beams of new light, of attention and friendship onto categories of beings considered once as means, seeing them also as ends, as beings which collaborate and have rights: this diminishes the extent of the empire of our arbitrariness. But a thousand things that we once obtained by command now we will get by cooperation. At the end of this road lies the ideal of a reality in which nothing is only a means, a thing, an instrument, but everything is a subject and an object of love. (1959: 5)[13]

In Capitini we find a "classic" theme of animal philosophy, *emancipation*, introduced in the nineteenth century by Henry Salt in *Animals' Rights: Considered in Relation to Social Progress* (1894), which inseparably links human civil progress and animal rights. The new element is that the revolt against all oppression and exploitation—which aims at liberation under-stood in the broadest sense, not only in society but also in nature—has a declared religious foundation, based on an open and progressive practice: nonviolence. In this way the initiative in favor of animals is part of the social-religious liberation that takes place—in Capitini's words—through:

1. the theoretical removal of the barrier that closes them within fixed and unchangeable limits;
2. prompting collaboration in various ways, including technical ones;
3. the progressive respect of their existence, gradually reducing violence and killing (non-violence is a value like music or goodness: these are not com-mands, and they do not run out; one can do one's best at that moment, aiming to do better). No one can say *you* in an exhaustive way. ("Lettera di religione n. 13," in Capitini 1969: 239–40)[14]

In this sense, Capitini can be considered an authentic precursor of con-temporary animal ethics for his anticipatory awareness of the ethical and political significance of the man-animal relationship and for having inau-gurated the debate on an ignored theme within Italian culture in the 1940s and 1950s: the case for animal rights—which today plays a signifi-cant role in bioethical reflection. His lesson proves to be of particular importance both for its opening up animal philosophy to nonviolence—and to the enhancement of an inseparable link between the two move-ments—as well as for the recovery of philosophical roots of different (existentialist-religious) inspiration. Establishing a precise analogy between oppressed groups of humans and animals, Capitini anticipates a theme that will be typical of liberationism in the 1970s, from Richard Ryder to Peter Singer and Tom Regan. If Ryder coined the term "speciesism," Capitini introduces in Italian philosophy the idea that a difference of species alone cannot provide an ethically defensible basis for giving the interest of one individual more weight than the interest of another. It is a remarkable transformation of thought, sufficiently radical to be aptly described as revolution. Convinced as Ryder that selfishness is the antithesis of moral-ity, Capitini anticipates Ryder's "painism" (2001; 2017) by stressing, as an antidote to this poison, the importance of others, of every other individual who can feel pain. We are, after all, all members of the same community of pain. Animals are in this sense persons and so deserve all the respect and rights of personhood.

This constitutes an authentic turning point in interspecies ethics which, in its effort to expand the circle of the moral community to the entire living world, has insisted on the theme of respect, interpreted in a utilitarian way—as in Peter Singer—or in a natural law way—as in Tom Regan. On the basis of the assumption that "all animals are equal," the former (1975) calls for the equal consideration of the interests of all sentient beings, regardless of the species to which they belong, and the latter (1983) argues for natural rights, applying both to humans and non-humans as "subjects-of-a-life" endowed with inherent value. In this reference to the demands of impartiality and universalizability, interspecies ethics takes an essential step toward the construction of a rigorous theory, a step that—let us not forget—was taken by Capitini as early as the 1950s—a true prophet unheard in Italian philosophy of what we now call "animal bioethics."

The novelty of Capitini's position lies in the religious inspiration of this tension toward enlargement that is linked to a palingenetic perspective of the moral regeneration of humanity, characterized by the end of violence and the exploitation of man over man and man over animals. The moral life inspired by vegetarianism is also animated by a perpetual tension that leads to extending the principle of respect for life in order to seek, once carnivorism has been refused, whether it is also possible to spare the life of plants: "Science progressively provides materials that will allow to avoid the destruction of plants [...]. If we cannot do everything, much can certainly be done, and we must act: we are indeed late" (Capitini 1990: 36).[15] The reason why the life of plants has remained mysterious for so long is perhaps simply because the way in which plant organisms perceive things is very different from that of animals. Can we speak—one begins to ask today—about "plant neurobiology"? The current debate between scholars of neurobiology, ecologists, and chemists focuses on an ancient question: which of the abilities and attributes that scientists have long considered an exclusive prerogative of animals, such as perception, learning, and memory, can be correctly transferred to plants? (cf., e.g., Mancuso 2017). Capitini anticipated and was sensitive to this attention toward the entire living world. In all his works it is possible to find a real manifesto, in which the strong condemnation of any exploitation opens up to themes of contemporary interspecific justice:

> One could ask the question whether animals have rights or have duties, as it is for any moral life. We have already seen that the problem of the relationship between humans and animals is evolving, and one step is certainly

vegetarianism. In many codes, some rights have been recognized to animals, for example the right of not being tormented except in certain conditions. The development will continue. But what we must note is that these steps are generally achieved on human initiative; and this also explains why it is so difficult to talk about *duties* on the part of animals, even though sometimes this has been done and even trials have been set up. This is the field where we will most have to work and experiment and create; and perhaps we will reach more precise results than we imagine. (1959: 6)[16]

3   An Ethics of Care

Here we have the intuition of the fundamentally asymmetric nature of our relationship with animals and hence the need for an ethic that reflects this relationship by assuming full responsibility for it, such as the ethic of care, which educates us to work not only when there is exchange in a society of equals but also when we must be at the center of a work done without receiving anything in return, as well as without calculating the results. For this reason, we might ask whether it is possible to interpret Capitini's concept of nonviolence—understood, as we have seen, in its positive sense of love for the entire living world—by placing it in relation to the idea of care. Situating Capitini's thought within the context of the contemporary reflection on care allows us, on the one hand, to deepen the concept of nonviolence, going beyond a purely negative determination in order to provide a positive meaning, and on the other, to enrich the area of the concept of care beyond a semantic point of view, opening it to new perspectives and dimensions:

> Nonviolence, in its essence, refers precisely to this principle of free personal decision, of loving kindness towards the other being, whatever their conduct. It is the practical proposal of a new way, in lieu of indifference or hatred; the offering of a *you* that can become deeper and deeper. It is useless to pompously cite from Eastern or Western thought in this regard. It is enough to insist on the character of this "openness" addressed to a being regardless of their condition, race, and also conduct. Love and help: the act of the Samaritan. Vegetarianism means addressing a group of non-human beings by taking the initiative to establish a relationship of openness and no longer of indifference or cruelty. (Capitini 1959: 4–5)[17]

The ethics of care is one of the most significant areas of philosophical and moral research of our time. In the ethics of care, fundamental religious perspectives, such as those of Emmanuel Levinas, as well as more secular approaches within the new theoretical concerns that have emerged in feminist thought in recent decades, intersect and, at least in part, overlap. Just think of Carol Gilligan's account of the importance of caring and the "different voice" in moral discourse (cf. Gilligan 1982; also Donovan and Adams 1996, 2007). In a general sense, the expression *taking care* refers to a plurality of meanings that all seem to refer to a fundamental attitude of availability toward the other, an attitude that arises from the recognition of an essential and constitutive interdependence and translates into a serious commitment to understand their real situation of need and to take responsibility for it. *Care* can therefore be defined as the solicitude for the fate of another individual (akin to Hans Jonas' idea of "altruistic fear," which he argued testifies to the apprehension of the vulnerability and fragility of other beings, the concern for their threatened existence; cf. Jonas 1984), which is supported by the most adequate possible knowledge of their reality, experiences, and needs. The different meanings of care therefore share a fundamental element: that of concern for the well-being of another being.

In Capitini we find this sense again when he proposes to concretely clarify the meaning of the term "love" by specifying: "As a profound interest of the soul, of the heart, towards a living being, it is connected with nonviolence, which is openness to the existence, freedom, and development of every being" (Capitini 1961: 175). According to Milton Mayeroff's definition of "care," "to take care of another person, in the most significant sense, is to help him or her to grow and to realize himself or herself" (1971, 1). I think this definition aligns with Capitini's vision of an interspecies ethics committed not only to ensuring the welfare but also to respecting the dignity of the individual animal, to making it grow by enhancing its capabilities. Once again, the reference to nonviolence is central, understood as a practice that translates into "openness to every being, desire, work and joy for its existence, freedom, and development, with the intention of always increasing it in intensity and breadth, from those near to us to those who are far away" (1961, 175).[18]

When we speak of "care" with reference to the animal world, it is appropriate to distinguish two fundamental senses of the term. The first, which should be understood in a more general sense, refers to an attitude of availability and openness toward non-human beings, considered in their vulnerability, and on this basis, elaborates a code of duties of a universal nature, even in the consideration of specific differences. The second sense of the term should be understood more specifically and favors the relations that bind us to particular animals, especially "familiar" ones. On this basis (that of an affective and sympathetic nature), the second sense identifies a series of direct responsibilities toward these particular animals. This is, for example, the meaning we find in the philosopher Raimond Gaita (2003), according to whom we feel deeply involved with animals that live in close contact with us, but much less with those that are far from our lives. Capitini's conception appears to be in line with the first—universalistic—meaning of care, in fact developing in the direction of a cosmic ethics. At its center lies the notion of vulnerability, which, in a broad and general sense, concerns the very precarious condition of all living beings, both human and non-human, who are exposed throughout their existence to the risk of being injured. One could therefore affirm that the basic rationale for Capitini's position—both on the religious and political front—can be identified in the theme of care: to take up the words he used in "La mia persuasione religiosa" (My Religious Persuasion), the affectionate revaluation of those suffering, of the least and the last, is translated, on the political and social level, into the affirmation of the need for justice (Capitini 1998: 34).

In fact, one cannot sensibly claim that one should take care of someone if one is not willing to understand them, to respect them, to worry about them, to commit oneself personally to their well-being, to work to reduce, as far as possible, their suffering, especially if the individual in question does not know or is not able to do so. Discovering what causes the suffering of a living being and how to respond to their needs requires, first of all, an exercise of *attention*. In revealing the other to me, attention also reveals the existence of an asymmetry of strength and power and, therefore, places me in front of responsibilities and duties that I did not see before but to which I now feel called to respond, without the depersonalized scheme of a role or an institution. In this way, a characteristic element of the ethic of care emerges, well present in Capitini's vision: *asymmetry*, that is, my being responsible for the other, for their good and their well-being, without expecting anything in return. *The other* asks me and forces me to take on an irrecoverable and asymmetrical responsibility—ethically, the highest one—since there is and can be no reciprocity.

In modernity, the tradition which has indelibly marked ethical-political thought is one whose essential terms are the social contract and individual rights and whose philosophical background is inspired, fundamentally, by a liberal and atomistic model. This model—which has historically acquired many merits and certainly possesses great qualities—seems, however, to neglect many dimensions of existence. In particular, as some authors have pointed out, it risks ignoring the variety of relationships with which people take responsibility and care of others. Reciprocity is valid in the exchanges between equals and corresponds to the philosophy of the contract, but the assumption of responsibility, in its deepest sense, disregards—and Capitini is well aware of this—any confirmation of this kind. If we assume non-reciprocity as a guiding thread of ethical reflection, we can recognize—going beyond the human sphere—the proximity of the silent gaze that questions us and that directly appeals to our moral conscience (it does not matter which species it belongs to), in an encounter with the other that verifies our justice toward them without any claim of reward. In its more general features, an ethics of care differs from an ethics of rights because it gives priority to relationships with others rather than the rights of individuals and is based on a league of interdependence rather than on a contract between equals.

Now, in an ethic extended to the community of living beings, *non-reciprocity* seems to be the rule. How, in fact, can animals reciprocate? Their impossibility to reciprocate does not seem, on the other hand, a valid argument to justify their exclusion from our moral universe. If non-humans live in the same ethical world as us, then we cannot use the old conceptual tools that provided the reciprocity of rights and duties proper to rational human subjects: "Why exclude non-human beings?" Capitini asks. "Because, it is said, they are not rational. But if we have something more than them, this just means more duties, more openness" ("La compresenza dei morti e dei viventi" [The Co-presence of the Dead and the Living], in 1998: 270).[19] Non-human access to ethical territory enables us to recover asymmetry as one of the essential dimensions of moral discourse, which is a dimension that has often been overlooked because of the prevalence of the contractual model. The asymmetry highlights the gratuitousness of ethical behavior, not waiting for anything in return, the overcoming of the logic of the transaction. The decisive point is that in this encounter the animal is not "anthropomorphized," through an always inappropriate, artificial process, on both an epistemological and an ethical level. It is not in fact a question of its conscience or self-awareness, but only of my conscience as a moral agent. Along this path, it seems possible not to be forced to limit to the human world the encounter with those others who present themselves to us as fragile and vulnerable, addressing a call for recognition to us.

For these reasons, care seems particularly suitable for constituting a bioethical paradigm of relations with the non-human world. It is a matter of elaborating, following Capitini's suggestions, a strong and constructive interpretation of the concept of care, not as a simple appeal to good feelings, but as a responsible commitment toward other non-human beings. The call to bioethical responsibilities cannot, in this way, disregard the concrete and resolute commitment that encounters more and more appropriate correspondence in positive regulation to defend animals. This is also Capitini's concern, and his originality lies, in my opinion, in the effort to show full compatibility between the theme of animal rights—of which he was a forerunner and a convinced supporter—and the theme of care, as a direct human responsibility in the struggle for their affirmation. He is well aware that affirming a right is equivalent to pointing out an injustice, often not provided for by the law: it is within a dynamic vision of justice that rights are an important, if not indispensable, instrument. It is not enough, however, to limit oneself to assigning rights to subjects who are, by their very status, at the mercy of others. "Taking care of the rights" of those subjects—such as animals—who do not have the possibility to claim them: this is still today the ethical and political challenge that Capitini's thought poses to us.

I believe that bioethics can take advantage of these new orientations of thought, particularly with regard to issues such as the relationship between humans and animals. I am thinking in particular of the "Barcelona Declaration"—written by 22 European scholars from different disciplines and philosophical horizons—which solemnly affirmed the new principles that bioethics should be inspired by: *autonomy*, *dignity*, *integrity*, and *vulnerability* (Kemp and Rendtorff 2008: 240). These four regulatory ideas are useful for the construction of a truly global bioethics, within the framework of an ethics of responsibility that clearly goes beyond the anthropocentrism of traditional morals. Integrity concerns the "coherence of life" of beings to whom an irreducible dignity is recognized and who cannot be offended. If the privileged reference is still the human person, the possibility of integrating the "coherence of animal and vegetable life" into this notion is nevertheless affirmed. In fact, as Capitini had guessed, integrity is a notion that can and must also be applied to other living beings. But it is in particular the principle of vulnerability that forms the basis, in the Declaration, of a public ethics of care that is intended to be built on the anthropological premise that we are all fundamentally vulnerable. This principle, which essentially expresses the idea of finiteness and the constitutive fragility of human existence, is the basis of the possibility and necessity of any moral discourse and any ethics that calls for responsibility and care. It specifically entails the duty to assist those who are incapable of realizing their human potential and who see their right to autonomy, integrity, and dignity threatened.

It seems to me worthy of the greatest attention that the final part of the Declaration states: "6. The application of guiding ideas should not be restricted to the human sphere; dignity, integrity and vulnerability might also be considered as a basis for legislation and legal practice in relation to animals, plants and the environment" (Kemp and Rendtorff 2008: 248). In this way bioethics would effectively become an ethics for the living world, realizing that vocation to which, from the beginning, its own etymology destined it. It should perhaps be said "bioethics returns to" instead, since this was the initial project of Van Rensselaer Potter (the inventor, it will be remembered, of the neologism "bioethics"), who explicitly referred to the ecological ethics of Aldo Leopold as a precious legacy to draw upon (cf., e.g., Potter 1988). The ethics of care should be considered extended beyond the human sphere, but as an expression of a fundamental "humanistic" vocation. The interspecies relationship—this is Capitini's fundamental lesson—is thus configured as an expanding circle beyond the human: man looks beyond himself; he opens himself to other worlds, but in order to realize his own humanity.[20]

NOTES

1. For some recent introductory literature (in Italian) on the life and work of Capitini, see, for example, Altieri (2003), Cavicchi (2005), Curzi (2004), Foppa Pedretti (2005), Polito (2001), and Tortoreto (2005).
2. "Finora si è considerato il campo animale come un campo libero dove uno potesse portare stragi; la nonviolenza inizia il piano di un accordo col campo animale che potrà arrivare molto lontano. […] Anche verso gli esseri non umani la nonviolenza ha un grande valore, appunto come ampliamento di amore e di collaborazione."
3. "Nel vedere un animale che si torce dal dolore e muore io posso tendermi alla vicinanza con lui […]. Non si tratta di semplice empatia che si può provare nel vedere un animale che soffre, c'è qualcosa di più, c'è una vicinanza dall'intimo che dà un contributo alla compresenza."
4. "Sarà da fare anzitutto opera costante e illuminata per alleviare ogni dolore dovunque lo si veda, per strappare gli esseri viventi alla sofferenza […]. Il dolore va affrontato universalmente, volgendoci agli esseri che lo soffrono,

cercando i modi di alleviarlo, ispirati dal proposito di non infliggerlo e con la convinzione che non è necessario che il dolore esista."

5. "Non mi interessa la specie cane (come è descritta nei libri di zoologia) quanto, molto, molto di più questo cane qui individuato, vivo e anche morto, come è ricordato da altri o da me."

6. "La diversità altrui non vince la somiglianza che scopro, somiglianza anche col corpo di ogni essere, animale, pianta, cosa."

7. "Col vegetarianesimo (cioè non nutrendosi della carne di animali macellati, ma di prodotti della terra e di derivati dagli animali, ma senza ucciderli) si realizza principalmente il riconoscimento del valore dell'esistenza di quegli esseri animali contro i quali si decide di non usare l'uccisione e, di riflesso, si realizza una maggior persuasione che non si debba usare violenza contro gli esseri umani."

8. Today a similar position has been (in part) developed by Donaldson and Kymlicka (2011).

9. "Così è avvenuto per la schiavitù giuridica [...] così è avvenuto per l'antropofagia e così potrà avvenire per il carnivorismo."

10. "Con l'alimentazione carnea noi sfruttiamo un vasto gruppo di esseri viventi e li sfruttiamo nel modo più radicale perché le azioni verso di loro sono tutte volte al fine di utilizzarli, stroncandone la vita. Ora, mentre stiamo operando per togliere lo sfruttamento sociale tra uomo e uomo, salendo di un grado verso il rispetto reciproco, è già qualche cosa salire di un grado per tanti esseri viventi e utilizzarne i prodotti (può essere una forma di collaborazione in attesa di altre ulteriori e meno costrittive) ma non distruggendone l'esistenza. Un socialista si sente indotto a essere vegetariano: non è una classe subalterna e oppressa anche quella degli animali?"

11. "Se si vedessero così, gli animali sarebbero chiusi in un destino inferiore, sorpassati da noi e incapaci di raggiungere il nostro livello e di liberarsi con noi. Ma di contro a questo modo *chiuso* di considerarli, c'è un modo *aperto*, che considera ogni essere come l'inizio di un'apertura e di ulteriori possibilità [...]. Da ciò deriva un diverso modo di condurre la lotta: o gli animali sono come un avanzo del passato o sono uniti a noi in un destino di liberazione. In questo secondo caso, noi prendiamo iniziative di speranza, di avvicinamento, di linguaggio, di concordia, con la persuasione che se anche nel momento attuale essi non sono consapevoli esattamente dell'orizzonte verso cui possono andare, lo saranno."

12. "Noi chiediamo loro di cooperare con noi e studieremo ciò che meglio possano fare, in modo che anche per essi la vita sia esplicazione ordinata di un lavoro, come è per noi."

13. "Gettare fasci di nuova luce, di attenzione e di amicizia su categorie di esseri considerati prima come mezzi, vederli anche come fini, come esseri collaboranti e aventi diritti, questo diminuisce l'estensione dell'impero del

nostro arbitrio. Ma mille cose che noi prima ottenevamo per comando, le otterremo per cooperazione. In fondo a questa strada sta l'ideale di una realtà in cui non ci sia più nulla, che sia soltanto mezzo, cosa, strumento ma tutto sia soggetto e oggetto di amore

14. "1. la rimozione teorica della barriera che li chiude in limiti fissi e immodificabili; 2. la sollecitazione, in vari modi anche tecnici, alla collaborazione; 3. il rispetto progressivo anche del loro esistere, riducendo gradatamente la violenza e l'uccisione (la nonviolenza è un valore come la musica, la bontà: non sono un comando, né si esauriscono; uno può fare meglio che può in quel momento, proponendosi di fare di più. Nessuno può dire un tu in modo esauriente."

15. "La scienza viene fornendo materiali che permetteranno di risparmiare la distruzione delle piante [...]. Se non si può far tutto, molto si può certamente fare e si deve: siamo anzi in ritardo."

16. "Si potrebbe porre il problema se gli animali abbiano diritti, abbiano doveri com'è di ogni vita morale. Noi abbiamo già visto che il problema del rapporto tra gli uomini e gli animali è in sviluppo e una tappa certamente è il vegetarianesimo. Alcuni diritti da molti codici sono stati riconosciuti agli animali, per esempio di non essere tormentati che in certe condizioni. Lo sviluppo continuerà. Ma il fatto da osservare è che questi punti vengano raggiunti, in genere, per iniziativa umana; il che spiega anche che difficilmente si possa parlare di *doveri* da parte degli animali, anche se talvolta questo discorso è stato fatto e si sono impiantati perfino processi. Questo è il campo dove bisognerà più lavorare e sperimentare e creare; e forse si arriverà a risultati più precisi di quanto immaginiamo."

17. "La non violenza, nella sua essenza, si riporta proprio a questo principio di libera decisione personale, di amorevolezza verso l'altro essere, quale che sia la sua condotta: è la proposta pratica di un modo nuovo, al posto dell'indifferenza o dell'odio; l'offerta di un *tu* che può diventare sempre più profondo. È inutile fare a questo proposito citazioni somme dal pensiero orientale o dal pensiero occidentale. Basti soltanto insistere sul carattere di questo 'fare aperto' rivolto a un essere indipendentemente dalla sua condizione, dalla sua razza, dalla sua condotta anche. Amore e aiuto: l'atto del samaritano. Il vegetarianesimo è il rivolgersi a un gruppo di esseri non umani prendendo l'iniziativa di stabilire un rapporto di apertura e non più di indifferenza o di crudeltà."

18. "apertura ad ogni essere, desiderio, lavoro e gioia per la sua esistenza, libertà e sviluppo, con il proposito di aumentarla sempre in intensità e in ampiezza, dal prossimo ai lontani."

19. "Perché escludere gli esseri non umani? – domanda Capitini –. Perché, si dice, non sono razionali. Ma se noi abbiamo qualcosa in più, ciò significa soltanto maggiori doveri, maggiore apertura."

20. Humanism is not necessarily anthropocentric, as I have tried to show in my book *Alle origini dell'etica ambientale* (*At the Origins of Environmental Ethics*, 2002), where I talk about "ecological humanism" and introduce the distinction between anthropocentrism and "anthopogenism."

Works Cited

Altieri, Rocco. 2003. *La rivoluzione nonviolenta. Biografia intellettuale di Aldo Capitini.* Pisa: BFS edizioni.

Battaglia, Luisella. 2002. *Alle origini dell'etica ambientale. Uomo, natura, animali in Voltaire, Michelet, Thoreau, Gandhi.* Bari: Dedalo.

Capitini, Aldo. 1950. *Nuova socialità e riforma religiosa.* Turin: Einaudi.

———. 1959. *Aspetti dell'educazione alla non violenza.* Pisa: Pacini-Mariotti.

———. 1961. *Battezzati non credenti.* Florence: Parenti.

———. 1969. *Il potere di tutti.* Florence: La Nuova Italia.

———. 1990. *Elementi di un'esperienza religiosa.* Bologna: Cappelli.

———. 1998. *Scritti filosofici e religiosi.* Ed. Mario Martini. Perugia: Fondazione Centro Studi "Aldo Capitini."

Cavicchi, Maurizio. 2005. *Aldo Capitini. Un itinerario di vita e di pensiero.* Bari: Piero Lacaita.

Curzi, Federica. 2004. *Vivere la nonviolenza. La filosofia di Aldo Capitini.* Assisi: Cittadella.

Donaldson, Sue, and Will Kymlicka. 2011. *Zoopolis: A Political Theory of Animal Rights.* Oxford: Oxford University Press.

Donovan, Josephine, and Carol Adams, eds. 1996. *Beyond Animal Rights: A Feminist Caring Ethic for the Treatment of Animals.* London: Continuum.

———, eds. 2007. *The Feminist Care Tradition in Animal Ethics.* New York: Columbia University Press.

FoppaPedretti, Caterina. 2005. Spirito profetico ed educazione in Aldo Capitini. In *Prospettive filosofiche, religiose e pedagogiche del post-umanesimo e della compresenza.* Milan: Vita e Pensiero.

Gaita, Raimond. 2003. *The Philosopher's Dog.* London: Routledge.

Gilligan, Carol. 1982. *In a Different Voice: Psychological Theory and Woman's Development.* Cambridge, MA: Harvard University Press.

Jonas, Hans. 1984. *The Imperative of Responsibility: In Search of an Ethics for the Technological Age.* Trans. Hans Jonas, with the collaboration of David Herr. Chicago: University of Chicago Press.

Kemp, Peter, and Jacob Dahl Rendtorff. 2008. The Barcelona Declaration: Towards an Integrated Approach to Basic Ethical Principles. *Synthesis Philosophica* 46 (2): 239–251.

Mancuso, Stefano. 2017. *Plant Revolution. Le piante hanno già inventato il nostro futuro.* Florence: Giunti.

Mayeroff, Milton. 1971. *On Caring.* New York: Harper & Row.

Michelet, Jules. 2018 (1868). *La Montaigne.* Sydney: Wentworth Press.

Paterson, David, and Richard D. Ryder, eds. 1979. *Animal Rights. A Symposium.* New York: Centaur Press.

Polito, Pietro. 2001. *L'eresia di Aldo Capitini.* Foreword by Norberto Bobbio. Aosta: Stylos.

Potter, Van Rensselaer. 1988. *Global Bioethics: Building on the Leopold Legacy.* East Lansing: Michigan State University Press.

Regan, Tom. 1983. *The Case for Animal Rights.* Berkeley: University of California Press.

Richard, Ryder. 2001. *Painism: A Modern Morality.* London: Open Gate Press.

———. 2011. *Speciesism, Painism and Happiness.* Upton Pyne: Imprint Academic.

Salt, Henry. 1894. *Animals' Rights: Considered in Relation to Social Progress.* New York: Macmillan & Co..

Singer, Peter. 1975. *Animal Liberation: A New Ethics for our Treatment of Animals.* New York: Random House.

———. 1981. *The Expanding Circle: Ethics, Evolution, and Moral Progress.* Princeton, NJ: Princeton University Press.

Tortoreto, Andrea. 2005. *La filosofia di Aldo Capitini. Dalla compresenza alla società aperta.* Florence: Clinamen.

CHAPTER 4

What Is Italian Antispeciesism? An Overview of Recent Tendencies in Animal Advocacy

Giorgio Losi and Niccolò Bertuzzi

1 Introduction

The publication of Peter Singer's *Animal Liberation* was a milestone in the history of animal advocacy worldwide. Although associations dedicated to vegetarianism and animal welfare had existed previously, after 1975 a consistent philosophical and social justice movement took shape, opposing the use of non-human animals in agriculture, research, and entertainment. Singer defined speciesism as "a prejudice or attitude of bias in favor of the interests of members of one's own species and against those of members of other species" (2009: 6) and believed it was the

---

Although authors vastly cooperated to this essay, Niccolò Bertuzzi originally wrote the sections *The Italian Rebus* and *Mainstream Positions*, while Giorgio Losi wrote *Defining Concepts* and *Radical Approaches*. We would like to thank Frank Brown Cloud and Amanda Vredenburgh for their excellent work of copyediting.

G. Losi (✉)
Indiana University Bloomington, Bloomington, IN, USA

N. Bertuzzi
Scuola Normale Superiore, Pisa, Italy

© The Author(s) 2020
F. Cimatti, C. Salzani (eds.), *Animality in Contemporary Italian Philosophy*, The Palgrave Macmillan Animal Ethics Series,
https://doi.org/10.1007/978-3-030-47507-9_4

moral foundation for all exploitative practices against other animals. In consequence, this social justice movement has been referred to as "anti-speciesism" or "animal liberation" (as in the title of the book). But does Singer's original definition encompass the complexities of the contemporary movement in its intellectual and activist variations in the Italian context? In this chapter, we outline the social and theoretical components of the antispeciesist movement in contemporary Italy by identifying two different trends: one leaning towards mainstream neoliberal positions, the other towards radical counter-hegemonic ones.

2   The Italian Rebus

Animal Liberation, Peter Singer's philosophical treatise which launched the term antispeciesism, was published in Italian translation in 1987.[1] At the moment, however, the word has acquired a range of meanings, to the point that some authors (in both Italy and other countries; see, e.g., Filippi and Trasatti 2013: 145–64) speak of several antispeciesisms, rather than referring to antispeciesism as a unitary, homogeneous phenomenon. As we will see for *Radical Approaches*, this distinction between forms of antispeciesism from a theoretical standpoint originated due to the application of continental, anti-humanist, and leftist philosophical interpretations to the analytic, utilitarian, or Kantian version of antispeciesism that liberal philosophers like Peter Singer and Tom Regan initially promoted. In Italy in recent years, a range of extremely different activists and groups self-define as antispeciesist, despite being motivated by very diverse perspectives and philosophical positions. The term antispeciesism is often an "empty signifier," to use the expression invented by Ernesto Laclau (2005) to define populism. This creates a methodological problem for sociologists—especially for critical sociologists—because we must take seriously the self-representations and auto-definitions of social actors (in this case, both animal advocacy groups and individual activists). At the same time, the range of groups that self-define as antispeciesist creates an epistemological issue because their use of the term can involve a total abdication of the initial meaning of antispeciesism itself, implying something more than a physiological reinterpretation of a philosophical frame, practice, and vision of the world.

We propose a general dichotomy (mainstream vs. radicalism), which we hope may begin to address this methodological concern, epistemological rebus, and the burgeoning political nihilism among this population of social actors. This approach will allow us to critically consider the current situation of both antispeciesist activism and more general contemporary social trends. For each of these two perspectives, we discuss examples in terms of both theory and practice. Because both theory and practice are generated by the same vibrant cultural and political environment—and mutually influence one another—the animal liberation movement can only be understood by considering both activism and intellectual work in parallel. Therefore, we will first review mainstream positions, considering forms of activism developed in the last decade by the most visible associations and groups, taking their modality of antispeciesism as a point of reference: this philosophical stance hopes to build a cultural and political dialogue with the current Western social structure, accepting its latent general assumptions and conditions, and seeks the improved treatment of non-human animals within such a system. Then we will address radical approaches, focusing on a set of scholars and activists which represent some of the most avant-garde and politically oriented elements: they propose drastically counter-hegemonic perspectives and are particularly appreciated in anti-capitalist political environments. Instead of reforming the status quo, this kind of approaches seeks radical change, using contentious practice and unconstrained philosophical thinking.

3   Defining the Concepts

Before proceeding with our analysis of mainstream positions, we give a brief account of the terminology currently in use in the larger field to which antispeciesism belongs. As in other countries, in Italy there is a fluid use of the terms animal advocacy, animal rights, veganism, animal liberation, and antispeciesism. These notions originated independently, and although the boundaries between their underlying ideas are often vague and porous, their usage does not completely overlap:

- The umbrella term *animalismo* (animal advocacy) was adopted during the late 1970s to identify a movement that arose from a single issue—opposition to animal testing—before incorporating several other issues related to animal advocacy.
- Another term, *diritti animali* (animal rights), was introduced to the Italian community in 1990, when the Italian translation of Tom Regan's *The Case for Animal Rights* was published.[2] Internationally acclaimed philosopher Paola Cavalieri, with her journal *Etica &*

Animali (1988–1998), contributed to the prevalence of this term describing animals as rights-holders, a terminology that became popular even among large institutional organizations.

- The vegan diet instead (a diet that intentionally excludes all animal products) began to be more common in the early 2000s as a radical form of solidarity among activists directly involved in campaigns against animal exploitation. Later, this diet was popularized, becoming a widespread,[3] vastly depoliticized commercial phenomenon, and in the process, the term *veganismo* absorbed a confusing array of meanings. In addition to food, many activists (or sensible consumers) would consider commodities and activities involving the use of other animals to be *non-vegan*, making veganism a broad philosophical stance (proscribing certain behaviours) rather than a simple diet. The all-embracing notion of veganism constitutes a new form of identity and tends to replace animal advocacy, deflecting attention from animal ethics to an obsession over the absence of animal ingredients in food and merchandise.[4] Veganism often goes with the belief that persuading single consumers, one by one, to change their shopping habits will be sufficient to bring a substantial change in the way other animals are treated and perceived. Even more confusion is generated by social actors who promote veganism as a universal solution for environmental concerns or personal health. In Italy, groups dedicated to supporting and expanding the vegan community are growing by the day, one of the best-known contemporary organizations being Progetto Vivere Vegan (The Vegan Life Project, 2001–current). As a dietary concept, veganism was preceded by vegetarianism (a less restrictive diet which excludes foods that require the killing of other animals; both a vegan and a vegetarian diet would exclude meat, whereas dairy and eggs are typically excluded by a vegan but not a vegetarian diet), which in Italy has long been practised by small groups guided by moral and religious motives.[5]
- The term *antispeciesism* describes instead what are considered to be the most—philosophically and sometimes politically—coherent elements in the varied landscape of animal advocacy. Antispeciesism is commonly treated as synonymic with *animal liberation* to describe unbending forms of activism, in both theory and practice, which demand a complete dismissal of animal use. Antispeciesism provides a rationale (if controversial) for animal advocacy campaigns and the practice of veganism. This philosophical stance is often referenced by

more theoretically prepared militants to distinguish themselves from other animal advocates whom they consider to be superficial, narrow-minded, or otherwise undesirable, yet the idea of antispeciesism is sometimes referenced even by these less cultured groups as a sort of ideological legitimization.

Compared to the relatively broad, equivocal, and imprecise notions of *animalismo*, those who ascribe their motivation to antispeciesism or animal liberation are typically against any use of other animals for human ends. A wide range of groups that fall under the generic term of *animalismo* would be likely to oppose cases of animal *abuse*, hoping to resolve these issues through the application of either existing laws or slightly modified variants. Only the smaller subset of antispeciesist groups opposes all instances of animal *use*. To provide a few examples, volunteers who dedicate their time to direct animal care and participate only in events in support of cats and dogs (like the annual, deviously racist protest against the Yulin Dog Meat Festival in China) also describe themselves as *animalisti*. Indeed, the definition of *animalismo* is broad enough to include groups whose propaganda is repressive and masculinist, like Animalisti Italiani (Italian Animal Advocates, 1998–current),[6] as well as more sophisticated and theoretically informed groups like Animalisti Friuli Venezia Giulia (Animal Advocates from Friuli Venezia Giulia, 2011–current), that can be fully considered antispeciesist. For this reason, some activists in antispeciesism recommend abandoning the umbrella term *animalismo*, which they believe to be irretrievably compromised.

4   Mainstream Positions

Among Italian animal advocates, mainstream positions are the most widespread, and these have proven to be generally effective in terms of enlarging the base of social actors who are engaged in animal advocacy as well as in the pursuit of some of their political goals. In this chapter we use the term "mainstream" to refer to those activists and groups that propose to find a space for the animal issues (and for veganism in particular, as a form of consumption) within the current neoliberal paradigm. Not always the theorists whose reception is more diffused in this area are themselves moderate or mainstream; however, they result particularly appreciated among this category of activists, sometimes also because of partial misunderstanding or simplification of their positions.

In recent years, the most visible groups among Italian animal advocates include Essere Animali (Being Animals, 2011–current), Animal Equality Italia (2012–current), Anonymous for the Voiceless (2016–current), Iene Vegane (Vegan Hyenas, 2016–current), and LAV. One indication of success for the mainstream position is that all of these groups have begun to issue an institutional endorsement of veganism. The considerable visibility and popular relevance of these organizations is sometimes a consequence of their large marketing budgets but is also an outcome of well-executed strategies, such as the attention-grabbing public actions undertaken by Anonymous for the Voiceless and Iene Vegane. Even among the more institutionalized groups, such as Essere Animali or Animal Equality, some key members were previously involved in grass-roots radical struggles, but then decided to professionalize their activism, explicitly stating that they wanted to work within the common rules of the Western market economy and social structure, trying to progressively veganize it from the inside.

Among these animal advocacy organizations, some claim to occupy apolitical positions, while others endorse reformist/moderate approaches. These are pragmatic philosophical stances: although they may recognize the philosophical contradictions inherent in compromising with speciesist institutions in particular and the overarching structure of the neoliberal economy in general, they have decided that a mainstream approach is the most effective. This is the basic situation if we consider the "supply side" (associations and key leaders, resource mobilizers), although the panorama seems to be even more complicated if we consider the movement base, which combines a range of peculiar elements: "ecumenical"[7] positions, transversal approaches, disinterest in frame-bridging discourses and operations, and a tendency to consider the animal question as the foremost issue among all contemporary social struggles. These generalizations do not represent the entirety of the animal advocacy base, but their prevalence has been documented by empirical survey data (e.g. Bertuzzi 2018) and by online debates that have occurred on some of the organizations' social media webpages (e.g. De Matteis and Bertuzzi 2019).

The philosophical variegation among the Italian antispeciesism community seems to have been caused by an institutional preference among mainstream organizations to enlarge their available base as much as possible, even at the expense of abdicating a strong theoretical support. In the short term, this approach seems to have been highly effective. These organizations have accrued large number of followers on social media; there has been a dramatic increase in attendance of vegan festivals across Italy, such as MiVeg (2013–current), Parma Etica (2014–current), and Sagra del Seitan (Seitan Festival, 2005–current); increasing numbers of people self-identify as vegans among the general population. This transversal approach, which popularized a vegan diet among the general public and seems to have increased both the visibility and favourable reception of animal-friendly positions, gained traction in the early 2010s.

As the mainstream approach has become more influential, it has also become less political; supporters are positively disposed to developments such as vegan offerings from fast-food chains and other major corporations. In the past, animal advocacy organizations endorsed companies like Slow Food or Eataly; now, animal advocacy organizations might endorse the vegan options even when proposed by McDonald's, Burger King, Granarolo, or other similar market actors, which are all notable not only as traditional symbols of the "worst" that neoliberal capitalism has to offer but also as massive exploiters (in both the past and present) of non-human animals. These companies seem to be implementing new lines of vegan products in order to attract and develop a new market niche (Evans and Miele 2012), not as evidence of a philosophical shift—none has concurrently reduced the exploitative nature of their traditional offerings. Clearly, these market actors have maintained their anthropocentric consumerist basis: it would be naive to expect for these market actors to eschew meat products, and the earnings made through their vegan offerings are reinvested in animal-exploitative facilities. And yet, despite the ongoing practice of these market actors, they have received the support of organizations that self-define as antispeciesist. The majority of the organizations that pursue a mainstream approach with big corporations are aware of this contradiction; however, they seem to believe that this attempt at *entryism* could be effective in the long term. Although we cannot predict which methods will prove most successful any better than the animal advocates who have chosen these strategies, the mainstream approach seems to have resulted in the corporatization of activism (Dauvergne and LeBaron 2014) and the hegemonic characterization of similar positions, all of which claim to want to change an omnivorous capitalist market with a vegan one, but without questioning the social structure and the economic system supporting it.

As anticipated, within this broad area of animal advocacy, only at the "top of the pyramid"—meaning among the leaders or the most visible figures—is the theoretical basis for the mainstream approach actively questioned; the "base of the pyramid" tends to passively receive the theoretical debate. We do not mean this as an elitist, normative commentary, but as an analytical observation (Bertuzzi 2018). Given this regime, the author who has been most influential in this area of Italian antispeciesism in recent years is Leonardo Caffo, thanks to his skills as a communicator along with his considerable visibility and adept use of social media. Caffo is not only the most influential scholar regarding the animal question in Italy in recent years but also the most divisive. It is not the aim of this text to critique Caffo's stance, but it is worth noting that some of his positions have been favourably received due to their simplification and vulgarization. It is unlikely that everybody in the group of Caffo's followers and admirers is an expert in his academic and published works. Indeed, the amenability of Caffo's work to reinterpretation or even partial misunderstanding seems crucial to his alignment with groups such as Essere Animali or Animal Equality—groups that, as we mentioned before, seem strategically and consciously biased towards enlarging their base, sometimes to the detriment of uniform reception of their discourses and campaigns.

Perhaps less transversal and less well known, surely more radical, Adriano Fragano has also been a decisive influence on this area of activism. Fragano works outside of academia and has typically adopted positions that are more political than Caffo's. His case is peculiar and needs further clarification: Fragano's positions could surely be defined as radical, but they have been positively welcomed also (and especially) among more moderate animal publics (Blue and Rock 2014). Fragano has built a reputation through his books, his public lectures, and especially his work on the website veganzetta.org. The claim introducing this website is paradigmatic of some radical positions of the author, stating: "Happiness is not finding vegan product at the supermarket, but not finding supermarkets." Some of his positions, however, are extremely appreciated among mainstream areas of animal advocacy, especially for the central role given to veganism and the primacy assigned at animal questions compared to other social justice issues. This pushed us to list Fragano among the mainstream authors, even if aware of the peculiarities of his approach: his production, in fact, is often mentioned by moderate and especially apolitical groups and activists (Bertuzzi 2018), namely, those who would like to acquire favourable position for non-human animals (and human consumers) within the current socio-economic structure. We are conscious that considering authors like Fragano mainstream could be controversial. However, the more positive diffusion his work had among moderate (or apolitical) activists and the generalist public induced us to locate it in the present section of the chapter.

Very different, and surely definable as completely mainstream, is the case of another commentator who has recently had a decisive impact on this area of antispeciesism: we refer to the former television personality Giulia Innocenzi, who has put in multiple appearances at festivals and on television shows.[8] Her stated positions are emblematic of some moral compromise inherent in the mainstream approach: for instance, the photograph that she posted on Facebook of herself alongside the well-known chef Gianfranco Vissani, followed by the comment: "We are the best couple of the world, and we are sorry for the others[9]... The battle against intensive farming creates unexpected alliances, even between an almost vegan and an extreme carnivore" (our translation).

With all the differences previously specified, Caffo, Fragano, and Innocenzi are among the most visible figureheads of the "new wave" of Italian antispeciesism, because of their favourable reception among transversal and depoliticized antispeciesist activists. This part of the animal advocacy movement is characterized by massive use of the Internet to promote petitions and legal initiatives, as well as to share animal-related content[10]: Animal Equality created a page on its website named "Animal Defenders" (*Difensori degli animali*), that mobilizes activists with the slogan "Stop animal cruelty with a click" and aims to "fight the cruelty of farms every day with simple and fast-paced actions that can be carried out directly from home to make a big difference for the animals" (our translation).[11] This huge emphasis on online activism seems to be related to the general trend of Western societies towards the development of so-called network societies (Castells 1996), characterized by narcissism (Lasch 1979), individualization (Giddens 1991), and the reduced time available for political activism (McAdam 1989). Online animal advocacy is a particularly contentious battlefield, both for the conflict between animal advocates and the proud defendants of human-centric speciesism and for the internal conflicts between various types of animal advocates themselves. "Clicktivism" is an emerging characteristic of this type of antispeciesism, and although this vein of activism has surely contributed to an expansion of the audience of possible supporters, it has also resulted in a progressive moderation on several issues. Large investments of energy and resources were made in an explicitly institutionalized manner to promote the European Citizens' Initiatives (ECIs), such as Stop Vivisection (2012)[12] or End the Cage Age (2018).[13] These pan-European initiatives are hallmarks of the shift in scale from the local movements typical of Italian antispeciesism in the early 2000s to the present-day internationalization of the movement. This shift inevitably resulted in a higher degree of legal action (delegating to a few representatives the decision-making) and a lower level of direct involvement.

The major European campaigns were also promoted by the long-standing Italian animal welfare association LAV, which now self-defines as an antispeciesist organization. This shift in self-designation—in its original conception, LAV did not self-define *antispecista* but *animalista*—is indicative of several concurrent phenomena among animal advocacy organizations. This shift is likely due not just to an increase in awareness among the traditional membership of LAV but also to an awareness among the directorship that such evolution is strategically necessary to maintain relevance and opportunities in the present day.

There are also grass-roots organizations that explicitly claim an apolitical stance. From our perspective, this represents an inherent contradiction: antispeciesism *is* political. Although Italy has both left-wing and right-wing antispeciesist organizations—among the latter, the explicitly xenophobic and racist Centopercento Animalisti (100% Animal Advocates, 2003–current)[14] is the best known—the basic definition of antispeciesism (the refusal to translate species differences into species hierarchies) should theoretically compel political change.

When the major animal advocacy organizations seek to be accepted within the current hegemonic structure, they do so by insisting on the distinctiveness and priority of antispeciesist struggle over other social justice conflicts. The mainstream approach to animal advocacy inevitably transforms what would seem to be a broadly counter-hegemonic philosophical stance into a single-issue cultural battle. Some such groups have employed strikingly visible demonstrations—for example, both Iene Vegane and Cani Sciolti (Loose Dogs = Mavericks, 2009–current) have bombarded supermarkets and public transportation with graphic images and slogans about animal exploitation and violence against non-human animals. Anonymous for the Voiceless has enacted similar spectacles. In general, though, the activists who belong to these groups do not seem to be particularly interested in the theoretical framework behind antispeciesism. Whereas the major institutional Italian animal advocacy organizations seem to be inspired by analytically inclined Italian authors (Caffo in particular but also the more political reflections of Roberto Marchesini or the articles in journals like *Animal Studies*, 2012–current), these grass-roots apolitical groups seem to be more influenced by foreign authors that stress the psychological and emotional aspects of animal issues, such as Melanie Joy or Jonathan Safran Foer.

5 RADICAL APPROACHES

Following several large animal advocacy campaigns (the most recent being Fermare Green Hill [Stop Green Hill] in 2012–2013), the last ideological developments of radical Italian antispeciesism occurred during a period of relative calm. Although some number of clandestine interventions continued to take place in Italy[15]—often without much influence on public opinion—the time from 2013 to 2019 stands in contrast to the prior decade, when the Italian tradition of public demonstrations against vivisection was revitalized by a wave of campaigns against animal testing, many of which started in Northern Europe.[16] Beginning in 2002, numerous demonstrations and sabotages targeted the firm Morini, in Reggio Emilia, which bred animals for laboratories in Europe and Israel. The campaign continued until Morini closed in 2010. Two years later, after a series of demonstrations in front of another breeding station called Green Hill (in Montichiari, close to Brescia), with thousands of participants, some of the protesters jumped over the barbed wire and stormed the facility, thereby forcing its closure. Big, moderate organizations like LAV and Legambiente were entrusted by a jury with the adoption of all 2639 dogs liberated from Green Hill (April 28, 2012). Finally, in 2013, activists of the same group that broke into Green Hill occupied the Pharmacology Department of the University of Milan, negotiating the release of several hundreds of transgenic mice and one rabbit. The Italian fur industry was also targeted by activists during this time period: the campaign Attacca la Pelliccia (Attack the Fur) targeted both farms in the countryside and clothing stores in city centres between 2004 and 2011. Between 2013 and 2019, however, long-lasting campaigns and ambitious acts of protest seem to have ceased, giving way to relatively generic marches in cities, often monopolized by the rhetoric of veganism as an individual practice capable of bringing substantial changes. During these years, veganism has often been represented as a personal choice and identity disconnected from its original ethical and political framework, both in the media (television, radio, and social networks) and at supermarkets (which began to provide a wide range of new products explicitly designed for vegan consumption). After 2013, animal advocates have perpetrated only one large-scale, illegal act of civil disobedience in Italy—on January 27, 2019, the French antispeciesist organization 269 led activists from several European countries in the occupation of a slaughterhouse in Turin.

Despite the relative paucity of radical action during the years between 2013 and 2019, this vein of antispeciesism continued to develop on a theoretical level, through intellectual debate and the publication of books and articles. This intellectual ferment has varied the theoretical landscape beyond the foundations of classical antispeciesism. As Carlo Salzani wrote:

> Both the utilitarianism of Peter Singer and the rights theory of Tom Regan – the two main philosophies that, in the late 1970s and early 1980s, redefined and gave new impulse to the modern animal protection movement – propose […] a thoroughly humanist enlargement of the moral community, whereby the criteria for inclusion and exclusion remain those of the humanist tradition. These philosophies are still founded on the traditional, humanist notion of subjectivity – which was precisely construed through the exclusion of animals. They necessarily reproduce, therefore, the same structure of exclusion, violence and sacrifice that characterized humanism, granting privileges to some groups while excluding others, as results evident from Peter Singer and Paola Cavalieri's *Great Ape Project*. (2017: 107)

In the last 10 years, alternative theories of antispeciesism have been developed in Italy which go in a different direction from the work of authors like Singer, Regan, or Cavalieri. Examples include the eco-Marxist approach promoted by Marco Maurizi (e.g. 2011) or Gianfranco Mormino's reading of speciesism through René Girard (Mormino et al. 2018).[17]

Here we will focus in particular on Massimo Filippi, because of his significant role in the reception and elaboration of the international debate on antispeciesism as well as his key position in between theory and activism. Filippi is the cofounder and editor of *Liberazioni – Rivista di critica antispecista* (*Liberations – Journal of Antispeciesist Critique*, 2010–current). He is also the author of many books on the question of animality, of which *Crimini in tempo di pace* (*Peacetime Crimes*, 2013, with Filippo Trasatti) and *L'invenzione della specie* (*The Invention of Species*, 2016a) are the most influential. Reformulating Matthew Calarco's idea of three main frameworks in critical animal studies (2011), Filippi interprets the history of antispeciesism as divided into three waves (analogous to contemporary understandings of the history of feminism). Antispeciesism began with a focus on identity, progressed to the focus on diversity, and only recently has started to encompass what Calarco calls "indistinction" and Filippi—echoing Antonio Negri—"the common." Antispeciesism of identity (i.e.

the theories espoused by Singer and Regan) is in fact a form of anthropo-centrism, extending human rights to some animals considered to be par-ticularly close to humans from an ethical perspective. Antispeciesism of diversity collects all those theoretical works (including Jacques Derrida's *The Animal That Therefore I Am*) where the line that traces the border between animals and humans is multiplied and complicated, but still not completely erased. With the third wave of antispeciesism, the very notions of animal and human are questioned, as are the mechanisms of species and speciation.

Following Judith Butler's critique of gender, Filippi—who considers himself to belong to the third wave of antispeciesism—rebuts the common assumption that classifying bodies into species is a neutral, innocent pro-cedure. Filippi highlights the biopolitical implications of assigning to someone a particular normative behaviour, essence, and predesigned place in the world based on its species. In so doing, Filippi draws upon queer theory to subvert species normativity and question the current relations between those whom we designate as humans or as non-human animals. Furthermore, Filippi adopts Giorgio Agamben's idea of the anthropologi-cal machine (Agamben 2004) to destabilize the idea that humanity or animality is something objective, natural, and stable by demonstrating the ways in which bodies, regardless of the biological species to which they are assigned, can shift from the status of humanity to animality through politi-cal mechanisms of privilege and oppression. As in Gilles Deleuze's concep-tion of *becoming animal*, Filippi tries to dissolve the species construct via an argument that he refuses to label as post-human, insofar as it is no more post-human than it is post-elephant or post-murine (2016b: 38). Filippi's critique leaves behind the ideological apparatus of rights, which attributes an abstract and universal ownership to individualized, autonomous sub-jects (2010: 286–89), and instead aims to deconstruct subjectivity, which works by focusing on the identity of the subject as human rather than on one of its many other characteristic features characterizing this subject (such as class, gender, race, and so on). Insofar as Filippi thinks of specie-sism as a machine that is both material and ideological, antispeciesism is then the liberating movement that interrupts the working of that machine and restores to bodies the fluid set of relations that constitutes them, rather than sectioning bodies into discretely identified, sacrificial individu-alities. In this logic, the central aim of antispeciesism is not expanding the circle of interests or rights, with humans at the centre measuring the dis-tance between themselves and others and establishing value according to this measurement, but rather to embrace a proliferating, intersecting togetherness of embodied and situated experiences—of sensual more than sensitive beings (in accordance with Filippi's preference of Butler over Singer).

Filippi as well as other antispeciesist philosophers who contribute to the journal *Liberazioni* have stimulated a lively debate around the question of animality that went well beyond the national borders, translating into Italian authors like Matthew Calarco, Ralph Acampora, and Carol J. Adams,[18] as well as interviewing leading intellectuals like Rosi Braidotti (2015), Judith Butler (2015), Michael Hardt (Filippi et al. 2016), and Jean-Luc Nancy (2019). Contributors to *Liberazioni* had a significant influence on the debate about the strategies of the animal liberation movement in Italy, both by questioning its direction and by proposing new challenging perspectives. For example, while he took part in several public debates arguing against animal testing,[19] Filippi rejected the use of arguments like the alleged uselessness of scientific research conducted on non-human subjects, recommending that animal advocates maintain an exclusively ethical and political stance, rather than taking shortcuts that are pseudoscientific and further enforce the centrality of the human subject.[20]

Although the past 6 years have seen a paucity of direct actions carried out by the radical antispeciesists, the national movement of radical antispeciesism has coalesced around a series of conferences and festivals. Incontro di Liberazione Animale (Encounter for Animal Liberation, 2004–current) is a meeting ground for those planning direct action—the organizers have emphasized this focus by scheduling their 2013 meeting in the Susa Valley, where the local populace has been fighting the construction of a new railroad for the last 20 years. The organizers of this conference lean towards John Zorzan's anarcho-primitivism, refusing the possibility that technology has any emancipatory potential and emphasizing the biological basis behind the notions of species and gender (in accordance with second-wave feminists and disagreement with ideologists of a queerer antispeciesism like Filippi). Consequently, the meeting recently changed its name to Incontro di Liberazione Animale e della Terra ("Encounter for Animal and Earth Liberation"). The Milan-area antispeciesist organization Oltre la Specie (Beyond Species, cofounded by Filippi, 2002–present) hosted the event Veganch'io (a pun for "I am coming too" and "I am going vegan too," 2006–2018), although the name was changed to Festa Antispecista

("Antispeciesist Festival") in 2016 because the term "vegan" had shifted in connotation, having lost its original radical significance and becoming instead both a commercial phenomenon and an overemphasized hallmark of ideological purity. For the intersection between trans-feminism and animal liberation, the two Liberazione Gener-ale ("General Liberation", but also "Gender Liberation") roundtables organized in Florence in 2013 and in Verona in 2016 by the collective Anguane (2012–current) were particularly remarkable.

Recent years have also seen the introduction of several projects belonging to the area of radical antispeciesism that are centred on a less paternalistic understanding of animal advocacy. Bioviolenza – Al mattatoio sani e felici (Bioviolence – Healthy and Happy at the Slaughterhouse, 2011–current)[21] opposes the rhetoric of organic farms where animals are given supposedly healthy, satisfactory lives until they are humanely put down. This rhetoric is used to relieve consumer anxieties about the slaughter of other animals by evoking an idealized notion of the good old rural life, where farmers and farmed animals live in symbiosis. Stressing the contradiction in thinking of non-human subjects as worth ethical consideration while still reaffirming that their destiny is the slaughterhouse, Bioviolenza appeals to the Foucauldian notion of biopolitics to reject the idea of benevolent control of others, given that humans are still delimiting their spaces and regulating their reproductive capabilities. The Bioviolenza collective exists mainly online, but its members also drew attention by interrupting public events like the gastronomy exhibition Salone del Gusto in Turin in 2011. Salone del Gusto is a gigantic event organized by Carlo Petrini's *Slow Food*. Alongside Oscar Farinetti's *Eataly*, Salone del Gusto is the most visible entrepreneurial initiative to establish an international brand for fair trade or eco-conscious traditional Italian foods, many of which are derived from animal products. Bioviolenza has maintained pressure against the drift towards welfarism in the antispeciesist movement (a welfarist approach would advocate for better conditions for farmed animals, rather than their liberation). For example, in 2015 members of Bioviolenza started a successful petition to oppose the growth of CIWF (Compassion in World Farming, 1967–current), a worldwide animal advocacy group that claims to fight industrial farming by giving prizes to corporations which introduce minimal improvements in their facilities.[22] Around the same time, in 2015 and 2016, Bioviolenza contested renowned antispeciesist intellectuals for taking part in a Summer School in veterinary science at the University of Milan that included the presence of CIWF and a visit to an organic farm.[23]

Resistenza Animale (Animal Resistance, 2013–current)[24] is a collaboratively maintained blog that collects stories from all over the world about acts of resistance by non-human animals, such as animals who retaliated against exploitation by escaping, attacking guardians or hunters, or refusing to perform in circuses or zoos. By overturning the notions of agency and resistance as exclusively human, Resistenza Animale highlights the manifest attempts of other animals to fight the systematic exploitation they suffer (despite thousands of years of genetic selection that should have made them completely harmless and docile). In the perspective promoted by Resistenza Animale, animal advocates are not heroic saviours but rather allies or accomplices in solidarity. Animals are not voiceless, contrary to the claims expressed by mainstream antispeciesist organizations like Anonymous for the Voiceless or Iene Vegane—La loro Voce (Their Voice). Rather, the work of Resistenza Animale argues that non-human animals have been forcefully muted by a multitude of factors: the physical displacement of their bodies, out of sight in barns and slaughterhouses; ridicule, like the folkloristic depictions of naughty animals who make inept attempts to live on their own, running away from the farm or the zoo; and speciesist propaganda that describes animals' pursuit of freedom as somehow horrid and monstrous, irrational beasts assaulting their affectionate masters. Members of these online radical antispeciesist organizations have also worked to expose Italian audiences to the writings of international scholars.

In 2017, Marco Reggio—a committed and intellectually influential activist who participates in Bioviolenza, Resistenza Animale, and also another controversial project he cofounded, Vegephobia (2009–current),[25]—collaborated with feminoska— a leading figure in both the queer, anti-ableist, antispeciesist group AH! SqueerTo! (2014–current) and the collective of militant translators Les Bitches (2016–current)—to edit and translate the work of Canadian author Sarat Colling, *Animal Without Borders*. Colling builds upon the work of authors like Jason Hribal, who describes domesticated animals as forced labour, and expands this theoretical framework by addressing animal resistance from a feminist, postcolonial perspective. Colling's master dissertation received more attention in Italy, where the efforts of Reggio and feminoska led to its publication as a book (2017),[26] than in North America.

Animal resistance is inevitably linked with the issue of animal sanctuaries. Animals who escape farms and slaughterhouses or were liberated by human activists need a place to stay, since they would otherwise be surrounded by a highly anthropic and often hostile environment or else simply unable to provide for themselves when suddenly returned to the wild. Sanctuaries for farmed animals (*rifugi per animali*)[27] are being developed in Italy, and these places are often run by antispeciesist organizations. Agripunk (2013–current)[28] was formerly an intensive farm for turkeys owned by the Amadori corporation in the hills of Tuscany but has since been occupied and turned into a hotbed for anarchist and antispeciesist initiatives, as well as a welcoming home for many formerly farmed animals. One of them is Scilla, a calf who in 2016 escaped a truck and swam across the Sicilian strait during his journey to the slaughterhouse (from France to Lebanon). Agripunk was finally able to adopt Scilla thanks to a mail-bombing campaign coordinated by Resistenza Animale. Sanctuaries like Agripunk allow for an actualization of the antispeciesist idea that human and non-human subjectivities could freely intermingle.

In Italy there is a vast public interest in the treatment of cats and dogs— the concern for these particular animals is generally apolitical and often disconnected from other branches of animal advocacy. This public support was indispensable to the success of campaigns like Fermare Green Hill in 2012, where the lives of dogs (rather than other, less popular critters) were at stake. At the same time, in a country where euthanizing cats and dogs in excess of the market's demand is against the law, many spend their lives in (degraded) kennels supported by the state. Activists who are part of the collective Resistenza Animale, like Davide Majocchi, are working to re-evaluate the condition of stray, wild, and community dogs whose lives, compared to pets, are less privileged but also less constrained, and can still receive medical and food assistance by supportive humans.[29]

Whereas first-wave antispeciesism (based on identity) recognized a kinship between animal liberation and other historical struggles to include marginalized subjects in the community of those deserving moral consideration, third-wave antispeciesism deconstructs dichotomies at the root of Western tradition (e.g. rich/poor, man/woman, white/black, able/disable), in this case noting flaws in the traditional distinction between human and animal, where all the negative, inferior poles in these binary distinctions are somehow associated with animality (beastlike poor, wild barbarians, emotional nonrational women, monstrous freaks, and so on). Not only the category of species can be fruitfully added to the set of traits canonically contemplated in the notion of intersectionality (like class, race, gender, and ability), but Federico Zappino (2015), a prominent scholar in queer studies, argued that the sacrificial norm (the material and symbolical creation of man at the expense, both material and symbolic, of other animals and of all traces of animality in "man") is foundational even to, for example, the heterosexual norm and can be seen as a common root of all other forms of oppression. Without establishing this sort of hierarchy or priority between antispeciesism and other struggles, many groups adopted a fully intersectional approach, criticizing symptoms of classism, racism, and sexism in mainstream animal advocacy but also trying to build alliances with other social justice movements. Oltre la Specie has organized many cultural events and demonstrations stressing the parallels between systemic exploitation of non-human animals and other forms of oppression directed at "animalized" subjects such as workers, migrants, Romani people, convicts, women, LGBTQI individuals, or disabled persons. In recent years antispeciesist activists have sought to work in solidarity with trans-feminist and queer organizations, drawing inspiration from published works Filippi and Reggio edited, such as Rasmus R. Simonsen's *A Queer Vegan Manifesto* (2012, 2014),[30] and the collection of essays *Corpi che non contano* (*Bodies That Do Not Matter*, 2015). Since 2017, the Italian branch of Ni Una Menos, Non Una di Meno (Not One Woman Less, 2015–current), includes one group specifically dedicated to ecofeminism and antispeciesism: Terra Corpi Territori e Spazi Urbani (Earth Bodies Territories and Urban Spaces, 2017–current). These initiatives continue the work of feminist and antispeciesist scholars like Agnese Pignataro, cofounder of the journal *Musi e Muse* (*Muzzles and Muses*, 2012–2014), and collectives like the above-mentioned Anguane, which already contributed to introduce in Italy the work of authors such as Carol J. Adams and pattrice jones.

As Filippi wrote about the history of antispeciesism, upon close scrutiny the notion of species dissolves as a fictional—yet lethal—construct (2016a: 55). Dehumanized subjects have always been exposed to dominion, regardless of the species assigned to them (2013: 163–164). The conflict between speciesism and antispeciesism resolves itself when we acknowledge the intrinsic vacuity of these notions, embracing instead a wider critique of anthropocentrism that would encompass all struggles for liberation, regardless of whether they are centred on subjects conventionally identified as "animal" or "human." When radical antispeciesist activists occupied a slaughterhouse in Turin in January 2019, the official statement released by the transnational protest organizers was rooted in anticapitalistic, intersectional philosophy. The organizers claimed that their antispeciesist struggle was one front in the battle against all forms of discrimination and domination (including racism, sexism, and xenophobia), expressing disgust for the anti-immigration Italian politician Matteo Salvini and denouncing the moral compromise of some mainstream animal advocacy organizations for allowing far-right ideologies to spread within their ranks. This combative, uncompromising attitude is indicative of the rift between Italian adherents of a moralistic, disembodied, allegedly apolitical mainstream antispeciesism and the subterranean persistence of a more politically oriented radical antispeciesism that aims for a radical revision of the common.

6   Conclusion

In this chapter we have provided an intellectual history of the recent debate within Italian antispeciesism, connecting theoretical proposals with activist's frames and action repertoires. This history is necessarily both partial and partisan: in the first part (*Mainstream Positions*), it would have been impossible to cover all the social agents active in the main field of animal advocacy, and in the second (*Radical Approaches*), we focused our attention only on a limited set of interrelated initiatives and projects that we consider to represent the most interesting and innovative tendencies in Italian radical antispeciesism. We think, however, that our account will resonate with other contributions of this book that look more in depth at some of the aspects we have only sketched. Focusing on organizations devoted to animal advocacy, we dichotomized the various perspectives as mainstream and radical. The recent history of antispeciesism has clearly perpetuated the classic contraposition of reformism versus radicalism. In our view, it is worth questioning the authenticity of certain reformist organizations, both because they still seem to be operating within an anthropocentric paradigm and because they may have been unwittingly instrumentalized by subtle marketing strategies. We have also tried to give more visibility to those organizations attempting to transcend cryptoanthropocentric discourses and practices, going beyond the traditional limits of antispeciesism and finding a more radical posture.

Notes

1. The first independent translation was provided by LAV, Lega Anti Vivisezione (Anti-vivisection League, 1977–current). Several years later the book was published by an official editor, Mondadori.
2. Original edition: 1983

3. According to Eurispes 2019, the percentage of Italian vegetarians and vegans is 7.3% of the population, meaning +0.2% compared to 2018, −0.3% compared to 2017, −0.7% compared to 2016, and +1.4% compared to 2015. Such oscillation indicates the instability of the phenomenon, but these fluctuations could be due to a limitation in data collection due to response bias, stigmatization issues, or distinct personal definitions of vegetarianism and veganism.

4. Consider the success of ruthless business operations like the label Vegan OK (2000–current), which identifies wholly vegan products on supermarket shelves.

5. The first Italian vegetarian association, founded in 1952 by Aldo Capitini, was called Società Vegetariana (Vegetarian Society). Capitini, an anti-fascist philosopher and politician, is known as the "Italian Gandhi" and initiated the famous Perugia-Assisi Peace March.

6. With strategies analogous to the American PETA (1980–current), Animalisti Italiani adopted as a sponsor porn star Rocco Siffredi, with slogans like "Pene più dure" ("Harsher penalties" but also "A harder penis") for those who abandon their pets.

7. By "ecumenical" we mean here that the approach aims at uniting all animal advocates, regardless of their political ideas and the tactics they use.

8. Much attention has been given recently to her book, *Tritacarne* (Innocenzi 2017).

9. This is a quote of a famous Italian song entitled *La coppia più bella del mondo* (Adriano Celentano and Claudia Mori 1968). Such quote sounds extremely weird referring to an animal advocate and a chef who supports culinary traditions and practices that are strongly meat-based, ferociously adverse to veganism.

10. Similar online initiatives have been promoted in the past years, for example, by the network Agire Ora (Act Now, 2003–current).

11. https://animalequality.it/news/2018/03/08/paluani-cede-e-dichiara-che-abbandonera-le-uova-di-galline-gabbia-entro-il-2019/

12. https://ec.europa.eu/citizens-initiative/public/initiatives/successful/details/2012/000007

13. https://ec.europa.eu/citizens-initiative/public/initiatives/open/details/2018/000004

14. The present analysis of *Mainstream Positions*, including self-defined antispeciesist activists and influencers whose political postures go from liberal to populist, excludes those figures affiliated with right-wing political movements that could hardly fit into the theoretical framework of antispeciesism. For an examination of this area of activism, we refer to the booklet *Conoscerli per isolarli* (2016) by Antispefa: https://antispefa.noblogs.org/files/2016/02/Conoscerli-per-isolarli-antispefa-2016.pdf. This text

precedes the formation of Movimento Animalista (2017–current) by Vittoria Brambilla, publicly endorsed by Silvio Berlusconi. For a more updated retrospective, see Bertuzzi and Reggio (2019).

15. As has been documented on websites like Bite Back http://www.directaction.info/

16. Particularly the campaign Stop Huntingdon Animal Cruelty (1999–2014) in the UK

17. In 2018 Professor Mormino launched the first official course on animal studies in Italy, at the University of Milan. Thanks to the efforts of Mormino and others, the University of Milan has recently become a centre for research, seminars, and conferences about the question of animality.

18. Although the first complete Italian translation of Adams' *The Sexual Politics of Meat* was published by Vanda in early 2020, excerpts of the book and other essays by the famous eco-feminist have been published before, for example, in 2010 in *Liberazioni* 1, 23–56, and in Filippi and Trasatti (2010: 23–38).

19. For example, in 2014, a conference at the Catholic University of Milan (https://www.youtube.com/watch?v=ia9ueHam3R0&t=6s%29) and one at the festival BergamoScienza (https://www.youtube.com/watch?v=8MK3xsGypYc)

20. http://www.liberazioni.eu/wp-content/uploads/2019/10/Filippi-07.pdf. This article has been published by *Liberazioni* in an unofficial issue, shortly before the publication of the first issue in 2010.

21. http://bioviolenza.blogspot.com/

22. http://bioviolenza.blogspot.com/2015/06/lettera-aperta-ciwf-compassion-in-world.html

23. http://bioviolenza.blogspot.com/2015/07/perche-collaborare-con-ciwf-e-compagnia.html, http://bioviolenza.blogspot.com/2016/09/animalistiche-organizzano-visite-agli.html

24. https://resistenzanimale.noblogs.org/

25. This is a project about the public stigmatization of veganism and vegetarianism: http://it.vegephobia.info/

26. Here is the original text in English: https://dr.library.brocku.ca/bitstream/handle/10464/5229/Brock_Colling_Sarat_2013.pdf.

27. In Italian, the definition of these places as shelters rather than sanctuaries is prevailing.

28. https://agripunkblog.blogspot.com/

29. See also the recent documentary by Davide Majocchi entitled *No Pet. Liberi e randagi* (*No Pets: Free and Stray*, 2018).

30. Originally published in English as a journal article, in Italy Simonsen's text was published as a book.

Works Cited

Adams, Carol. 2020. *Carne da macello. La politica sessuale della carne. Una teoria critica femminista vegetariana.* Milan: Vanda.

Agamben, Giorgio. 2004. *The Open. Man and Animal.* Trans. Kevin Attell. Stanford: Stanford University Press.

Bertuzzi, Niccolò. 2018. *I movimenti animalisti in Italia. Strategie, politiche e pratiche di movimento.* Milan: Meltemi.

Bertuzzi, Niccolò, and Marco Reggio, eds. 2019. Destre e liberazione animale. Fra qualunquismo e strumentalizzazione. In *Smontare la gabbia. Anticapitalismo e movimento di liberazione animale*, ed. Niccolò Bertuzzi and Marco Reggio, 43–63. Milan: Mimesis.

Blue, Gwendolyn, and Melanie Rock. 2014. Animal Publics: Accounting for Heterogeneity in Political Life. *Society and Animals* 22 (5): 503–519.

Braidotti, Rosi. 2015. Per amore di Zoe. *Liberazioni* 21: 6–14.

Butler, Judith. 2015. Una molteplicità di animali sensuali. In Filippi and Reggio 2015: 23–26.

Calarco, Matthew. 2011. Identity, Difference, Indistinction. *The New Centennial Review* 21 (2): 41–60.

Castells, Manuel. 1996. *The Rise of Network Society: The Information Age: Economy, Society, and Culture.* Cambridge: Blackwell.

Colling, Sarat. 2017. *Animali in rivolta. Confini, resistenza e solidarietà umana.* Trans. Les Bitches, and Ed. Feminoska and Marco Reggio. Milan: Mimesis.

Dauvergne, Peter, and Genevieve LeBaron. 2014. *Protest Inc.: The Corporatization of Activism.* Cambridge: Polity.

De Matteis, Francesca, and Niccolò Bertuzzi. 2019. Attivisti nella rete? L'influenza della comunicazione web sulle nuove forme di protesta. In Bertuzzi and Reggio 2019: 27–42.

Evans, Adrian B., and Mara Miele. 2012. Between Food and Flesh: How Animals Are Made to Matter (and Not to Matter) Within Food Consumption Practices. *Environment and Planning D- Society and Space* 30 (2): 298–314.

Filippi, Massimo. 2010. Not in My Name. In Filippi and Trasatti 2010: 277–313.

———. 2016a. *L'invenzione della specie. Sovvertire la norma, divenire mostri.* Verona: Ombre corte.

———. 2016b. I quattro concetti fondamentali dell'antispecismo. *Liberazioni* 27: 29–42.

Filippi, Massimo, and Marco Reggio, eds. 2015. *Corpi che non contano. Judith Butler e gli animali.* Milan: Mimesis.

Filippi, Massimo, and Filippo Trasatti, eds. 2010. *Nell'albergo di Adamo. Gli animali, la questione animale e la filosofia.* Milan: Mimesis.

———, eds. 2013. *Crimini in tempo di pace. La questione animale e l'ideologia del dominio.* Milan: Eleuthera.

Filippi, Massimo, Michael Hardt, and Marco Maurizi. 2016. *Altre specie di politica*. Milan: Mimesis.

Giddens, Anthony. 1991. *Modernity and Self-identity: Self and Society in the Late Modern Age*. Cambridge: Polity.

Innocenzi, Giulia. 2017. *Tritacarne*. Milan: Rizzoli.

Laclau, Ernesto. 2005. *On Populist Reason*. New York: Verso.

Lasch, Christopher. 1979. *The Culture of Narcissism: American Life in an Age of Diminishing Expectations*. New York: Norton.

Maurizi, Marco. 2011. *Al di là della natura. Gli animali, il capitale e la libertà*. Aprilia: Novalogos.

McAdam, Doug. 1989. The Biographical Consequences of Activism. *American Sociological Review* 54: 744–760.

Mormino, Gianfranco, Raffaella Colombo, and Benedetta Piazzesi. 2018. *Dalla predazione al dominio. La guerra contro gli animali*. Milano: Cortina.

Nancy, Jean-Luc. 2019. *La sofferenza è animale*. Ed. Massimo Filippi, and Antonio Volpe. Milan: Mimesis.

Regan, Tom. 1990. *I diritti animali*. Trans. Rodolfo Rini. Milan: Garzanti.

Salzani, Carlo. 2017. From Post-Human to Post-Animal: Posthumanism and the 'Animal Turn.'. *Lo Sguardo* 24 (2): 97–109.

Simonsen, Rasmus R. 2012. A Queer Vegan Manifesto. *Journal for Critical Animal Studies* 10 (3): 51–81.

———. 2014. *Manifesto Queer Vegan*. Trans. Filippo Trasatti, and Ed. Massimo Filippi and Marco Reggio. Verona: Ortica.

Singer, Peter. 2009. *Animal Liberation*. New York: Harper.

Zappino, Federico. 2015. Postfazione. In Filippi and Reggio 2015: 75–90.

PART II

Animality in Perspective

CHAPTER 5

Beyond Human and Animal: Giorgio Agamben and Life as Potential

Carlo Salzani

1 Deferral

Among the major representatives of so-called Italian Theory, Giorgio Agamben explicitly addressed the question of animality in his book *The Open: Man and Animal* (published in Italian in 2002), linking it to his theorization of biopolitics. *The Open* does not explicitly belong to his celebrated series *Homo Sacer* and, unlike many others of his works, has not sparked much debate or controversies—with the exception of the field of Animal Studies. Here this book, as very often happens with Agamben's work, polarized the readers: on the one hand, his archaeology of the human-animal divide, his theorization of the "anthropological machine," and his call for the dissolution of the metaphysical difference separating human and non-human animals have been hailed by some as a major contribution to the field and have become part of the basic theoretical

C. Salzani (✉)
Messerli Research Institute, Vienna, Austria

© The Author(s) 2020
F. Cimatti, C. Salzani (eds.), *Animality in Contemporary Italian Philosophy*, The Palgrave Macmillan Animal Ethics Series,
https://doi.org/10.1007/978-3-030-47507-9_5

toolbox; on the other, critics have emphasized his lingering anthropocentrism and the lack of a real and concrete commitment to animal suffering and domination. It is possible to argue that both positions are (partially) true, and if this chapter will be mainly devoted to defending the importance of Agamben's take on the animal question, it is perhaps convenient to start off with the criticisms.

The main critique to *The Open* is that non-human animals are absent from the book and from Agamben's reflections on animality: *The Open*, writes Dominick LaCapra, "has virtually nothing specific to say about other-than-human animals or their lives" (2009: 168), and Matthew Calarco notes that it focuses "entirely and exclusively on the effects of the anthropological machine *on human beings* and never explore[s] the impact the machine has on various forms of animal life" (2008: 102, emphasis in the original). This inattention, for Fayaz Chagani (2018), amounts to a "deferral" or "suspension" of the question of the animal outside the human and is therefore guilty of a "performative anthropocentrism" (Calarco 2008: 98; Chagani 2018). In fact, the only animal mentioned in *The Open*, notes again Chagani (2018), is the tick, whose relationship with its environment stands in for that of all animals—a "flattening" of all animal individuals into the macro-category "the animal" so typical of traditional anthropocentrism and humanism (cf. Wolfe 2013: 27). Animals remain here a mere figure of the limit of the human. Moreover, insofar as animals are included only by way of exclusion, Agamben "reproduces the very logic he seeks to undermine" (Chagani 2018). For Leonardo Caffo and Ernesto Sferrazza Papa, Agamben "does not talk about animals; he uses them": in his discourse, animality is merely a "political-ontological operator" used to problematize the categories of humanity and inhumanity (2015: 135), or, in LaCapra's words, it only functions as an "abstracted philosophical topos" (2009: 166). In the end, Agamben's critique of traditional humanism would not really question anthropocentrism as such but would only aim at renegotiating a certain form of anthropocentrism and would tend towards a "local or weak anthropocentrism." That is, Agamben's redemptive proposal in fact aims at horizontalizing hierarchies within the human, leaving however untouched the power relations with the other living beings (Caffo and Sferrazza Papa 2015: 137–38). This seems confirmed by the fact that, as Krzysztof Ziarek points out, after its deactivation the anthropological machine would stop working, but the terms constituting it would remain unchanged, prolonging thereby human domination over animals (2008: 197; cf. also Chrulew 2012: 58). This all means, for Cary Wolfe, that, within Agamben's system, the "political" (or biopolitical) dimension of the plight of real, concrete non-human animals in contemporary society remains unthought, or even "unthinkable" (2013: 27).

These criticisms are well-founded and difficult to counter. However, they mostly read *The Open* as a stand-alone book, isolating it from Agamben's corpus of writings, and especially approach it from theoretical perspectives extraneous to it, such as posthumanism, antispeciesism, or animal rights theory. Agamben's interest in writing this book lies elsewhere (cf. Attell 2015: 168; Colebrook and Maxwell 2016: 103), and, to really assess and appreciate it, one needs to situate it within his peculiar philosophical project. From this perspective, Claire Colebrook and Jason Maxwell even argue that "there is a sense in which all of Agamben's work concerns animality" (even though he differs from contemporary vitalisms and materialisms in that "it still grants language supreme importance") (Colebrook and Maxwell 2016: 103, 35), which is also the central claim of a recent book by Ermanno Castanò (2018). This chapter will adopt this latter perspective and put forward the argument that *The Open* and the animal question are pivotal to the comprehension of Agamben's whole philosophical project. The rift dividing human and animal represents and constitutes in fact the main structure of Western metaphysics, which always presupposes an unknowable and unnameable substrate (from time to time *physis*, animal "voice," nature, natural life, animality, and so on) supporting a knowable and nameable "substance" (*nomos*, *logos*, culture, politics, humanity). This presuppositional structure always leads to the subjection and dominion of one part over the other and to the deadly production of "bare life." The overcoming of this structure in what Agamben calls "form of life," a life of "potential," that would therefore deactivate the caesura dividing human and animal and open both to a new "use," that is, to a new understanding and a new relationship. In what follows, I will first analyse Agamben's "early" production (previous to his so-called biopolitical turn) to highlight his lingering anthropomorphism, and then I will read *The Open* as a constitutive part of the *Homo Sacer* project and will finally conclude with an overview of the notion of "potentiality" in his work.

2   LINGERING ANTHROPOCENTRISM

It can be and has been argued (Calarco 2008: 79; Prozorov 2014: 151) that a certain discontinuity characterizes Agamben's thought in regard to animality and anthropocentrism, analogous (and related, as we will see) to that between his so-called pre-political and biopolitical phases: whereas from the 1970s to the early 1990s his (Heideggerian) critique of humanism (precisely like Heidegger's) never questioned anthropocentrism as such, his "biopolitical turn" of the mid-1990s led him to see the traditional human-animal divide as ultimately untenable. His early phase can be characterized as the quest for a "post-metaphysical definition of the human" (Calarco 2008: 79), basically still articulated around the traditional (and metaphysical) formula "unlike the (other) animals, man is/has/can..."[1] As, for example, already in his first book, *The Man Without Content*, where Agamben rehearses (in passing) the Western metaphysical *vulgata* that non-human animals are submitted to necessity and only human beings are free and capable of free action (1999a: 69, 79–80; admittedly, he was only quoting Aristotle and Marx).

It is however in his third book, *Infancy and History*, where the metaphysical divide becomes a founding part of the argument: what opens up the human to the experience of history and culture, Agamben argues here, is, as for the whole Western tradition, language; but, unlike for this tradition, for Agamben language does not separate human and non-human animals because the former possess it and the latter don't. Quite the opposite: "Animals are not in fact denied language; on the contrary, they are always and totally language," whereas "man" (sic) "is not the 'animal possessing language', but instead the animal deprived of language and obliged, therefore, to receive it from outside himself" (1993a: 52, 57). Man must learn language, must receive it from the outside, and is therefore split in his nature ("the split between language and speech, between semiotic and semantic") (1993a: 52).[2] This founding split is what distinguishes animal "voice" (animal vocalizations and communication) from human language and allows thereby for human historicity and culture; Agamben names it, etymologically, *in-fancy* (from the Latin *in-* that negates the verb *fari*, to speak). So Agamben seems to be attuned to the recent findings in animal cognition when affirming that animal communication is fully linguistic and that animals are linguistic beings, but is empirically wrong when categorically denying some forms of language acquisition at least in some animal species (cf. Calarco 85–86; Watkin 2014: 262).[3] And he is certainly wrong when extracting "man" from the evolutionary continuum and singling him out as *deprived* of language. The point he wants to make, however, is that animals communicate *immediately*, that is, without the mediation of the sign, and that as such animal language is one with nature, just like the chirp of the cricket or the braying of the donkey (cf. 1993a: 3); human language is instead conceptualized as a *lack*—a lack that, in a move typical of the whole Western tradition, is then turned into the very condition of possibility of freedom.[4]

"Experimentum linguae," the preface added to the 1989 French translation (and also to the 1993 English translation) of *Infancy and History*, adds a fundamental point, which will return also at the very incipit of *Homo Sacer*: Agamben quotes the famous (and central for his whole project) passage from Aristotle's *Politics* (1253a 10–18) about the difference between animal "voice," mere indication of pleasure and pain, and human speech, able to indicate "what is useful and what is harmful, and so also what is right and what is wrong" (Agamben 1993a: 7–8). This difference, which is that of in-fancy, is what opens up for humans the space of ethics and politics (besides the already mentioned historicity and culture). This structure becomes central to the seminars collected 4 years later in *Language and Death*, where Aristotle's passage is again quoted and commented on (cf. 1991: 87) and "voice" becomes the mark of the very workings of Western metaphysics: here the voice epitomizes the negative structure of Western metaphysics insofar as it is "*ground* [*fondamento*], but in the sense that it goes *to the ground* [*va a fondo*] and disappears in order for being and language to take place" (1991: 35). For human language (and all that comes with it) to emerge, the argument goes, the animal voice (and animality with it) must disappear and become a negative ground or foundation. Animal voice is therefore the very place of negativity that marks our entire tradition with a negative presupposition: the structure of Western metaphysics is for Agamben that of a caesura, of a separation between an unknowable and unnameable substrate which goes to the ground in order for a knowable and nameable "substance" to emerge. The suppression of the animal (voice) is therefore the condition of possibility for the emergence of the human (language), and this structure will recur in many a founding paradigm throughout Agamben's career, as in the couplets *physis/nomos*, nature/culture, silence/witness, *zoé/bios*, and so on. More precisely, the suppression of the animal *constitutes and establishes* this very structure: the book and its title are in fact based on a quotation from Heidegger that "Mortals are they who can experience death as death. Animals cannot do this. But neither can they speak" (qtd. in 1991: xi). Language, negativity, and anthropogenesis are tightly intertwined—in a deadly embrace.

That is why, for Castanò (2018), the question of the animal is central to, and founding for, Agamben's whole philosophy, way beyond the limited and partial intervention of *The Open*. And this is also why, as Prozorov argues (2014: 151–52), the discontinuity between his anthropocentric and non-(or less) anthropocentric phases should not be overstated: the overcoming of Western metaphysical negativity, which starts with the exclusion of the animal (voice), has always constituted the core of his soteriology. That is, to make the divide between human and animal (or between voice and *logos* and so on) inoperative *has always been* (though perhaps only implicitly) the main item on his agenda. The negative structure of Western metaphysics is what makes it so deadly and so doomed, and the overcoming of the division has always been the goal of his messianic philosophy. The primacy of language itself, so fundamental for anthropocentrism's self-image and self-justification, is for Agamben not much a presupposition as the very problem to face and analyse, as concisely shown by a short, poetic text, *The End of Thought* (*La fine del pensiero*, 1982)[5]: here the traditional human exceptionalism is reaffirmed in the thesis that only human beings are not one with their animal voice (properly speaking, *there is no human voice*) and that this is precisely the origin of thought, but this disunity calls precisely for a recomposition, for a healing of the fracture, whereby the fulfilment of thought would therefore mean simultaneously *the end of thought*.[6] By the same token, and in a very consistent way throughout his entire career, Agamben sees in poetry the messianic *désœuvrement* of the communicative and informative function of language (of the caesura between semiotic and semantic) and thus the fulfilment (and the end) of language itself.[7]

Agamben's "biopolitical turn" of the early 1990s will move the focus from language to life, with important repercussions on his anthropocentrism which will bring the question of the animal explicitly to the foreground. But the structure of the analysis, and most of all the advocated way out from the deadly metaphysical tangle of the West, will remain the same.

3 ANIMAL AGAIN

The whole *Homo Sacer* project rests on the Foucauldian biopolitical premise that "For millennia [...] man remained what he was for Aristotle: a living animal with the additional capacity for political existence; modern man is an animal whose politics calls his existence as a living being into question" (*The Will to Knowledge* qtd. in Agamben 1998: 3). This means that (human) animality takes a radically new political role, since political life becomes increasingly indistinguishable from the (animal) life of the body. And it also means therefore that, in modernity, the ontological abyss[8] separating humans and animals has narrowed to a babbling brook, or, in Chagani's words (2018), that "inasmuch as the 'total management' of biological life has become the 'last historical task' [...], humanity has already become animal again." Under biopolitics, Wolfe rightly remarks, the human/animal distinction becomes a "discursive resource, not a zoological designation" (2013: 10), a "floating signifier in a second-order operation, one that can be deployed as needed to supplement the first-order political work" of managing life (2003: 7). The animal question, therefore, necessarily imposes itself to a biopolitical thought "from within" (Calarco 2008: 87), and that is precisely what happened to Agamben's philosophy—even though this recalibration never removed the human from the centre of his thought (cf. Colebrook and Maxwell 2016: 167).

The continuity in his thought is however marked by the fact that Agamben begins *Homo Sacer* with the very passage from Aristotle's *Politics* (1253a 10–18) that was so central to *Infancy and History* and *Language and Death*: now the "going to the ground" and disappearing of animal voice is related to (and made to coincide with) the exclusion of (animal) mere life from political life that constitutes the very foundation of the *polis* and of Western politics (cf. Agamben 1998: 7–8). Here, however, the Heideggerian logic of the *Ab-Grund* that structured Agamben's previous readings of the voice is subsumed in, and practically replaced by, the Schmittian logic of the exception[9]: the "going to the ground" of voice and life is here an exclusion that simultaneously includes them, an "inclusionary exclusion," which is for Agamben the very "logic of sovereignty" at the base of Western metaphysic. The exclusion of animal voice and animal life (i.e. of animality as such) is what allows for the birth of the human *polis*, but also of humanity as such, and thus coincides with what Agamben will later identify as the process of "anthropogenesis."

Politics remains here a prerogative of humanity and a mark of its difference from animality. As Prozorov notes (2014: 152–53), the distinction between *zoé* and *bios* makes sense only for human life, and the same holds for the notion of "bare life," which is precisely the product of the inclusionary exclusion of *zoé* from *bios* and is thus "species-specific" (Shukin 2009: 10). It is true that in *Homo Sacer* the "werewolf" is made to represent bare life as the outside of sovereign protection and thus as a zone of indistinction between human and animal (cf. Agamben 1998: 104–111), but as a stand-in for the *Homo Sacer*, the werewolf is still human—albeit reduced to an animal. The true paradigm of bare life is in fact the Muselmann, whose animalization is so horrific precisely because it exposes humans *as humans* "in their constitutive capacity not to arrive at their (supposedly) essential humanity" (Colebrook and Maxwell 2016: 44). The infinite de-humanization of the Muselmann never reaches the animal, whose a priori exclusion is precisely what founds the human as such, and that is why in *Remnants of Auschwitz* (1999b) animals enter only figuratively, as the "lice" to which Hitler compared the Jews or the "stray dogs" to which the guards compared the Muselmann (Prozorov 2014: 153). Nevertheless, Castanò (2018: 194–97) proposes here an interesting thesis: he sees *Remnants of Auschwitz*, which is labelled as "volume III" of the *Homo Sacer* series, as a sort of threshold between the *pars destruens* of the volumes I and II of the project and the *pars construens* of volume IV. By showing the workings of the "anthropological machine" in all its deadly purity, Auschwitz represents the apex of metaphysics and as such unveils the nature of the originary metaphysical rift. The book would thereby be a preparatory work that opens up a space for the messianic proposal of *The Open*.

What changes in *The Open* is indeed not only the fact that the animal question is here tackled directly and explicitly as a problem (or rather as *the very problem* founding Western metaphysics[10]) but also that the messianic way out is identified in the de-position of the human-animal divide itself, that is, in the overcoming of the very anthropocentrism that structured not only the Western tradition but also Agamben's thought so far. The shift is minimal but important: the issue is here, again, that of investigating the "practical and political mystery of separation" (2004: 16) that structured Agamben's previous analyses of metaphysics, but the Aristotelian passage around which this investigation is construed this time is not that about the voice (still quite—and too—logocentric) but a passage from *De Anima* (413a, 20 – 413b, 8) in which Aristotle identifies and defines "nutritive life" as separated from, and articulated with, the other levels of life. Nutritive life is identified as the "foundation" of any form of life, but, as we know by now, as such it must "go to the ground" and disappear in an inclusive exclusion (2004: 14). This apparatus is what structures the workings of what Agamben calls "anthropological machine," which from time to time "construes" the human by separating its vegetative and animal parts as excluded-included foundation from its superior and proper "human" part. The first half of the book is devoted to an archaeological description of the workings of the anthropological machine aimed at showing how its final product is, every time, "neither an animal life nor a human life, but only a life that is separated and excluded from itself – only a *bare life*" (2004: 38).[11] In fact, the collapsing of the difference between animal and human in biopolitical modernity—when the "total humanization of the animal coincides with a total animalization of man" (2004: 77)—and the consequent "running idle" of the anthropological machine is no "salvation" from its deadly workings; rather, it signifies on the contrary the universalization of the state of exception that the machine always produces—even when it is working "well." The only salvation, the only way out, is therefore to stop the machine itself.

The second part of *The Open* is constituted by a series of chapters proposing an intense reading of Heidegger's take on animality, in particular of the 1929–1930 lecture course later published as *The Fundamental Concepts of Metaphysics*. These chapters reiterate Agamben's Heideggerian understanding of Western metaphysics; however, they ultimately show how also Heidegger's *Dasein* is predicated on the inclusionary exclusion of the animal's particular mode of relation (Calarco 2008: 99) and thus that, despite Heidegger's uncompromising rejection of humanism, it is the culmination and the ultimate expression of the anthropological machine of humanism. Ziarek, in turn, accuses this interpretation of misreading Heidegger's core intention: Heidegger decisively refuses to inscribe *Dasein* and the human into the horizon of living beings—and thus of animality—as the humanist tradition has done for centuries (where *anthropos* has always been understood as *zoon* + *x*), and, by framing once again the definition of the human around the concepts of "life" and "living being," Agamben would remain *humanist* and fully *metaphysical* (2008: 189–90, 198). Agamben's reading of Heidegger here is in fact *biopolitical*—a stance that Heidegger himself would never have taken—and cannot but turn on its head Heidegger's hyper-anthropocentric antihumanism.

Be as it may, in a move typical of his *modus operandi*, Agamben turns at the end from Heidegger to Benjamin in order to find a messianic way out from the deadlock of metaphysics. Benjamin provides him with the possibility of thinking humanity and animality beyond the dominant logic of traditional metaphysical categories and also beyond the traditional *oikonomia* of "salvation." The way out is not the Heideggerian exasperation of the human-animal divide, but rather its de-position, in a messianic "jamming" of the metaphysical apparatus of humanism:

> To render inoperative the machine that governs our conception of man will therefore mean no longer to seek new – more effective or more authentic – articulations, but rather to show the central emptiness, the hiatus that – within man – separates man and animal, and to risk ourselves in this emptiness: the suspension of the suspension, Shabbat of both animal and man. (2004: 92)

Only this deposition—beyond any possible reconciliation (Abbott 2011: 94, 96), and unlike the deconstructive multiplying of differences[12]—can undermine the logic of inclusive exclusion. Only the deposition of this logic will overcome the presuppositional structure of metaphysics that relates two terms (*bios* and *zoé*, language and voice, human and animal, actual and potential, and so on) only by positing one as the submerged and negative foundation of the other. Despite all the unquestionable limits of his discourse, this is definitely Agamben's most important contribution to the contemporary debate on animality.

4   Life and/as Potential

The phrase Agamben uses to identify the messianic overcoming of the anthropological machine (and of metaphysics as a whole) is "outside of being"—which is the title of the last chapter of *The Open*. This concept is an evolution that goes far beyond that of "whatever being" Agamben proposed in *The Coming Community* (1993b), which, insofar as it was modelled on the experience of language, was restricted to humanity (Prozorov 2014: 171). And it also goes beyond the Heideggerian concept of potentiality that Agamben embraced early on and made all along the core of his philosophical proposal. If potentiality can be said to be the "central term of his philosophy" (de la Durantaye 2009: 3), it must be pointed out that it also underwent an evolution parallel to, and intertwined with, that of his critique of humanism. Potentiality is the mode in which Heidegger's *Dasein* exists, and as such it informed from the very beginning Agamben's philosophy, though implicitly at first. If the term entered Agamben's vocabulary only in the mid-1980s, it constitutes nevertheless already the logical structure of the experience of infancy, which is in fact not the actuality but the potentiality of speech (cf. Prozorov 2014: 71–72). And it already marked, in Heideggerian fashion, human exceptionality: if only human beings have infancy, it is because only humans have the potentiality *not to* speak, that is, to remain in *in-fancy*. This is, for Agamben, the very structure of potentiality—not only the potentiality of something but that *not to* do or be something—and it is what gives humans a freedom denied to non-human animals.

The link between infancy and potentiality is made explicit in a text of the mid-1980s included in *Idea of Prose*: "The Idea of Infancy." Here Agamben takes the axolotl, a neotenic salamander native of Mexico, as a paradigm of infancy and as a key to interpret human evolution: the axolotl is characterized by neoteny (a term coined by the German zoologist Julius Kollmann at the end of the nineteenth century precisely to describe the axolotl), that is, the retention of larval (or infantile) traits in adulthood that according to some theories played a pivotal role also in human evolution. Human beings, it is argued, evolved not from individual adults, but from the young of primates who had acquired, like the axolotl, the capacity for reproduction. What interests Agamben is that this "eternal child" is "so little specialized and so *totipotent* that it rejects any specific destiny and any determined environment"; unlike the other animals, who "develop only the infinitely repeatable possibilities fixed in the genetic code" and thus "attend only to the Law," humans—and only humans, though many other animal species display neotenic traits and behaviours—is "able to pay attention precisely to what has not been written, to somatic possibilities that are arbitrary and uncodified" and is thus "free from any genetic prescription" (1995: 95, emphasis added).

Prozorov (2014: 73) rightly notes how paradoxical it is that the example Agamben chooses to illustrate the exclusively human phenomenon of infancy belongs to the animal realm[13]: the axolotl is seen as the exception among animals (though this is empirically untrue), while this exceptionality is considered the rule for human beings. Or, in Chrulew's words, "Agamben's anti-biologism here relies, in fact, on a biologism of animals" (2012: 56). This exceptionalist pattern returns in a seminal 1986 lecture, published only in 1999 as "On Potentiality":

> *Other living beings – reiterates here Agamben – are capable only of their specific*
> *potentiality; they can only do this or that. But human beings are the animals*
> *who are capable of their own impotentiality. The greatness of human potential-*
> *ity is measured by the abyss of human impotentiality.* (1999c: 182, emphasis in
> the original)

Human exceptionality consists in the potential *not to* adhere to the species'
limited set of possibilities, and this "not to" is what bestows on humans an
exceptional freedom. After all, this is a very traditional and unoriginal
argument, which Agamben merely adopts from the Western philosophical
tradition and which he uncritically repeats without giving it too much
thought, even after the publication of *The Open*.[14]

In *The Open*, however, this argumentative line constitutes the core of
Heidegger's take on the human-animal divide (or rather "abyss"), which
Agamben reads precisely as the culmination of the metaphysical tradition. In *The Fundamental Concepts of Metaphysics*, Heidegger famously
adopted Jakob von Uexküll's notion of the *Umwelt* as the species-specific, spatiotemporal, subjective reference frame of animal life, which
ultimately cages animality within a limited set of possibilities, determined
by what Uexküll called "carriers of significance" and Heidegger renamed
"disinhibitors." Heidegger called "captivation" (*Benommenheit*) the animal's limited and deterministic relation with its disinhibitors, and it is
the impossibility to escape the limits of its captivation that constitutes
the animal's "poverty in world." The *Dasein*, to the contrary, experiences in profound boredom "the disconcealing of the originary possibil-
itization (that is, pure potentiality) in the suspension and withholding of
all concrete and specific possibilities" (Agamben 2004: 67). This is how
Agamben formulates Heidegger's thesis:

> What appears for the first time as such in the deactivation [...] of possibility,
> then, is *the very origin of potentiality* – and with it, of Dasein, that is, the
> being which exists in the form of potentiality-for-being [*poter-essere*]. But
> precisely for this reason, this potentiality or originary possibilitization con-
> stitutively has the form of a potential-not-to [*potenza-di-non*], of an impo-
> tentiality, insofar as it is *able to* [*può*] only in beginning from a *being able not*
> *to* [*poter non*], that is, from a deactivation of single, specific, factical possibili-
> ties. (Agamben 2004: 67)

But if, therefore, the human is "simply an animal that has learned to
become bored" (Agamben 2004: 70) and has become human through a

suspension of animality's captivation, then this whole structure does nothing but replicate the inclusive exclusion of animality at the very core of Western metaphysics. Excluding animal life from potentialities and freedom represents the core workings of the anthropological machine.

"Outside of being" means the de-activation and de-position of this very structure and of all its historical declinations, from the "open" to language, law, history, nature, biology, genetic code, and any biological or historical destiny. It means to let both human and animal "be outside of being, saved precisely in their being unsavable" (Agamben 2004: 92). If Agamben's biopolitical project started from the thesis that "[t]his biopolitical body that is bare life must itself [...] be transformed into the site for the constitution and installation of a form of life that is wholly exhausted in bare life" (Agamben 1998: 188), then *this body is the animal body* unmarked by any historical and metaphysical signature—first and foremost the signature "human." And bare life becomes thus a life of pure potentiality insofar as it is completely unmarked and finally free from any determination.

Admittedly Agamben never pursued this bigger picture, but he certainly opened up the field, when finally "abandoning" the *Homo Sacer* project, for others to continue and expand it (cf. Agamben 2016: xiii). One of his core methodological principles is to focus, in every work he approaches, on what Feuerbach called *Entwicklungsfähigkeit*, the capacity for elaboration (cf. Agamben 2009: 7–8); what remains "unsaid" and needs to be taken up and elaborated in Agamben's philosophy is a fuller thought of animality, and this is the task he has assigned to the coming philosophy.

א *Grass.* Agamben concluded the essay "Bartleby, or On Contingency," with an uplifting image of Bartleby the scrivener (a "reversal" in respect to traditional interpretations), where life is returned to pure (im)potentiality: "There is sky and there is grass. And the creature knows perfectly well 'where it is'" (1999c: 271). But between sky and grass, Agamben choses grass. His recent "autobiography," *Self-portrait in the Studio* (*Autoritratto nello studio*), ends in fact with these words: "But if I had to tell now in what I finally invested my hopes and my faith, I could only confess in an undertone: not in the sky – in grass. [...] Grass, grass is God. In grass – in God – are all those I have loved. For the grass and in grass and like grass I have lived and I will live" (2017: 166–67). Like Deleuze and Guattari, who counterpoised the transcendent, metaphysical tree—firmly rooted, vertical, and sky-bound—to antimetaphysical, horizontal, earth-bound grass (1987: 7–8), Agamben choses grass as the final image of life as pure immanence, of a life finally released even from the ultimate signature of "animality" and returned to pure potential.

NOTES

1. That I write "man" and not "human being" is no oversight: the use of the neutral universal "man" instead of (the politically correct) "human being" is not only still very common and widespread in Italian academia and society at large, but also marks a certain "gender blindness" characteristic of Agamben's writings.
2. "It is not language in general that marks out the human from other living beings – according to the Western metaphysical tradition that sees man as a *zoon logon echon* (an animal endowed with speech) – but the split between language and speech, between semiotic and semantic (in Benveniste's sense), between sign system and discourse" (Agamben 1993a: 51–52).
3. In a "gloss" (1993a: 56), Agamben cursorily cites the ethologist William Thorpe and his findings that in certain birds the "song" is not entirely written in the genetic code and must therefore be "learned," but for him this is definitely the proverbial exception proving the rule.
4. The founding myth in this sense is that of Epimetheus, who, according to Plato (*Protagoras* 320d–322a), gave all natural gifts to non-human animals forgetting human beings and forced his brother Prometheus to steal fire (= arts and technology) from the gods and give it to them, making thereby humans *free*. As, among others, Derrida notes, it is "paradoxically on the basis of a fault or failing in man that the latter will be made a subject who is master of nature and of the animal" (2008: 20). Roberto Marchesini has analysed and demystified this myth in a number of books (cf., e.g., 2014: 84–112).
5. This short text was initially published in a bilingual edition (Italian-French) the same year of the first Italian edition of *Language and Death* (1982) and was then included as "Epilogue" in its 2008 Italian reissue, but is not included in the English translation.
6. For an overview of this topic in Agamben's later work, and especially in *The Use of Bodies* (2016), see Cimatti (2015).
7. This long itinerary of *désœuvrement* climaxes in the conclusion of *The Sacrament of Language* (2011: 71): "It is perhaps time to call into question the prestige that language has enjoyed and continues to enjoy in our culture, as a tool of incomparable potency, efficacy, and beauty. And yet, considered in itself, it is no more beautiful than birdsong, no more efficacious than the signals insects exchange, no more powerful than the roar with which the lion asserts his dominion."

8. The term "abyss" is obviously the one used by Heidegger (*Abgrund*) to characterize the human-animal divide and that epitomizes the view of the whole Western tradition (cf., e.g. Heidegger 1995: 264 and passim).

9. In German *Abgrund* means simply "abyss," but the term is construed by Heidegger as the absence (negative *ab-*) of a foundation (*Grund*): that its logic characterizes Western metaphysics means for Heidegger (and also for Agamben) that metaphysics lacks a positive foundation in Being. The logic of the exception is instead the inclusion of something via its exclusion, as in Carl Schmitt's state of exception which includes the law by suspending (i.e. excluding) it.

10. Agamben writes in fact (2004: 16): "What is man, if he is always the place – and, at the same time, the result – of ceaseless divisions and caesurae? It is more urgent to work on these divisions, to ask in what way – within man – has man been separated from non-man, and the animal from the human, than it is to ta positions on the great issues, on so-called human rights and values."

11. An "antispeciesist" corollary to the anthropological machine appears a few years later in the short text "Special Being" of *Profanations*, where Agamben laments the hypostatization of the (human) species, where the term originally only means "appearance, aspect, or vision" (2007a: 56): "The transformation of the *species* into a principle of identity and classification is the original sin of our culture, its most implacable apparatus [*dispositivo*]" (2007a: 59). Species is the apparatus used to establish an identity by distinguishing between one's own appearance from that of other beings; self-recognition depends on a (fictitious because only imaginary, in the sense of depending on one's image) separation between humans and non-humans.

12. Derrida argues in fact that our "war" against the animal cannot be resolved through a deposition of the human-animal difference; rather, this difference—that he still calls, in Heideggerian fashion, a "rupture or abyss"—is necessary and to call it into question would be "asinine" [*bête*]. Derrida's strategy consists instead in "multiplying its figures, in complicating, thickening, delinearizing, folding, and dividing the line precisely by making it increase and multiply" (2008: 29–30; see also Chagani 2018).

13. By the same token, in "In Praise of Profanation," the cat who plays with a ball of yarn as if it were a mouse is taken as a paradigm of profanation, which deactivates the usual behaviour and opens it up to a new, possible use (Agamben 2007a: 85).

14. For example, in another important essay from 2004, "The Work of Man," he states again (admittedly citing Dante): "While the intelligence of the angels is perpetually in act without interruption (*sine interpolatione*) and that of the animals is inscribed naturally in each individual, human thought

is constitutively exposed to the possibility of its own lack and inactivity: that is to say, it is, in the terms of the Aristotelian tradition, *nous dunatos*, *intellectus possibilis*" (2007b: 9). That is why man (sic) has no proper "work", no *opera*, and is thus, in his essence, *inoperative*, that is, a *potential being*, open to all possibilities.

WORKS CITED

Abbott, Mathew. 2011. The Animal for Which Animality Is an Issue. *Angelaki* 16 (4): 87–99.

Agamben, Giorgio. 1982. *La fine del pensiero/La fin de la pensée*. Bilingual edition with French Trans. Gérard Macé. Paris: Le Nouveau Commerce.

———. 1991. *Language and Death: The Place of Negativity*. Trans. Karen Pinkus, and Michael Hardt. Minneapolis: University of Minnesota Press.

———. 1993a. *Infancy and History: On the Destruction of Experience*. Trans. Liz Heron. London: Verso.

———. 1993b. *The Coming Community*. Trans. Michael Hardt. Minneapolis: University of Minnesota Press.

———. 1995. *Idea of Prose*. Trans. Michael Sullivan and Sam Whitsitt. Albany: Suny Press.

———. 1998. *Homo Sacer: Sovereign Power and Bare Life*. Trans. Daniel Heller-Roazen. Stanford: Stanford University Press.

———. 1999a. *The Man Without Content*. Trans. Georgia Albert. Stanford: Stanford University Press.

———. 1999b. *Remnants of Auschwitz: The Witness and the Archive*. Trans. Daniel Heller-Roazen. New York: Zone Books.

———. 1999c. *Potentialities: Collected Essays in Philosophy*. Ed. and Trans. Daniel Heller-Roazen. Stanford: Stanford University Press.

———. 2004. *The Open: Man and Animal*. Trans. Kevin Attell. Stanford: Stanford University Press.

———. 2007a. *Profanations*. Translated by Jeff Fort. New York: Zone Books.

———. 2007b. The Work of Man. Trans. Kevin Attell. In Giorgio Agamben: Sovereignty and Life, eds. Matthew Calarco and Steven DeCaroli, 1–10. Stanford: Stanford University Press.

———. 2009. *The Signature of All Things: On Method*. Trans. Luca D'Isanto with Kevin Attell. New York: Zone Books.

———. 2011. *The Sacrament of Language: An Archaeology of Oath*. Trans. Adam Kotsko. Stanford: Stanford University Press.

———. 2016. *The Use of Bodies*. Trans. Adam Kotsko. Stanford: Stanford University Press.

———. 2017. *Autoritratto nello studio*. Milan: Nottetempo.

Attell, Kevin. 2015. *Giorgio Agamben: Beyond the Threshold of Deconstruction*. New York: Fordham University Press.

Caffo, Leonardo, and Ernesto C. Sferrazza Papa. 2015. Nuda vita come animalità: Un argomento di ontologia sociale contro Giorgio Agamben. *Lo Sguardo – rivista di filosofia* 18: 129–138.

Calarco, Matthew. 2008. *Zoographies: The Question of the Animal From Heidegger to Derrida*. New York: Columbia University Press.

Castanò, Ermanno. 2018. *Agamben e l'animale: La politica dalla norma all'eccezione*. Aprilia: Novalogos.

Chagani, Fayaz. 2018. The Animal That Therefore I Am Not: Agamben, Derrida and the Limits of the Human. Unpublished Paper.

Chrulew, Matthew. 2012. Animals in Biopolitical Theory: Between Agamben and Negri. *New Formations* 76: 53–67.

Cimatti, Felice. 2015. Agamben and Lacan (By Way of Joyce). A Short Note About *L'uso dei corpi* (2014). *Rivista Italiana di Filosofia del Linguaggio* 9 (1): 344–350.

Colebrook, Claire, and Jason Maxwell. 2016. *Agamben*. Cambridge: Polity Press.

de la Durantaye, Leland. 2009. *Giorgio Agamben: A Critical Introduction*. Stanford: Stanford University Press.

Deleuze, Gilles, and Felix Guattari. 1987. *A Thousand Plateaus: Capitalism and Schizophrenia*. Trans. Brian Massumi. Minneapolis: University of Minnesota Press.

Derrida, Jacques. 2008. *The Animal That Therefore I Am*. Trans. David Wills. New York: Fordham University Press.

Heidegger, Martin. 1995. *The Fundamental Concepts of Metaphysics: World, Finitude, Solitude*. Trans. William McNeill and Nicholas Walker. Bloomington: Indiana University Press.

LaCapra, Dominick. 2009. *History and Its Limits: Human, Animal, Violence*. Ithaca: Cornell University Press.

Marchesini, Roberto. 2014. *Contro i diritti degli animali? Proposta per un antispecismo postumanista*. Casale Monferrato: Sonda.

Prozorov, Sergei. 2014. *Agamben and Politics: A Critical Introduction*. Edinburgh: Edinburgh University Press.

Shukin, Nicole. 2009. *Animal Capital: Rendering Life in Biopolitical Times*. Minneapolis: University of Minnesota Press.

Watkin, William. 2014. *Agamben and Indifference: A Critical Overview*. New York: Rowman & Littlefield.

Wolfe, Cary. 2003. *Animal Rites: American Culture, the Discourse of Species, and Posthumanist Theory*. Chicago: The University of Chicago Press.

———. 2013. *Before the Law: Humans and Other Animals in a Biopolitical Frame*. Chicago: The University of Chicago Press.

Ziarek, Krzysztof. 2008. After Humanism: Agamben and Heidegger. *The South Atlantic Quarterly* 107 (1): 187–209.

CHAPTER 6

Deconstructing the *Dispositif* of the Person: Animality and the Politics of Life in the Philosophy of Roberto Esposito

Matías Saidel and Diego Rossello

1   INTRODUCTION

Roberto Esposito has become one of the most influential Italian political philosophers of the last three decades. His writings on topics such as the impolitical, community, immunity, biopolitics, personhood, the impersonal, politics, and negativity, as well as the specificity of Italian philosophy in the history of modern and contemporary political thought, have been

Matías Saidel wishes to acknowledge the financial support of CONICET/Universidad Católica de Santa Fe, Argentina. He also wishes to thank Professor Constanza Serratore for collaborating with important references from Esposito's *Ten Thoughts on Politics* (2011).

Diego Rossello wishes to acknowledge the financial support of FONDECYT, Regular Project 1171154.

M. Saidel (✉)
Catholic University of Santa Fe/CONICET, Santa Fe, Argentina

D. Rossello
Adolfo Ibáñez University, Santiago, Chile

© The Author(s) 2020
F. Cimatti, C. Salzani (eds.), *Animality in Contemporary Italian Philosophy*, The Palgrave Macmillan Animal Ethics Series,
https://doi.org/10.1007/978-3-030-47507-9_6

115

widely read, discussed, and translated. In this context, animality is a key element in his more general reflection on the relationship between politics, life, historicity, language, and thought. Rather than confronting the question of the animal head on, Esposito introduces this *problématique* as an element that traditional metaphysics and political philosophy have not been able to properly reflect upon. In fact, the tradition of metaphysics and political philosophy, widely understood, often misrecognizes, and at times even denigrates, the potentialities of animality in and beyond humanity.

On the one hand, animality appears in Esposito's early writings somewhat obliquely, as a feature, for example, of the Centaur: the mythical figure, half human and half animal, introduced by Machiavelli to characterize the skills a virtuous prince should have in order to remain in power. The figure of the Centaur suggests the necessary connection between human action and animality and seeks to account for the corporeality, vitality, and contingency of politics. The Centaur as a figure is also linked to Esposito's reflections on the impolitical as a realistic and agonistic perspective on politics, aimed at deconstructing the neutralizing function of the modern theory of sovereignty. In this sense, and as an alternative to the Hobbesian tradition, the Machiavellian republic can be the place of the immanentization of antagonism instead of the realm that excludes conflict as such. At the same time, the Centaur is also a figure of contamination and continuity between animality and humanity; of de-hierarchization of different forms of life, leaving behind the Hobbesian renunciation of our animality in order to become political subjects. In this sense, beyond the traditional hierarchy between *bios* and *zoé*, inspired by Aristotle and criticized by Agamben, Machiavelli and other Italian philosophers analyzed by Esposito in *Living Thought* (2012b) like Bruno, Vico, Leonardo, and Leopardi will help us see that both politics and philosophy are not the realms of pure *logos*, in which our animality must be excluded. On the contrary, politics and philosophy itself cannot be fully understood without taking into consideration animal drives, passions, life, and their place in historicity.

On the other hand, animality is a crucial element in order to deconstruct a tradition in the philosophy of law and metaphysics that invests in the notion of the *person*, making ample room for a politics of depersonalization and of animalization, therefore reifying certain aspects of human life (reason) while denigrating and objectifying others (drives, affects, bodies, etc.). Thus, if immunitary philosophy makes use of animality to deny personhood and humanity to certain groups of people, which can therefore be reduced to things and are excluded from the community (depersonalization), Esposito will introduce a critical philosophy of *impersonal life* and *flesh* that recognizes our animality in terms of *living corporeality* and its positive relation toward things. This deconstruction of personhood implies the deactivation of both the theologico-political and the anthropological machines that parcel living beings in terms of value or worth, by introducing metaphysical, juridical, scientific, and political hierarchies between humans, animals, and things. These hierarchies have legitimized different historical *dispositifs* of power (Christian, totalitarian, and/or liberal) as well as the domination over our own bodies; over other human beings considered to be less than human, including colonized peoples, women, Jews, Gypsies, and people with disabilities; and also—and this seems to be beyond Esposito's explicit concerns—over other non-human living species.

In that sense, Esposito's philosophy suggests that if we want to leave behind an immunitarian politics over life (*politica sulla vita*) that has produced subjection, suffering, and death of different groups of humans, animality should neither be thought as what can be destroyed with impunity nor as that part of humanity that must be rejected and controlled by higher spiritual and mental human abilities. On the contrary, Esposito shows—reading together Italian "living thought" and French post-structuralism—that it is only by establishing a new political and philosophical framing of existence in which corporeality and spirituality, animality and humanity, and *zoé, techné*, and *bios* are not severed, that a politics of life (*politica della vita*) can be both thought and experienced.

In what follows, and drawing on the work by Esposito, we will focus our discussion on two alternative paradigms for thinking about the relation between the personal, the animal, and the political. These alternative paradigms are represented precisely by the two mythical figures discussed above: Machiavelli's Centaur and Hobbes's Leviathan. According to our reading, the Centaur and Leviathan show alternative ways of thinking and organizing what counts as human and what counts as animal in the realm of the political. We will defend the thesis that whereas Machiavelli, as read by Esposito, is hospitable to the animality of the human being for thinking the political, Hobbes is concerned with keeping animality at bay in the state of nature, at a distance from a pacified and well-ordered political society. Finally, we will also explore Esposito's deconstruction of the notion of the person, as a kind of apparatus that controls and domesticates the animality of the human being. Drawing on Esposito's notion of the impersonal, we will examine the potential of such category for reconceiving the relation the human can have not only with his or her own animality but also with non-human animals and nature in the realm of the political.

2   Politics, Life, and Historicity: The Centaur and the Leviathan

Although it would be difficult to assert that animality as such is a central issue in Esposito's philosophy, his reflections on the topic can be traced back to his early writings on Machiavelli where he discusses, as it was stated above, the mythical figure of the Centaur (1984). Esposito comes back to the Centaur later on in his corpus, in particular when he addresses the relationship between philosophy and politics, politics and life, as well as humanity and animality (2011, 2012b, 2015a). Indeed, Machiavelli is a key author in Esposito's genealogy of a specific Italian tradition of political thought in which antagonism, bodily existence, and historicity can neither be excluded nor transcended. The immanent relationship between politics and life as seen through antagonism, passions, and corporeality is unavoidable in contemporary biopolitical thought but was already theorized by Machiavelli, and other Italian philosophers, well before the early modern turn toward rationalism exemplified by the philosophy of René Descartes. Although different from Descartes in many relevant ways, the political philosophy of Hobbes takes part in such turn by means of the substitution of the *Ordo naturalis* with an artificial machine whose main goal is to exclude any expression of conflict within civil society.

In this sense, Hobbes inaugurates mainstream modern political thought (and hence modern political theology) in at least three ways. Firstly, he construes a clockwork mechanism that excludes from civil society not only conflict but also any contact between humanity and animality. Leviathan is, at the same time, an artificial person and a mortal god that leaves the chaos of the state of nature behind only to be threatened by external or civil war. Secondly, Hobbes establishes Leviathan as the condition sine qua non to institute private property and to create a labor market which will become, from Locke onwards, the main goal of liberal political philosophy and institutions.[1] Thirdly, Esposito finds in Hobbes a new relationship between life and politics that gives birth to the modern immunitarian paradigm. Indeed, from Hobbes onwards, the main goal of the state (and hence of politics) is not *eudaimonia*, happiness, self-sufficiency, virtue, justice, or even power, but *conservatio vitae*. According to Esposito, the conservation of life is achieved through an immunitarian mechanism by which in order to protect life, politics must negate alternative forms of relationality (Esposito 2010). Put differently, according to Esposito reading Hobbes, the only legitimate social relationship is that of protection and obedience between sovereign and subject.[2]

These three levels of Hobbesian thought are intermingled, and Esposito will deal with them in different parts of his work, opposing Italian living thought—in which Machiavelli has a key role—to the cold machinery put forward by Hobbes. Indeed, the exclusion of conflict will be thoroughly reflected upon in his early texts on Machiavelli, especially in *Ordine e conflitto* (*Order and Conflict*, 1984), and will be revisited in his two important books dedicated to the impolitical—*Categories of the Impolitical* (2015a, originally published in 1988) and *Nove pensieri sulla politica* (*Nine Thoughts on Politics*, originally published in 1993 and now republished with an extended version as Esposito, 2011)—as well as in *Living Thought* (2012b).

According to Esposito, after Hobbes political philosophy will be structurally incapable of thinking conflict, and hence politics, because it will only reflect upon an order that excludes conflict as such. Furthermore, Hobbes theorizes modern *political representation* through the trope of theatrical representation—in which the people authorizes any action made by the sovereign actor. Thus, political representation understood in this way tips the balance in favor of the representative excluding any possibility of dissent, based on the argument that the people cannot question actions that they have themselves approved and authorized in the first place.[3] Needless to say, this way of understanding political representation reinforces the structural impossibility of grasping conflict in political philosophy:

> There is a close and binding connection between order and representation, in the sense that representation [...] is always of the order. Even when it encounters conflict, it does it from and within the presupposition of order, possible if not actual. [...] Yet the conflict, in all its vast range of expressions, is nothing but the reality of politics, its *factum*, its facticity [...]. This real facticity, this factual reality, does not enter into the representative schemes of political philosophy, it is not pronounceable in its conceptual language. (2011: 30-31)[4]

On the contrary, the realistic tradition in which Machiavelli's thought is inscribed cannot conceive a world outside conflict and hence outside politics, coinciding in that sense with the impolitical perspective proposed by Esposito. If political philosophy understands representation as that of an order, the impolitical perspective will necessarily think beyond representation, since conflict is unrepresentable. In that sense, Esposito (2011) maintains that it is not surprising that Machiavelli was rejected as a nonphilosopher by thinkers of the Counter-Reformation, as well as by modern Hobbesian and Cartesian trends. The latter accused him of subtracting politics from a truly scientific study by ignoring the assumptions of unity and stability, as well as the (normative but also scientific) inquiry on the best regime; the former read Machiavelli as a humanist pagan and a teacher of evil. Thus, while political philosophy tries to reduce multiplicity to unity, and difference to sameness, the very figure of the Centaur introduces difference and multiplicity in the apparent notion of a unitary subject:

> In this sense Machiavelli's thought is eminently anti-representative, anti-symbolic. Or "diabolical" [...] recognized and fixed in an eternal *diaballein* that breaks every possible *synballein*, every philosophical symbol, that is, every presupposed unity, of the political. No coincidence that the only symbol used by Machiavelli to characterize it – the Centaur – incorporates, without excluding it, the split and indeed it "symbolizes" it. Divided, broken, between man and beast, reason and force, order and conflict, the political subject, in order to "consist," to "last" in its own inconsistency, must "finish" as a whole, die as a subject-human, bring within itself his own difference, his own other, his own inhumanity: "To have as teacher a half-beast, half-man means nothing other than that a prince needs to know how to use both natures; and the one without the other is not lasting" (Machiavelli, *The Prince*). What only lasts, what "resists" the "wheel" of fortune, is *the unity of the divergent*: the breaking of the political subject both in its ancient humanistic typology and in the new individualistic-universal connotation told by the myth of the contract. (2011: 40)

From this quotation, in which Esposito evokes his previous book *Order and Conflict*, we can extract different elements for our subject. On the one hand, with regard to the relationship between animality and humanity, *zoé* and *bios*, classical political philosophy maintains that outside the political realm human beings are transformed into either beasts or gods. Hobbes reinforces this idea, since Leviathan, which is depicted both as an artificial animal and as a mortal god, wants to exclude any form of animality—understood as irrational behavior, voracity, aggressive instincts, and so on—from the civil state and relegates animality to the state of nature. Indeed, even though Hobbes's animalization of the human being in the state of nature, which has upset his contemporaries from different schools of thought, "relaxed the human-animal divide" (Rossello 2012: 259), it also implied that we become fully human only in the civil state, where animality has no place:

> Hobbes's animalization of the human being came with crucial qualifications. His political theory tempered or tamed his apparent demotion of the human to the animal by, at the very same time, redrawing the lines that distinguish between the two, and calling attention to the contrast between the nonpolitical gregariousness of nonhuman animals and the political character specific to human associations [...]. According to Hobbes, human animals depend on the artifice of the commonwealth to be what they are; if they fall outside of it, they do not just exit a political order, they also default on their humanity – and return to the brutish existence proper to the "liberty of the beasts." (Rossello 2012: 259)

Hence, although Hobbes recognizes that "man surpasseth in rapacity and cruelty the wolves, bears, and snakes that are not rapacious unless hungry and not cruel unless provoked" (1998: 40), he wants to oppose the voraciousness of animals to the speech of human beings who can establish a covenant (Rossello 2012: 260; Derrida 2011: 23, 65).[5]

On the contrary, the Machiavellian figure of the Centaur shows that politics is a realm from which animality cannot be excluded. Rather, the prince needs both humanity and animality and law and force, in order to rule and remain in power (Rossello 2013). As Esposito underlines, in this unity of opposites the political subject explodes, questioning not only the prior humanist assumptions from Aristotle to Cicero but also the universal, ahistorical, and self-interested individual that the figure of the social contract from Hobbes onwards needs to presuppose in order to work:

> In this sense, the figure of the Centaur not only enables to put into question order as origin and becoming of the political, and hence the categories that organize political thought since modernity. It also enables to question the place of the human using this image, since in order to govern, to keep power, one must behave as a Centaur, i.e., from a dual nature in which reason and animality differ and complement. (Soto García 2015: 78)

Hobbes's exclusion of animality from politics has also to do with a different notion of temporality and origins. While Hobbes wants to leave the state of nature behind after the contract, reducing politics to a neutralization of the originary conflict, Machiavelli—and Italian living thought—conceives politics and violence as originary and inescapable, since they are rooted in human life and passions.[6] These irrational aspects of human behavior have, according to Esposito, no place in Hobbes's Leviathan, which ultimately hinges upon the negation of the natural human being and its drives toward aggression and on human being's rationality and its capacity to create artificial political constructs (Esposito 1984: 194). As Esposito maintains:

> The Potency of the machine-state, unlike the Machiavellian state-body, is proportional to the control-exhaustion of every passional impulse […] to the break with any biologically rooted element in the natural life of the body […] it is the renunciation of one's own bodily origin that makes the operating speed of the machine perfectly functional […]. The irreversibility of the transition from the natural to the civil state marks the sharpness of this break. We are at the antipodes of the Machiavellian state, which draws the strength to resist the ineluctability of its "civil" degeneration only from the relationship with its origin […] on the Hobbesian scene […]. Foundation means first of all cancellation of the origins. (Esposito 1984: 194)

Furthermore, according to Esposito, Hobbesian philosophy excludes both natural and political histories since they are based either on experience or authority, but not on reason (Esposito 1984: 208). While, as Esposito reminds us, Hobbes maintains that we shall not seek political examples from the Greeks and the Romans,[7] and characterizes the ancient Roman people as ferocious beasts,[8] Machiavelli links innovation to the return to beginnings (*ritorno ai principii*) from which Esposito maintains that "real innovation is precisely to reconquer the roots, finding the beginning again, and coming back to the origin" (Esposito 1984: 200). Needless to say, this origin does not need to be fully left behind.[9] On the contrary, reflecting on Vico's notion of *ricorso* (recurrence or reflux), Esposito maintains that the origin can be reactivated as an energetic resource for politics—or, on the contrary, suffered as a phantasmatic return (Esposito 2012b).[10]

Hobbes is also critical of the Republican idea of a mixed state or constitution. According to him, there is a need of absolute power in order to avoid social conflict that he identifies with civil war (Esposito 1984: 184). On the contrary, Machiavelli's originality lies in his appreciation of a mixed state precisely taking into account the social conflictivity it enables. If in most humanist Aristotelian tradition the golden mean guarantees order, Machiavelli rejects this idea as it avoids the productive tension of the extremes (Esposito 1984: 185). Hence, while Machiavelli bets on the connection or co-implication between order and conflict, Hobbes maintains that there can only be either one or the other (Esposito 1984: 186). While Machiavelli shows that conflict is the essence of politics, Hobbes theorizes a mutual exclusion between politics and conflict. If for Machiavelli conflict is not only unavoidable but also creative, for Hobbes it is only artificially neutralizable (Esposito 1984: 188). But in that case, according to Esposito, Hobbes's politics consists only in an act of de-politicization: outside the state politics is war (and hence non-politics); within the state politics is civil law (again, non-politics). Politics, in the Machiavellian sense, does not exist neither before nor after the state (Esposito 1984: 188).

As Daniela Calabrò (2012) reminds us, following Esposito, the novelty of Machiavelli's reflection on politics in contrast with the Christian-humanist anthropology is situated precisely in this double dimension: Machiavelli proposes a virtuous oscillation between the political and the state, subjectivity and institution, decision and machine. The figure of the Centaur is a key element in Esposito's reading of Machiavelli, since these oppositions cannot be pacified and modern politics is "suspended in an antagonistic dynamic between sovereignty and institution, in the balanced game of their conflictual representability" (Esposito 1984: 15). The passage from the Centaur to Leviathan can thus be read as an attempt to neutralize an emerging conflictivity. In other words, the objective is to hide a gap in the history of political thought, represented by Machiavelli, toward a sovereign apparatus that conceives sovereignty as belonging exclusively to a compact and unitary subject (the State) (Esposito 2011).

Precisely this figure of the Centaur gives the name to the journal in which Esposito took part during the 1980s (*Il Centauro*) to reflect upon the notions of negativity and crisis of the political in a non-dialectical way. Far from the figure of the organic intellectual, this academic review goes back to the origins of modern political philosophy in order to better understand both its categories and their crisis. In this context, the figure of the Centaur marks a threshold of discontinuity between humanism and political modernity and alerts us of a crisis, a scission even, in the origins of modern political thought (Calabrò 2012). Esposito's critical reflection on the notion of subject, origin, sovereignty, politics, and historicity, which will be systematized later in *Categories of the Impolitical,* emerges clearly in this period. In that sense, Esposito maintains that the impolitical attitude coincides with a political realism that, after Machiavelli, was censored by modern political philosophy. Such realism conceives the political beyond ethical and organicist presuppositions, identifying dissent as the a priori of human co-existence (2011).

In that sense, one of the key elements that Esposito traces in Italian thought is an alternative to modern rationalism that considers animality as irrationality and instinctiveness that corrupts our higher spiritual faculties, including animality as something that must be excluded or governed by personhood. In fact, even though Hobbes's one-time recovery of the metaphor of *homo homini lupus* has been amplified well beyond its original scope,[11] the beast in his theory is always related to ferocity in human beings themselves, suggesting that when there is no sovereign power, humans are naturally inclined to wage war reciprocally.

In *Living Thought* Esposito shows that, like Machiavelli, Leonardo da Vinci has a very different understanding of this issue. In fact, the figure of the Centaur reappears in the depiction of horsemen in the battle of Anghiari, in which we can not only see an expression of "conflict as such" but also assist to an animalization of humans that parallels the humanization of animals—like the horses that seem to fight with the same passion of their riders. This ambivalence is present in Leonardo's idea on the aggressiveness of human beings "who have always lived by the death of others: by the death of the animals they eat, but also by the death of the other human beings." He also maintains that animals "do not eat specimens of their own kind, unless they are mad" and that "Man has much power of discourse which for the most part is vain and false; animals have but little but it is useful and true and a small truth is better than a great lie"(Leonardo qtd. in Esposito 2012b: 103). As we can see, these reflections are strictly opposite to the Hobbesian idea that the ability to speak, ratiocinate, and make oaths is what makes us political subjects capable of leaving behind ferine rapacity and voracity. From Leonardo's thoughts and paintings, Esposito concludes that

> one cannot help thinking that, for Leonardo, instead of the animal corrupting the human [...] it is rather humans who have managed to corrupt animals by transmitting our homocidal [sic!] fury to them. If this is the case, then the "monster" is not the man with the bestial features that critics have

also seen in the *Battle*, but the beast with the human features. (Esposito 2012b: 103)

According to Leonardo, we cannot blame animals for our ferocity. Irrational and unnecessary violence, which can ultimately derive in the annihilation of other living beings, is an all-too human trait. At the same time, humans as political subjects cannot leave animality behind. In that sense, as Soto García argues, the image of the Centaur helps us depart from rationalist conceptions of the human and interrogate the relationship between biological and political life, since such mythical figure represents the conjunction of *zoé* and *bios* within human life.[12] Hence, human life cannot be reduced to reason. If Hobbesian sovereignty institutes the protection of life as the main task of politics through immunitarian mechanisms, Italian living thought can be recovered in order to think another version of biopolitics. This alternative version recognizes that life, passions, bodily needs, and desires are at stake in politics and hence our ferine agonism cannot be fully domesticated.

3   PERSONHOOD, ANIMALIZATION, AND POLITICS OVER LIFE

As said above, Hobbes's theory of sovereignty has been read by Esposito not only as symptomatic of political philosophy's incapacity to think politics as such but also as the beginning of the modern immunitary paradigm in (bio)politics. As the reader may know, the notion of biopolitics became prominent in political philosophy after Foucault's theorization of the productivity of modern biopower and its difference with, and relation to, sovereign power.[13] "The right of sovereignty was the right to take life or let live. And then this new right is established: the right to make live and to let die" (Foucault 2003: 241). Biopolitics is a modern way of exerting power that takes care of phenomena that become relevant when considered statistically. Hence its main referent is the population.[14] Indeed, new sciences that rely on statistical knowledge, like biology, public health, demography, and political economy, provide governments with information that allows them to manage their populations.

Far from opposing biopolitics to sovereign power (as in Hardt and Negri) or assimilating the former to the latter (as in Agamben), Esposito articulates biopolitics and sovereignty in what he calls the immunitary paradigm. Through this articulation Esposito characterizes not only the *temporality* of biopolitics but also its *meaning*. Immunization involves the protection of life through its own negation. While in early modernity this immunization was mediated by different conceptual apparatuses, like sovereignty, property, freedom, and security, among others, in the last century, the relationship between politics and life becomes intimate and unmediated. In fact, once immunization has been developed, the dose of immunizing agents becomes a strategic decision. Thus, although there is an originary relationship between life, technics, and politics, we now live in a biopolitical and immunitarian era in which the main goal of politics is to protect and enhance *human* life, and this leads to the subordination of certain forms of life that are considered not fully human. Modern racism, sexism, and speciesism are clear examples of a politics that protects certain forms of life at the expense of others, a process that led to the elimination of certain groups of people in order to make life "healthier and purer" (Foucault 2003: 255) as well as the continued exploitation of non-human animals.

In that framework, as Esposito (2015c) and Agamben (2004) have pointed out, animality has functioned in western philosophical and scientific thought as a way to introduce differential thresholds within the human, dividing zones of different value or worth (Saidel 2013, 2014). Accordingly, Cary Wolfe maintains that "to live under biopolitics is to live in a situation in which we are all always (potential) 'animals' before the law" (Wolfe 2013: 10). If animality is what can be exploited to death or even killed with impunity, it is not surprising that biopolitics needs to animalize human beings in order to kill them. In fact, serial production invented by Ford in the automobile industry was inspired by the slaughterhouses of animals and will be tragically applied to massive production of death in Nazi extermination camps a few decades later. Hence, Wolfe maintains that it is "ironic, then, that the main line of biopolitical thought has had little or nothing to say about how this logic effects non-human beings – a cruel irony indeed, given how 'animalization' has been one of its main resources" (Wolfe 2013: 10).

Even if genocide as a juridical and political concept was born in the twentieth century, throughout history different political regimes have found ways to depersonalize human beings in order to treat them like animals—that is to say to make them available, disposable, and killable. During the colonial enterprise, modern racism was used to justify the enslavement and slaughter of millions of human beings that were not seen as fully human. At the same time, witch hunts presupposed the dehumanization of women in order to assassinate them—through fire or other means. In a similar vein, Nazism has been considered paradigmatic of how biopolitics, through the dispositive of racism, can turn into a systemic thanatopolitics, a politics of death oriented to erase certain human groups from the face of the earth. In Esposito's terms, Nazism led the immunization to a point in which, like an auto-immunitarian disease, the mechanism that was supposed to protect the national body became so aggressive that began to attack it instead (Esposito 2008). But in order to do so, the Nazis had first to convince themselves that Jews or Gypsies were *Untermenschen*, creatures that are not fully human—or even parasites, bacteria, or insects. Once again, the destruction of different layers of personhood and the animalization of human beings appeared as the condition of possibility of their annihilation.

In this context, it is not surprising that after World War II the *Universal Declaration of Human Rights* (1948) and the national and international law derived from it have tried to protect human life by establishing the legal sacredness of the person. Nowadays, animal rights groups claim that animals can also be considered as persons and that it follows that nonhuman creatures also have rights. However, according to Esposito, what Christian personalism, human rights law, and most of liberal political philosophy fail to acknowledge is that personhood presupposes the possibility of depersonalization and animalization, since the very dispositive of personhood produces the distinction between the person and what is subordinated and subjected to such person. In this sense, his thesis is that "the essential failure of human rights, their inability to restore the broken connection between rights and life, does not take place in spite of the affirmation of the ideology of the person but rather *because* of it" (2015c: 5). This dispositive of the person operates "separating what it purports to join and by unifying what it divides, by submitting one part to the domination of the other" (2015d: 3).

We can see this, according to Esposito, in nineteenth-century biological knowledge which establishes a separation between a relational life and merely organic one; this separation enables, in turn, the construction of degrees of humanity as well as the domination of one form of life over others. Accordingly, those who are closer to animality must be subjugated to rational and hence personal forms of life. Esposito also addresses the role of human sciences such as linguistics and anthropology in the general biologization of politics, as well as the role played by animality within that framework. According to him, the idea of a "double biological layer within every living being – one vegetative and unconscious, and the other cerebral and relational – [...] initiated a process of desubjectivization, which was destined to drastically change the framework of the modern concept of the political" (2015c: 6). If our will and passions are determined by the blind force of vegetative life, it means that we cannot rule ourselves completely. Therefore, the political depends on a biological fact that precedes our subjectivity and cannot be modified. When these kinds of theorizations are transferred from the individual to the human species, the latter is divided into zones of different value: some human beings are considered to be closer to animality than to humanity, initiating a process of depersonalization that arrives at its apex in the concentration camps.[15]

In juridical terms, according to Esposito, the dispositive of personhood is based on the presupposed separation between person, body, and thing. The person is conceived as an artificial entity different from the "human as a natural being, whom the status of person may or may not befit" (2015c: 9). Indeed, at least since Roman Law, the juridical and theologico-political institution of personhood presupposes that some human beings, who are not persons, can be treated as things and that even those who get the status of a person remain in the vicinity of, and can eventually fall into, the status of a thing. In the terms of Timothy Campbell, "[i]n other words, only those who have sufficiently been personalized, which is to say those who have successfully subjugated the animal, are considered to be fully human" (Campbell 2012: 38). According to Esposito, this separation between *persona* and *homo* penetrates modern juridical, philosophical, and political thought. For instance, according to Esposito, both Hugo Engelhardt and Peter Singer, considered the greatest exponents of liberal bioethics:

> are linked together by the ontological distance between personal life and biological life, which they both take for granted. With the failure of the idea of body as an inseparable substrate of personhood, this latter becomes (or is re-established as) a qualifying condition dependent on a series of attributes – reason, will, moral sense – that not all human beings possess, or that they possess only partially. It is precisely the presence, or the extent, of these "indicators of humanhood," as Singer describes them, that divides what we commonly call human beings into two distinct categories: those who can be considered simple "members of the species Homo sapiens" and those who deserve to be called real and proper "persons." (Esposito 2015c: 97)

On the contrary, from his impolitical stage onwards, Esposito's work deconstructs personhood as a dividing *dispositif* that introduces the possibility of depersonalization and which remains a key element of the theologico-political tradition since antiquity—being also at the basis of the idea of sovereignty (from *pater familias dominium* to the modern theory of the state).

However, the crisis of modern sovereignty does not lead to the end of the *dispositif* or apparatus that creates and recreates the person. Nowadays, personalization through human capital's competitiveness goes together with the depersonalization and animalization of disposable or surplus populations (Campbell 2012: 42ff). In this vein, Esposito shows how liberal-utilitarian bioethicists, like Singer and Engelhardt, justify the suppression of "defective" or "handicapped" children that could take the place of "normal" ones, raising the costs in suffering and the efforts of their parents (and society): "In neoliberal society, those who cannot develop their human capital and hence their employability; those who do not adapt to the requirements of market economy – willingness to compete, self-entrepreneurship, flexibility, etc. –, are condemned to a lower degree of personhood: less property, less freedom, less rights and less humanness" (Saidel 2014: 122).

As we can see, Esposito is well aware of the thanatological consequences of the depersonalization and animalization of human beings. However, it seems that, like other thinkers of biopolitics, Esposito does not explore the issue of how late capitalism affects the fate of non-human animals. In that framework, Wolfe asks whether is it possible "to rethink the common fate of human beings and non-human beings as having a shared subjection to these technologies and mechanisms that take the body as a new kind of political resource, to be in some cases killed, and in other cases, maximized and manipulated?" (qtd. in Glaser 2011). Esposito gets close to this pressing issue, perhaps, is his deconstruction of personhood and the proposition of a philosophy of the impersonal, which, at least potentially, can be conceived as an extension of the *communitas* to non-human animals and nature.

4   DECONSTRUCTING PERSONHOOD: TOWARD AN IMPERSONAL COMMUNITY

So far we have seen that immunitary apparatuses impose norms over life producing different forms of exclusion and that personhood separates life into zones of different worth or value. In order to contest such apparatuses, Esposito tries to think an affirmative biopolitics linked to the notion of *impersonal life*. In this notion, norm and life coincide, and *communitas* is understood as a form of commonality that is not based on any kind of property. In both cases, there is a deconstruction of both the person and the modern subject, a demotion of the boundaries that secure the separation between man and animal, persons and things, and at the same time a theorization of life as the insurmountable horizon of contemporary politics and thought.

This undoing of the rigid frontier between person and thing, man and animal, can be found in the notion of the *impersonal*, which traverses Esposito's philosophy from the impolitical stage to our days and enables him to think the relationship between commonality and singularity beyond modern subjectivity. Esposito repeatedly refers to Simone Weil's philosophy of the impersonal to deconstruct the metaphysical, juridical, theological, and political notion of personhood. From *Categories of the Impolitical* to *Third Person*, and also in *Persons and Things*, Esposito takes Weil's assertion that the most sacred in us is not the person but the impersonal. Indeed, Weil maintains that since ancient Rome, the juridical tradition has turned human beings into things, since any right always implies a kind of privilege that can be maintained by violence. According to Weil, if the notion of personhood implies the separation and subordination of human beings, the only possibility of avoiding coercion is through the impersonal: "Just as rights belong to the person, justice pertains to the impersonal, the anonymous" (Esposito 2015c: 101). Interrupting the immunitary mechanism that "introduces the 'I' into the simultaneously inclusive and exclusive circle of the 'we'" (Esposito 2015c: 102), Weil links the impersonal to the singular. Only by defusing personhood, human beings can be thought as such, for what they have as most unique and also as most common with each other exceeds the boundaries of the person. In this sense, Weil wants to overturn the particularism of the juridical form into the aporetic figure of a common right:

This is what lies behind the intention of re-establishing – against personalism – the primacy of obligations over rights: the obligation of each, added to that of all the others, corresponds in a global count to the rights of the entire human community. Only the community – conceived of in its most radical signification – can rebuild the connection between rights and human beings that was severed by the ancient blade of the person. (Esposito 2015c: 103)

After analyzing the **third person** in the realm of linguistics, Esposito retrieves this coincidence of right and justice and of singularity and commonality beyond personhood in Kojève's suggestion that, at the end of history, humans will become animal again. In fact, following Hegel, Kojève sees human history as a process of de-animalization through which the juridical subject emerges in its difference from the animal species *Homo sapiens*. Law must emancipate from animality in order to give life its authentic human dimension. However, Kojève overturns these conclusions when referring to the universal and homogeneous state that will emerge at the end of history, in which no one will have a private interest that does not coincide with the common one, and juridical civilization will return to the very animal dimension from which it had strived to depart: "at the end of time, when the general will coincide with the singular and the proper with the common, the third person will rediscover, or recognize, its own impersonal substance" (Esposito 2015c: 114). According to Esposito, the fact that human beings will become animals at the end of history implies that "this is not a pure return to a primitive condition, but the achievement of a state never before experienced: rather than a simple re-animalization of the now humanized human, it is a way of being human that is no longer defined in terms of alterity from our animal origins" (Esposito 2015c: 114).

According to Esposito, commonality and singularity also coincide in the notion of *impersonal life* developed by Gilles Deleuze. Esposito identifies three sources of attack to the dualistic model of the person in *A Thousand Plateaus*, which are related to the notion of life: first, the substitution of possibility with Bergson's virtuality; second, the notion of individuation that displaces the horizon from subject to life and where the notion of *haecceity* refers to the temporality of the *event*; third, the notion of *becoming animal*:

> Deleuze's becoming-animal takes on its full constitutive and countering force when we recall that the animalization of man was the most devastating outcome of the dispositif of the person, but also of the thanatopolitical powers that imagined they were opposing it while actually enhancing its coercive power. In a theological, philosophical, and political tradition that has always defined the human through opposition to the animal – to that part of the human, or that area of humanity, that was bestialized as a prophylactic measure – the vindication of animality as our most intimate nature breaks with a fundamental interdiction that has always ruled over us. (Esposito 2015c: 149-150)

As we can see, in this philosophy of life, the ontological difference that metaphysics wanted to establish between man and animal crumbles. Animality is our most intimate nature, what we have in common with other human beings and other non-human animals. Hence, becoming animal brings back humankind to its natural alteration and also opens humanity to

> plurivocity, metamorphosis, contamination – and preventive critique of any claim to hereditary, ethnic, or racial purity. In opposition to purity, against its supposed immune effects, the becoming-animal becomes "propagation by epidemic, by contagion, [which] has nothing to do with filiation by heredity." (Esposito 2015c: 150)

According to Esposito, this concept by Deleuze and Guattari not only puts at stake the relationship with the animal but also, and more importantly, the becoming of a life that individuates itself by breaking the chains and prohibitions that man has imposed on it. Thus, becoming animal unties the metaphysical knot we have inherited and refers to a form of being human that does not coincide with the person nor with the thing.

As we can see, following Deleuze, Esposito defines as impersonal singularity the possibility to restitute normativity to the becoming of life in its different forms. A singular impersonality that traverses man, plants, animals "independently of the matter of their individuation and the form of their personality" (Esposito 2008: 194). According to this philosophy, all the living is part of the same process, and no portion of life can be sacrificed to another. This conception of *bíos* escapes sacrifice by subtracting itself from the theologico-political *dispositif* of western metaphysics. This philosophy of impersonal singularity has nothing to do with the relationship between universal or transcendental and the subject or person but with the pre-individual and trans-individual (Simondon), which can also be connected to the notion of *flesh* (Merleau-Ponty), as an originary biological pulp in which man and animal become indiscernible (as in Leonardo's *Battle of Anghiari*).

Calabrò (2012) maintains that this relationality toward other forms of life, like plants, bacteria, viruses, and animals, which does not take place according to instituted hierarchical divisions and is not defined by external sovereign mediations or any other transcendence, gives ecological relevance to the notion of "affirmative biopolitics" (Esposito 2008: 10). If the metaphysical and theologico-political dispositive of personhood is founded on the separation between different forms of life and the subordination of some of them to the person, impersonal life seeks to think a "*synolon* of form and force, interior and exterior, *bíos* and *zoé*" (Esposito 2015c: 151). At the same time, the idea that becoming animal implies a propagation not by inheritance but by contagion is linked to the notion of *communitas* as a form of commonality beyond identity.

Some scholars have seen in Esposito's affirmative biopolitics a kind of neovitalism that leads to aporias that stem from the generic dedifferentiation between forms of life. For instance, following Tim Luke, Cary Wolfe states: "Will we allow anthrax or cholera microbes to attain self-realization in wiping out sheep herds or human kindergartens? Will we continue to deny salmonella or botulism micro-organisms their equal rights when we process the dead carcasses of animals and plants that we eat?" (Wolfe 2013: 59). At the same time, Claire Colebrook's challenges Esposito's recourse to life that only seem to be anti-humanist by making "all forms of life equal since all express it" (qtd. in Wolfe 2013: 63).

Other scholars have pointed out that Esposito is not posing life as *hypokeimenon*. Rather, his concern is to think a politics that potentiates life without enclosing it in biologically assigned identities. Needless to say, Esposito would accept that, at the ontological level, life is something we have in common with the rest of living beings. However, life is not a transcendental magma from which all individuals stem. Life is always already formed; it is always singular, traversed by power, contingency, and historicity. In fact, if we attend to Foucault and others, life *itself* did not even exist before modernity (Foucault 2001; Tarizzo 2010).

Esposito finds an anticipation of this thought on biopolitics and impersonality in an Italian tradition that has reflected permanently on life, politics, and historicity and that, at the same time, has always been an anti-statist tradition, in tension with the theologico-political thought of sovereignty.

Hence, he finds the possibility to think an *immanentization of antagonism* and a subject not separated from its bodily existence, open to the community and to the world. In fact, the body is, at the same time, the target of immunitary apparatuses of modern power and the *locus* of resistance and connection. The body can close into itself and reject what is extraneous to it but can also open itself to the other, as in the philosophy of the *flesh*. This flesh can potentially deconstruct any form of anthropocentrism and subjectivism, since it exposes every-body to alterity and contamination with other bodies and the world. Like life, *flesh* is impersonal and common to a plurality of singular beings. It is not the opposite but the other of the body, its extroversion (Esposito 2008; Saidel 2016). That is why Esposito seeks to elaborate a philosophy that takes the living body, and not the exclusive relationship between persons and things, as its starting point. The body is that materiality in which passions and affections that force us to think reside. According to Esposito, once personhood is deconstructed, the inviolability of the body does not refer to a religious transcendence, but to its commonality:

> Because the body does not coincide with the mask of the person, and yet cannot be reduced to the appropriability of the thing, it falls under the third genus consisting of the *res sacra*. Belonging neither to the state nor to the Church, nor exclusively to the person that dwells inside it, the body owes its inviolability to the fact that it is eminently common. (2015b: 107)

A philosophy of the impersonal conceives the body not merely as subordinated to a person and deconstructs the metaphysical and juridical dualism between the person and its biological substratum. In this sense, Esposito follows both Bruno and Spinoza who reject the distinction between a bodily substance and a spiritual part that governs it. However, here we see the limits of Esposito's thought regarding animality. He deconstructs personhood rather than humanness, and that is why philosophical reflection (*pensiero*) can be a key moment in order to develop the potentialities of the impersonal:

> The full ownership of thought by a single individual, rather than its autonomy, became the noetic assumption of the subjection of the individual to a legal order that is always able to impute to individual people the responsibility of their acts. But what the principle of the impersonality of thought calls into question, perhaps even prior to this, is the set of exclusionary thresholds

that cut the human race into overlapping segments based on the amount of reason attributed to them [...]. To see intelligence not as a property of the few, to the detriment of others, but as a resource for all, through which one can pass without appropriating it for oneself, means to assign it a collective power that only the human species as a whole can fully actualize. (Esposito 2015d: 12)

These passages on body and thought suggest that maybe the *Impersonal*, in Esposito's own terms, is not enough in order to think animality as such. Esposito shows that his concern is always the possibility of animalization of certain human beings to make them disposable, but animality as such seems to escape his framework of analysis. The impersonal opens a gap in a philosophy based on subjectivity, property, and identity. At the same time, it shows the arbitrariness of the hierarchies on which our political and metaphysical thoughts are founded. In this sense, even though Esposito does not thematize it as such, a new conception of the common and of rights beyond personhood seems promising in order to put into question the way we treat animals as things.

In this sense, even though Esposito does not seem to be specifically concerned with the inclusion of animality in the community, we would like to suggest that his notion of *communitas* as reciprocal donation that alters the subject can include, beyond Esposito's account, non-human animals as well. Indeed, Esposito retrieves the original meaning of *communitas* as a mandatory gift, as an obligation to donate that expropriates the subject and exposes it to the outside. This shared *munus* that opens boundaries exposes subjects to a contagion that immunitary apparatuses will try to avoid at any cost. So why can't we think this contagion or this contamination beyond the realm of the human?

In many ways, a contamination between man and animal can be seen at work in the pages at the end of *Communitas*, where Esposito discusses the work of Bataille. Countering Heidegger's anthropocentric view of an insurmountable discontinuity between a *weltbildend Dasein* and a *weltarm* animal, in which the hand appears as the vehicle of *munus*, and drawing on Bataille's notion of originary community, Esposito states:

Again, Bataille will reverse the argument. Against every originary logos, what the grottos of Lascaux reveal is the unbreakable interweaving of humanity and animality produced precisely by the civilizing hand of man. In the design in which the superiority of the human species takes form, "man ceases to be animal giving to the animal (and not to himself) a poetic image," and indeed depicting himself with an animal mask. What comes to light in this text is certainly another and extreme mode for breaking the identity of the subject through its violent rootedness in that animal that at one time men "loved and killed"; something as well in close proximity of friendship and death refers to that same *communitas* that constitutes us without belonging to us. (Esposito 2010: 134)

This interweaving can also be found in Vico's aforementioned conception of an originary community, the *ingens sylva* (massive forest), in which man and beast are not yet differentiated:

In order for a specifically human history to begin, this world in common, made indistinct by the superimposition of bodies, must give way to a differ-ent form of life gradually governed by the mind and by the immunitary dispositifs that make possible a fully historical humanity. But what is crucial to our discussion is that this originary element, which is nonhistorical [...] never completely fades away, but rather, moves in a covert fashion, so to speak, into history itself. (Esposito 2012b: 27)

In both passages we can grasp how Esposito's notion of the originary *com-munitas* undermines the rigid separation between man and animal upon which western metaphysics and law are founded. Once again, as in the distinction between Hobbes and Machiavelli, the notion of origin as the place of contagion between man and animal, *mythos* and *logos*, passions and rationality, violence and institution, and its continuity within history is a key element in Esposito's deconstruction of modern thought.

Maybe, what Esposito's impersonal community enables us to think is that in order to establish another kind of relationship to animality we can-not simply include the animal within the existing legal framework. According to Gabriel Giorgi, therein lies an important difference between animal studies and biopolitical thought: while the former stresses the dimension of animals as subjects of rights, the latter tends to see the rela-tion between politics and life as a possibility for reinventing the common (Giorgi 2014: 28n15). And yet, a productive point of intersection between animal studies and biopolitics can be found in the literature on *critical* animal studies, where the focus is no longer on the mere inclusion of non-human animals into pre-existing philosophical and institutional frame-works but on the impact the question of the animal can have in re-thinking such frameworks, even for human animals (Derrida 2009; Oliver 2009;

Massumi 2014). Therefore, deconstructing personhood and affirming the common as impersonal may be a necessary step in order to think something like a right in common; a non-anthropocentric framework for conceiving rights that does not treat living beings as mere things to be exploited or killed with impunity, as well as a non-anthropocentric community that recognizes our ontological and political duties toward those living beings "whose value and expectation are no less than ours" (Canetti qtd. in Esposito 2015a: 113).

Thus, Esposito's impersonal community may not be able to give an account of animal's suffering, but it shows how our connection with animality beyond personhood makes us understand better who we are and, in the process, by deactivating immunitary apparatuses, helps us glimpse a better life for both human and non-human animals.

5   Concluding Remarks

In this chapter we tried to highlight Esposito's contribution to the discussion of the question of animality. Through a series of close readings of his works, we showed that Esposito introduces a productive contrast between Machiavelli and Hobbes that hinges upon the role played by animality in their theories. Thus, according to Esposito, Machiavelli is a major representative of the Italian living thought tradition whose main characteristic is to incorporate drives, affects, the body, and the animal dimension of the human being in his agonistic account of the political. In contrast, Hobbes is read by Esposito as a representative of an alternative trend in political theory that, unlike Machiavelli, seeks to domesticate, secure, and control animality. Hence, in Hobbes's political theory, animality is seen as a vehicle of dangerous, irrational, and anarchic drives that should be confined to the non-political state of nature. In this context, and drawing on Esposito's argument, we argued that the Centaur and Leviathan stand as allegories of alternative modalities of thinking the political. Whereas Machiavelli's Centaur is hospitable to animality by drawing on a mythical figure that is, at the same time, human and animal, Leviathan's polysemy (a clock, a mortal God, a sea creature) is articulated in such a way that ultimately serves as a barrier or frontier between the state of nature, where man is a wolf to man, and the pacified and well-ordered civil state. Accordingly, the political as conceived by Hobbes can only exclude and negate conflict and can never see the latter as a vehicle for institutional innovation or creative political imagination.

In accordance with the tradition of Italian living thought, Esposito follows Machiavelli's advice and takes issue with the notion of person and personhood, understood as an apparatus of domestication and control of the animality of the human. Needless to say, the person is a key notion in Hobbes's project as he famously discusses not only how to personate things but also, and mainly, how to personate sovereign authority. But Esposito goes beyond Hobbes to track iterations of the person even in contemporary approaches to personal dignity and human rights. In order to question a growing consensus around the notion of person and personal dignity, Esposito explores the possibilities opened by the notion of the impersonal understood as a deconstruction of the proper: subjectivity, property, rights, and so on. We tried to show the potential that the notion of the impersonal can have not only for rethinking the "subject" of the political but also for recuperating the common in such a way that the scope of the political can be reconfigured beyond the exclusive realm of the human.

NOTES

1. It is important also to stress the difference between Locke's theorization on the naturality of labor/property as something ontologically primordial that the state must recognize and Hobbes's recognition of the Leviathan as the condition of possibility of property and labor market. In *Leviathan* Hobbes points out that in the state of nature "there be no Propriety, no Dominion, no Mine and Thine distinct; but onely that to be every mans that he can get; and for so long, as he can keep it" (Hobbes 2012: 196). Later he adds: "Seeing therefore the Introduction of Propriety is an effect of Common-wealth; which can do nothing but by the Person that Represents it, it is the act onely of the Soveraign; and consisteth in the Lawes, which none can make that have not the Soveraign Power" and that "mans Labour also, is a commodity exchangeable for benefit, as well as any other thing" (Hobbes 2012: 388). See also Macpherson (1962).
2. As Carl Schmitt reminds us, *protego ergo obligo* is the cogito ergo sum of the state (2008: 52).
3. According to Esposito, the idea of an institution through covenant implies an aporetic relationship for the subjects involved: "they are subjects of sovereignty to the extent to which they have voluntarily instituted it through a free contract. But they are subjects to sovereignty because, once it has been instituted, they cannot resist it, for precisely the same reason: otherwise they would be resisting themselves" (Esposito 2008: 59–60).

4. All translations of not translated works are our own.
5. Diego Rossello (2012) shows how this divide can be deconstructed in Hobbes's texts through lycanthropy and melancholy, but, for reasons of space, we cannot develop the issue here.
6. In *Living Thought* Esposito maintains that "Both for Machiavelli and Vico the origin is characterized by violence and lethal conflict" (2012b: 260).
7. "And as to Rebellion in particular against Monarchy; one of the most frequent causes of it, is the Reading of the books of Policy, and Histories of the antient Greeks, and Romans [...]. From the reading, I say, of such books, men have undertaken to kill their Kings, because the Greek and Latine writers, in their books, and discourses of Policy, make it lawfull, and laudable, for any man so to do; provided before he do it, he call him Tyrant. [...] From the same books, they that live under a Monarch conceive an opinion, that the Subjects in a Popular Common-wealth enjoy Liberty; but that in a Monarchy they are all Slaves" (Hobbes 2012: 508).
8. "It was the speech of the Roman people [...] that all Kings are to be reckon'd amongst ravenous Beasts. But what a Beast of prey was the Roman people, whilst with its conquering Eagles it erected its proud Trophees so far and wide over the world" (Hobbes 2002: 23).
9. This nonmetaphysical use of the notion of origin is explained in Esposito (2012a, b).
10. Esposito reminds us that "Nothing is more deadly, for Vico, than the typically modern idea that we can sever the knot that binds history to its nonhistorical beginning, unraveling it through a process that fully temporalizes life" (2012b: 27).
11. See Johnson (1987), Derrida (2011), and Torrano (2016).
12. On the contrary, sovereign power has always tried to separate *zoé* from *bíos* and subordinate the former to the latter (Agamben 1998).
13. As Foucault puts it in the last chapter of *La volonté de savoir*: "for millennia, man remained what he was for Aristotle: a living animal with the additional capacity for a political existence; modern man is an animal whose politics places his existence as a Living being in question" (1978: 143).
14. "Biopolitics deals with the population, with the population as a political problem, as a problem that is at once scientific and political, as a biological problem and as power's problem" (Foucault 2003: 245).
15. As Arendt (1951) pointed out, the individual that arrived in the Nazi concentration camp had to be previously depersonalized in juridical, moral, and even individual terms in order to be annihilated with total impunity.

Works Cited

Agamben, Giorgio. 1998. *Homo Sacer: Sovereign Power and Bare Life*. Trans. Daniel Heller-Roazen. Stanford: Stanford University Press.

———. 2004. *The Open: Man and Animal*. Trans. Kevin Attell. Stanford: Stanford University Press.

Arendt, Hannah. 1951. *The Origins of Totalitarianism*. New York: Schocken Books.

Calabrò, Daniela. 2012. *Les détours d'une pensée vivante: Transitions et changements de paradigme dans la réflexion de Roberto Esposito*. Paris: Mimesis.

Campbell, Timothy. 2012. "Enough of a Self": Esposito's Impersonal Biopolitics. *Law, Culture & the Humanities* 8 (1): 31–46.

Derrida, Jacques. 2009. *The Animal That Therefore I Am*. Trans. David Wills. New York: Fordham University Press.

———. 2011. *The Beast and the Sovereign. Volume 1*. Trans. Geoffrey Bennington. Chicago: University of Chicago Press.

Esposito, Roberto. 1984. *Ordine e conflitto. Machiavelli e la letteratura politica del Rinascimento italiano*. Naples: Liguori.

———. 2008. *Bíos: Biopolitics and Philosophy*. Trans. Timothy Campbell. Minneapolis: University of Minnesota Press.

———. 2010. *Communitas: The Origin and Destiny of Community*. Trans. Timothy Campbell. Stanford: Stanford University Press.

———. 2011. *Dieci pensieri sulla politica*. Bologna: Il mulino.

———. 2012a. *Dall'impolitico all'impersonale: Conversazioni filosofiche*, ed. Matías Saidel and Gonzalo Velasco. Milan: Mimesis.

———. 2012b. *Living Thought: The Origins and Actuality of Italian Philosophy*. Trans. Zakiya Hanafi. Stanford: Stanford University Press.

———. 2015a. *Categories of the Impolitical*. Trans. Connal Parsley. New York: Fordham University Press.

———. 2015b. *Persons and Things: From the Body's Point of View*. Trans. Zakiya Hanafi. Cambridge, MA: Polity Press.

———. 2015c. *Third Person: Politics of Life and Philosophy of the Impersonal*. Trans. Zakiya Hanafi. Cambridge, MA: Polity Press.

———. 2015d. *Two: The Machine of Political Theology and the Place of Thought*. Trans. Zakiya Hanafi. New York: Fordham University Press.

Foucault, Michel. 1978. *The History of Sexuality: An Introduction*. Trans. Robert Hurley. New York: Pantheon Books.

———. 2001. *The Order of Things: An Archaeology of the Human Sciences*. London: Routledge.

———. 2003. *Society Must Be Defended: Lectures at the Collège De France, 1975–76*. Trans. David Macey. New York: Picador.

Giorgi, Gabriel. 2014. *Formas comunes: Animalidad, cultura, biopolítica*. Buenos Aires: Eterna Cadencia Editora.

Glaser, Linda. 2011. Biopolitics Views Humans as Animals Before the Law. *Cornell Chronicle.* http://news.cornell.edu/stories/2011/10/lecturer-biopolitics-views-humans-animals-under-law. Accessed on 10 Nov 2019.

Hobbes, Thomas. 1998. On Man. In *Man and Citizen (De Homine and De Cive)*, ed. Bernard Gert, 33–85. Indianapolis/Cambridge: Hackett Publishing.

———. 2002. *De cive: The English Version Entitled in the First Edition "Philosophicall Rudiments Concerning Government and Society".* The Clarendon Edition of the Philosophical Works of Thomas Hobbes, vol. 3. Oxford: Clarendon Press.

———. 2012. *Leviathan.* The Clarendon edition of the works of Thomas Hobbes, vol. 3–5. Edited by Noel Malcolm. Oxford: Oxford University Press.

Johnson, Paul. 1987. Hobbes and the Wolf Man. In *Hobbes's Science of Natural Justice*, ed. Craig Walton and Paul J. Johnson, 139–151. Dordrecht: Springer.

Macpherson, Crawford B. 1962. *The Political Theory of Possessive Individualism.* Oxford: Oxford University Press.

Massumi, Brian. 2014. *What Animals Teach Us About Politics.* Durham: Duke University Press.

Oliver, Kelly. 2009. *Animal Lessons: How They Teach Us to Be Human.* New York: Columbia University Press.

Rossello, Diego. 2012. Hobbes and the Wolf-Man: Melancholy and Animality in Modern Sovereignty. *New Literary History* 43 (2): 255–279.

———. 2013. La wertud maquiaveliana: el príncipe como centauro. In *La revolución de Maquiavelo: El príncipe 500 años después*, ed. Diego Sazo, 269–286. Santiago: RIL.

Saidel, Matías. 2013. Más allá de la persona: lo impersonal en el pensamiento de Roberto Esposito y Giorgio Agamben. *Eikasia. Revista de filosofía* 51: 159–176.

———. 2014. Biopolitics and Its Paradoxes. An Approach to Life and Politics in R. Esposito. *Lo Sguardo – Rivista di Filosofia* 15: 109–131.

———. 2016. Roberto Esposito y la deconstrucción de la teología política: hacia una política de los cuerpos vivientes. *Filosofía Italiana.* http://www.filosofiaitaliana.net/wp-content/uploads/2018/04/Saidel.pdf. Accessed on 10 Nov 2019.

Schmitt, Carl. 2008. *The Concept of the Political.* Trans. George Schwab. Chicago: University of Chicago Press.

Soto García. 2015. Conflicto y vida: La recuperación del Centauro en el pensamiento de Roberto Esposito. *Revista de la Academia* 20: 67–84.

Tarizzo, Davide. 2010. *La Vita, un'invenzione recente.* Rome: Laterza.

Torrano, Andrea. 2016. Werewolves in the Immunitary Paradigm. *Philosophy Today* 60 (1): 153–173.

Wolfe, Cary. 2013. *Before the Law: Humans and Other Animals in a Biopolitical Frame.* Chicago: The University of Chicago Press.

CHAPTER 7

Animality Between Italian Theory and Posthumanism

Giovanni Leghissa

1   ANIMALITY AND POSTHUMANISM

The reason why animality is an important philosophical issue is twofold. First, there is an ontological commitment regarding the position of human beings within the realm of existing things. Long after the establishment of Darwinism within the sciences of the living being, the idea that human beings are endowed with properties that make them different from other living beings still continued to persist within philosophy. The reasons for resisting against the naturalization of the ontological frame within which the philosophical discourse takes into consideration the human might have been various and of different provenance. Equally various have been both the motives and the strategic moves that led, in the recent decades, to a shift from an ontological perspective that had the function to underpin the unicity of human beings within the set of the things existing on earth toward an ontology that allows for describing human beings as living beings among other living beings, whereas the belonging to the

G. Leghissa (✉)
University of Turin, Turin, Italy

© The Author(s) 2020
F. Cimatti, C. Salzani (Eds.), *Animality in Contemporary Italian Philosophy*, The Palgrave Macmillan Animal Ethics Series,
https://doi.org/10.1007/978-3-030-47507-9_7

animal realm is the only relevant ontological trait that the former have in common with the latter. What matters is the fact that, at present, a heavy burden of proof rests on the shoulders of those who want to single out those peculiarities of the human that are supposed to disclose an ontological difference between humans and other living beings.[1]

Second, there is an ethical commitment toward animal life, not in the sense that the life of the living being must be preserved at any rate, but in the sense that the articulation of any ethical stance must begin with observing how emerges and works the system of differences among living beings.[2] It is this glance cast on hierarchies that allows for the critical distance needed to evaluate the degree of violence that imbues a given social formation. In other words, the perception of how animals belonging to other species are bashed about or ill-treated is supposed to open the way to a more refined and acute perception of how various forms of violence among humans—including the symbolic violence Bourdieu made us attentive to—are exerted across different social contexts and historical periods. In this sense, the exclusion of the animal other not only foreruns the exclusion of the feminine, the enemy, the stranger—just to evoke the most striking figures of the excluded other—but constitutes the anthropological frame within which the self-affirmation of any strong identity structure takes place.

That said, there could be some objections to the supposed primacy of the question of animality with regard to how hierarchical structures arise and unfold whenever a subject defines its own identity in opposition to this or that instantiation of otherness; one could say, for example, that the perception of this primacy became the starting point of a fruitful critical inquiry only after the widespread reception of a philosophical discourse centered on the concept of difference within the Humanities. It is true, in fact, that precisely the theoretical devices made available by authors like Derrida, Deleuze, or Foucault, for whom the notion of difference plays a fundamental role, could make room for the awareness that the exclusion of the animal other is strictly intertwined both with the subordination of the feminine to the masculine order and with the exclusion of the stranger, or the foreigner.[3] Nevertheless, as Cary Wolfe has persuasively pointed out,[4] the posthumanist project is not to be considered as the mere continuation of postmodernist philosophies with other means.[5] Surely, posthumanist thought is interested in the issue of embodiment, namely in the idea that the humanist way of considering the subject as disembodied must be deconstructed and put in question for its philosophical as well as political consequences. But it is further interested both in recasting the concept of humanity within the broader epistemic frame offered by evolutionary theory and in laying bare how the prosthetic character of *Homo sapiens* is not the result of a process of hybridization that began in the era of the cyborg, but constitutes its hallmark since the emergence of our species. Thus, what makes the posture of posthumanism so peculiar is to be seen in the claim that the emphasis put on the question of difference, which provides, on its part, the starting point for undertaking a radical critique of violence, is alone insufficient to result in a paradigm change; the posthumanist project aims at recasting the epistemological and ontological frameworks—as well as the relationship between the two—that allow for articulating the relation of the human with its own animality, the relation of the human with other living beings, and, finally, the entanglement between humans and the artifacts they create.[6] The human disappears as a privileged object of investigation, whereas the chain of relations in which the human is involved acquires visibility. The shift is toward an observation of the different networks within which human beings can be defined as human. What makes "human" a human being is not a property that can be isolated once forever—a property, thus, that would be equal to the essence of the human in the traditional sense of the term (i.e., "that what does not change across time"). It is rather a set of cognitive and behavioral patterns that the observer identifies time after time by looking carefully at the embeddedness of human beings into networks that include living beings, objects that can be recognized as belonging to natural kinds, and artifacts (from the Acheulean hand axe to Google's Quantum Computer).

2   CALVINO BETWEEN LITERATURE AND CYBERNETICS

In 1967—long before the term "posthumanism" became established—Italo Calvino wrote an essay about literature meant as a combinatory process. In his essay he pointed out how deep has been the break that the shift from an ontology of the continuous to an ontology of the discrete caused within both the realm of physics and the realm of biology.[7] By referring to the pathbreaking works of authors like Shannon, Wiener, and von Neumann, whose role as forerunners of the posthuman stance has been meanwhile widely recognized,[8] Calvino caught with precision to which extent the cybernetic revolution succeeded in impinging upon the idea that a specific set of traits possessed by humans—above all the capability to speak and articulate complex chains of thought through language—were sufficient to vest them with a privileged position in the realm of being. If the research program of cybernetics holds, then language reveals to be a combinatorial game. It is a game that can be played in various forms and can have different outcomes—myths, novels, and poems, among others. Seen in this way, the use of language in all its heterogeneous expressions is nothing but the result of the evolutionary history of our species, and a long chain of narrative products connects the mythical storytelling of our ancestors during the Paleolithic and the most sophisticated works of contemporary literature.[9] As Calvino does not fail to understand, such a way of looking at the function of language starting from the perspective disclosed by cybernetics affects the very idea of subjectivity. The subject is no more the external observer of a given set of rules that can be managed at will in order to achieve an *opus* closed in itself and meant as the result of a conscious effort. It is true that Calvino addresses the act of writing, but his way of considering the latter has a far broader relevance. The experience of writing, in which the author is dissolved, instantiates what the subject in general goes through whenever it is affected by language. To be a subject, thus, does no longer mean to make the experience of a complete self-possession, which is supposed to be the presupposition for the conscious management of each act of sensemaking. It means, rather, to occupy a site within the chain of signifiers, whereas this positioning can never be unmovable and permanent.

Calvino's attempt to recast the concept of subjectivity in accordance with the epistemic revolution ushered by the scientific program of cybernetics was not the outcome of a solitary intellectual undertaking. In the same period (i.e., the second half of the 1960s) an author like Silvio Ceccato, for example, contributed as well to acclimatize the discourse of cybernetics to the Italian cultural landscape. However, the resistance against such an effort, whose unrelatedness to the humanistic attitude that has always deeply marked the tradition of Italian philosophy could not go unperceived, has been very strong since the beginning. Notwithstanding the bulk of translations into Italian of the main works that contributed to the spread of the cybernetic thought, Italian philosophers, generally speaking, have always preferred to stick to the traditional frame of mind imbued with idea that human subjectivity is to be conceptualized as a matter separated both from animality and from the realm of the mechanical. When the long wave of Marxism faded, it became possible to take advantage of the decentralization of the subject that French authors like Derrida or

Foucault have successfully carried out in their work. Nevertheless, if observed retrospectively, the Italian reception of both Foucauldian genealogy and Derridean deconstruction seems to have been strongly biased by the heritage of Italian historicism, in the sense that the critical work philosophy is called to carry on in order to reshape the concept of subject has been put in service of a political project the scope of which still remained the fight against capitalism. In this way, the real change of mind that the decentralization of the subject implies risks to get lost. The history along which subjectivity can be explored by identifying its different features and deployments is a "deep history,"[10] namely a history that goes far back before the rise of modern capitalism. If subjectivity becomes a relevant issue only when the modern subject is at stake, that is, only when the philosophical gaze is directed to how the world order issued from modernity—by looking at the whole gamut of its cultural and even metaphysical implications—shaped a specific set of subjective self-positionings, then what has been achieved by introducing the concept of difference falls apart completely. On the contrary, the identity of the Self arises from an unstable and precarious making and re-making of systems of differences between living and non-living entities, whereas the culturally coded character of these systems can never conceal its own rootedness in those processes of self-differentiation along which living beings unfold.[11]

The present picture would be incomplete—it must be added—if one overlooked the fact that an effective rupture with the past tradition of Italian philosophy took place when issues, authors, and conceptual frameworks coming from analytical philosophy began to be incorporated in the Italian philosophical landscape. Analytical philosophers are used to work together with both neuroscientists and those scientists that contribute to realize the scientific program of AI—a program which in many respects issues from cybernetics.[12] Yet, even if the whole stuff of concepts concerning the human and its relations to the world it inhabits—including the "internal world" of its mind—receives a completely new definition thanks to this strict intertwinement between philosophy and the natural sciences, the research program of analytic philosophy is, generally speaking, not interested in focusing on the anthropological changes that occur when the human is posited between animality and the mechanical. This lack of interest is responsible for the fact that analytic philosophy cannot provide any sort of critical observation upon those power relations which, on the contrary, the philosophical project of posthumanism is willing to focus on. In other words, it is not enough to naturalize the question of subjectivity and, consequently, to recast or even to abandon the bulk of concepts that served to define the specificity of the human (consciousness, intentionality, and the like) in favor of a conceptual map that upholds the representation of those positions that a possible subject—not necessarily a human one—may occupy from time to time.

Whatsoever philosophical project aimed at decentralizing the subject must, in fact, comply with the following condition in order to acquire—as in the case of posthumanism—both a theoretical and a political relevance: the action oriented toward a goal is to be seen with regard to the environmental context within which it takes place, and it is at this level of analysis that questions concerning the political—in its broadest sense—can arise. When the observer looks at how human beings interact with each other, other living beings, and artifacts, what emerges is the system of differences that codes and socially embedded scripts convey in order to underpin or simply make obvious and unquestionable specific power arrangements at organizational level. When the observer looks at how the cognitive system of the actor performs any possible goal-oriented course of action, an inquiry of such kind has to lay bare to which extent what philosophy used to call "subjectivity" is but a multilayered structure of occurrences that can be described both as mental events (this happens when the observer tends to ascribe whatsoever form of intentionality to an agent) and as conducts (meant as the supposed effects of the intention to undertake a certain course of actions). The point to be made here is that a comprehensive explanatory theory of mental causation (which is still lacking, despite the efforts that both analytic philosophy and the cognitive sciences make together in this direction) is as important as a critical theory that allows for understanding how performative are those rules and norms that shape institutions and organizations. According to a posthumanist perspective, even the latter have to be seen as tools, as artifacts, namely as objects that did not came to earth because rational agents want them to exist in order to achieve some goals, but because they help agents to adapt to their environment.

3   THE IDENTIFICATION WITH THE ANIMAL OTHER

Thus, it is not surprising that, in Italy, a first complete assessment of what is really at stake with regard to the posthuman stance could take place only within the work of Roberto Marchesini,[13] who is not a philosopher. This elucidation is given surely not to diminish the significance of his theoretical contribution, but rather to highlight the fact that the Italian philosophical community, which is still inclined to show a deaf but steadfast opposition toward the issues raised by posthumanism, could be affected by the latter only thanks to contributions coming from other disciplinary fields. Marchesini is an ethologist, and his scientific provenance guarantees for the scientific accuracy that characterizes his way of recasting the difference between the human and its animal other. Starting from an ethological perspective, in fact, it becomes quite impossible to overlook the bulk of commonalities that tie together humans and animals.[14] The consequences that must be derived from these commonalities are important, first of all, for understanding those animal traits of the human that the posthuman perspective is willing to underscore. But not less important is the ethical commitment that arises when humans look at themselves as a part of the animal realm. According to Marchesini, this commitment should not arise only from the acknowledgment of the fact that the human species is an animal species among others; much more important is the empathy that humans can feel for the animal other. The interaction between *Homo sapiens* and other animals cannot be reduced, in fact, to the utilitarian gains that could be achieved by holding some "useful" species among the human group.

The point is twofold. On the one hand, with regard to the practice of hunting, the idea that humans have exploited other species just for the benefits deriving from a diet rich in meat does not hold.[15] On the other, it is no less controversial to represent even the friendship with animals as a form of advantage contributing to human wellness: if animals like our pets are cozy, charming, and funny companions, the personal affection that fills the relation with them goes far beyond the desire to make up for one's own feelings of loneliness and/or sense of despair. According to Marchesini, the animal other is the only possible mirror of the human—a mirror that the latter needs in order to achieve its own identity. In this sense, the presence of animals among human beings, which dates back to the beginning of our evolutionary history, rests on an anthropologically grounded necessity. The presence of the animal other is not only the epiphany of alterity, but it is as well the epiphany of a constitutive element of human identity itself. And it would be no objection to this statement to observe that humans, within their relationship with animals, simply project upon them some traits or behaviors that can be actually only human. What appears to be, prima facie, a sort of naïve projection aimed at "humanizing" animals expresses an ontological commonality, whose signs can be at best found in those widespread myths and narratives that talk about how animals and humans have been living together since the origin of the world.[16] Levinas' face of the other was a human face—and it would have been impossible, for Levinas, to represent the face of the other in a way that differs from the human one, as Derrida convincingly pointed out in his analysis of the humanistic and even theological presuppositions of Levinas' thought.[17] In Marchesini's posthumanism, on the contrary, the other in whom the human subject reflects itself assumes precisely the features of the animal's face—a face which is perceived as such not in force of a projection, but in force of an ontological necessity.

The ethical commitment that imbues Marchesini's thought seems, thus, to rest on premises that have a philosophical consistency which is not always to be found within the discourse on animality that characterizes the claims of the antispecist movement.[18] The latter aims at giving voice to the feelings of uncomfortableness, indignation, or even disgust that arise—or should arise—as soon as one considers the dreadful and excruciating way of treating those animals that belong to the species humans consider obvious to exploit for their own purposes reducing them to food or using them as test subjects. It is worth pointing out, however, that the project of posthuman philosophy can well include among its presuppositions the desire to give voice to the silent cry of sufferance that comes from the animals killed in a lab or in a slaughterhouse. The question about how to ground philosophically the ethical commitment toward the animal other only by considering the pity we as humans can feel for the physical and emotional pain they suffer is controversial. If it is undisputed to acknowledge that animals can suffer, less self-evident is the set of criteria needed to shore up those actions that can buttress both the public discourse and the concrete tools—including the legal ones—that are necessary to hinder the violence against the animal other.[19] Nevertheless, if one assumes that feelings and emotions play a fundamental role within the process that culminates in the formulation of moral judgments,[20] then it is not at all out of place to hold that feelings like empathy for the animal suffering caused by humans can sustain the burden of being the starting point for arguments against violence to animals.[21]

4   AGAMBEN AND THE POLITICS OF THE LIVING

Among those Italian authors that work within the disciplinary field of philosophy the only one who devoted a specific attention to the question of animality is Giorgio Agamben. The question of life meant as a target of political interventions has been playing a pivotal role since the beginning of his philosophical career.[22] In this sense, he shares with other authors who are now well known in the United States under the label "Italian Theory"[23] an acute interest toward the Foucauldian legacy. Yet, while Foucault specifies very clearly the extension—both at the conceptual and at the historical level—that the concept of biopolitics can cover, Agamben tends to use the latter in order to describe the telos that has been animating the whole of the political project of Western civilization since its ancient beginning. This twisting from the original sense of the term is of course not worth of disapproval for philological reasons. The point is that it is rather questionable to say that each and every form of power targets the body of the individual. Foucault had in mind a specific series of events and specific historical changes that occurred in the fading of classical liberalism as he introduced the term biopolitics.[24] Agamben, on the contrary, is inclined to term as biopolitical not specific instantiations of governmental interventions on individuals and their social setting, but the general form that all possible manifestations of power and control have in common. The outcome of Agamben's conception of power is a truism: of course the scope of those agents that govern institutions and organizations (no matter whether they are human beings or algorithms) is to manage the relation between the necessity to keep alive the basic structure of institutions and organizations and individual needs, desires, skills, levels of aspiration, and capability to make conscious choices. What needs explaining is, rather, the specific and historically determined forms that take the various attempts to combine together what constitutes the life of an individual and what constitutes the "life" of an organization or an institution. It would be out of place to expose with more details why Agamben's theory of the (bio)political results to be rather unsatisfactory, but it is necessary to keep in mind the major traits of it in order to better understand his way of dealing with the question of animality.

The latter finds its place in *The Open*, a book originally published in Italian in 2002.[25] Here Agamben investigates how philosophers (Aristotle and Heidegger), theologians (Thomas Aquinas), and natural scientists (Linnaeus and von Uexküll) have articulated the difference between the human and the animal in order to extract from their discourse—a discourse whose coherence and persistence along time Agamben correctly underscores—the hidden essence of those biopolitical traits that characterizes the government of the living within the Western tradition. There can be no doubt that it is worth inquiring how and to which extent the difference between the human and the animal sheds light on the sources of the discourse that has the function to underpin the power of the sovereign.[26] And Agamben is surely right in saying that the understanding of how works any possible biopolitical dispositive presupposes the understanding of how works what he calls the "anthropological machine," that is, the discourse that allows for the production of what is not human *within* the human in order to produce the human as something radically different from the animal (Agamben 2004: 37–38). He is right in showing, in other words, what a tremendous effort has been undertaken in order to arrange a political space whose internal hierarchies among humans are based on specific power mechanisms that are responsible for including some living beings and excluding others—to the point that some humans are excluded (and even murdered) precisely because they are not to be differentiated from animals.

The limit that affects Agamben's reconstruction, however, is given by his willingness to offer a tetragon and allover reconstruction of how arose the political space. Such a reconstruction may be convincing, at least to some extent, as far as some moments of the history of Western civilization are concerned, but its internal differentiation needs being articulated with much more precision. First, Agamben is not inclined to underline the gap that separates Christianity and modernity—a philosophical gesture that characterizes the whole of his philosophy.[27] Second, Agamben does not seem aware of the fact that the process of anthropogenesis, meant as a chain of events occurred along with the evolutionary history of *Homo sapiens*, cannot simply coincide with the history of mankind. Truly, a bulk of ethnographic data shows that it is an anthropological invariant to produce a difference between the human and the animal in order to define the proper of the human. But Agamben's reconstruction seems to confer a universal character to some categories and narratives that are typically Western in their provenance and origin.

What seems to be rather persuasive within the perspective disclosed by *The Open* is, nonetheless, the idea that a sort of suspension of the difference between humanity and animality can inaugurate a new way of articulating the political space, that is the power relations among humans. As if the inactivity of the animal, which does not want to do anything to eschew the *ennui* that arises when the living being abandons itself to the sheer fact of being alive, could per se anticipate the utopia of a world without violence and exclusion (Agamben 2004, 87–92). The hint at this utopic dimension is surely an important element of Agamben's argument, but I wonder whether it were more fruitful to spend more theoretical energy in pursuing the project of melting philosophy with other disciplines in order not only to suspend the difference between the human and the animal but to ground the impossibility to hold on this difference.[28] The result of this effort would be, perhaps, that the question of how it is like to be a bat can be translated into a much more radical version, namely how it is like to be an octopus.[29]

NOTES

1. de Fontenay (1998) gives the most complete account of how the Western philosophical tradition has since its inception upheld the idea that animals are inferior to humans—and it is needless to say that the philosophical discourse on animality is only a strand among others within the whole history of how animals have been placed in a subordinated position with respect to humans.
2. Haraway (1991) still remains a pathbreaking work on this matter.
3. An exception that is worth mentioning is Adams (2010), whose first edition dates back to 1990, that is, to a period when the main attention was devoted to discussing issues related mainly to the question of gender and race.
4. See especially Wolfe (2010).
5. In this sense, Braidotti (2013) is far from convincing in her attempt to build a bridge between the critical discourse issued from postmodernism and the posthumanist stance.
6. On this point, see in particular Hodder (2012).
7. "Cybernetics and Ghosts," in Calvino (1986).
8. On this, see Hayles (1999: 50–112).
9. For a recent and exhaustive account of the anthropological significance of human storytelling—and, consequently, of the anthropological significance of literature as such—see Cometa (2017).
10. The notion of "deep history" is taken from Smail (2008).
11. Vitale (2018) provides a fine analysis of the relationship between Derrida's reflection upon life and death and the most recent conceptions of the living issued from biology by showing that in both cases what is at stake is the metastable state of the identity of the living.

12. On this, see Dupuy (2009).
13. Marchesini (2001), (2009), (2014), (2016a), and (2016b).
14. On this, see also de Waal (2006) and (2009).
15. The literature on this issue is wide; for a good assessment, see Speth (2010).
16. Daston (2005) makes acute and convincing considerations on this point.
17. See Derrida (2008).
18. As otherwise, in Italy as well the movement has its supporters. See, among others, Rivera (2010), Caffo (2011), Filippi and Trasatti (2015), and Filippi (2017).
19. The literature on the topic is huge, and some works have reached a relevance that goes far beyond the realm in which they arose as, for example, in the case of Rollin (1981), Regan (1983), and Singer (1984). On the current debate about the opportunity to confer equal rights to animals, see Zuolo (2017) and (2019).
20. See, among many others, Nussbaum (2001)—in the present context, it is worth remembering that Nussbaum (2006) deals specifically also with the question of animality.
21. Wolfe (2012) brings about more detailed arguments on this topic by showing how the attempt to confer some rights to some animals is only a part of the problem.
22. See Agamben (1998), which is perhaps the book that contributed at most to the establishment of the author's fame in Italy and abroad.
23. See Hardt and Virno (1997), Chiesa and Toscano (2009), Gentili (2012), and Gentili et al. (2018).
24. See Foucault (2008).
25. See Agamben (2004).
26. Derrida (2009–2011) offers the most complete and accurate reconstruction of this issue.
27. It seems to me highly problematic to adhere to the paradigm of secularization, as Agamben does, after Blumenberg (1985). But this is not an issue to be discussed here. What is to be noticed, rather, is that Derrida (2009–2011) seems more correct in showing that modernity introduced a new way in making the difference between the animal and the human productive in order to define the function of the sovereign power.
28. In this direction, see, for example, Vallortigara (2005).
29. See Godfrey-Smith (2016).

Works Cited

Adams, Carol J. 2010. *The Sexual Politics of Meat: A Feminist-Vegetarian Critical Theory*. New York/London: Continuum.

Agamben, Giorgio. 1998. *Homo Sacer: Sovereign Power and Bare Life*. Trans. Daniel Heller-Roazen. Stanford: Stanford University Press.

———. 2004. *The Open: Man and Animal*. Trans. Kevin Attel. Stanford: Stanford University Press.

Blumenberg, Hans. 1985. *The Legitimacy of the Modern Age*. Trans. Robert M. Wallace. Cambridge, MA: The MIT Press.

Braidotti, Rosi. 2013. *The Posthuman*. Cambridge: Polity Press.

Caffo, Leonardo. 2011. *Soltanto per loro. Un manifesto per l'animalità attraverso la politica e la filosofia*. Roma: Aracne.

Calvino, Italo. 1986. *The Uses of Literature. Essays*. Trans. Patrick Creagh. New York: Harcourt Brace & Company.

Chiesa, Lorenzo, and Alberto Toscano, eds. 2009. *The Italian Difference. Between Nihilism and Biopolitics*. Melbourne: re.press.

Cometa, Michele. 2017. *Perché le storie ci aiutano a vivere. La letteratura necessaria*. Milan: Raffaello Cortina Editore.

Daston, Lorraine. 2005. Intelligences: Angelic, Animal, Human. In *Thinking with Animals: New Perspectives on Anthropomorphism*, ed. Lorraine Daston and Gregg Mitman, 37–58. New York: Columbia University Press.

de Fontenay, Élisabeth. 1998. *Le silence des bêtes: la philosphie à l'épreuve de l'animalité*. Paris: Fayard.

de Waal, Frans. 2006. *Primates and Philosophers: How Morality Evolved*. Princeton: Princeton University Press.

———. 2009. *The Age of Empathy*. New York: Three River Press.

Derrida, Jacques. 2008. *The Animal That Therefore I Am*. Trans. David Wills. New York: Fordham University Press.

———. 2009–2011. *The Beast and the Sovereign*, 2 vols. Trans. Geoffrey Bennington. Chicago: University of Chicago Press.

Dupuy, Jean Pierre. 2009. *On the Origins of Cognitive Science: The Mechanization of the Mind*. Cambridge, MA: The MIT Press.

Filippi, Massimo. 2017. *Questioni di specie*. Milan: Elèuthera.

Filippi, Massimo, and Filippo Trasatti. 2015. *Crimini in tempo di pace. La questione animale e l'ideologia del dominio*. Milan: Elèuthera.

Foucault, Michel. 2008. *The Birth of Biopolitics. Lectures at the Collège de France, 1978–1979*. Trans. Graham Burchell. Basingstoke: Palgrave Macmillan.

Gentili, Dario. 2012. *Italian Theory. Dall'operaismo alla biopolitica*. Bologna: Il Mulino.

Gentili, Dario, Elettra Stimilli, and Glenda Garelli, eds. 2018. *Italian Critical Thought. Genealogies and Categories*. Trans. Glenda Garelli. London: Rowman & Littlefield.

Godfrey-Smith, Peter. 2016. *Other Minds: The Octopus and the Evolution of Intelligent Life*. London: HarperCollins.

Haraway, Donna. 1991. *Simians, Cyborgs, and Women: The Reinvention of Nature.* London: Free Association Books.

Hardt, Michael, and Paolo Virno, eds. 1997. *Radical Thought in Italy: A Potential Politics.* Minneapolis: University of Minnesota Press.

Hayles, Katherine N. 1999. *How We Became Posthuman: Virtual Bodies in Cybernetics, Literature, and Informatics.* Chicago: The University of Chicago Press.

Hodder, Ian. 2012. *Entangled: An Archeology of the Relationship between Humans and Things.* Malden: Wiley-Blackwell.

Marchesini, Roberto. 2001. *Posthuman. Verso nuovi modelli di esistenza.* Turin: Boringhieri.

———. 2009. *Il tramonto dell'uomo. La prospettiva post-umanista.* Bari: Dedalo.

———. 2014. *Epifania animale. L'oltreuomo come rivelazione.* Sesto San Giovanni: Mimesis.

———. 2016a. *Etologia filosofica. Alla ricercar della soggettività animale.* Sesto San Giovanni: Mimesis.

———. 2016b. *Alterità. L'identità come relazione.* Modena: Mucchi.

Nussbaum, Martha. 2001. *Upheaval of Thoughts: The Intelligence of Emotions.* Cambridge: Cambridge University Press.

Nussbaum, Martha. 2006. *Frontiers of Justice. Disability, Nationality, Species, Membership.* Cambridge (MA) – London: The Belknap Press of Harvard University Press.

Regan, Tom. 1983. *The Case for Animal Rights.* London: Routledge.

Rivera, Annamaria. 2010. *La bella, la bestia e l'umano: sessismo e razzismo senza escludere lo specismo.* Rome: Ediesse.

Rollin, Bernard E. 1981. *Animal Rights and Human Morality.* New York: Prometheus Book.

Singer, Peter. 1984. *Animal Liberation. Toward an End to Man's Inhumanity to Animals.* Wellingborough: Thorson Publishers.

Smail, Daniel L. 2008. *On Deep History and the Brain.* Berkeley: University of California Press.

Speth, John D. 2010. *The Paleoanthropology and Archeology of Big-Game Hunting: Protein, Fat, or Politics?* New York: Springer.

Vallortigara, Giorgio. 2005. *Cervello di gallina. Visite (guidate) tra etologia e neuroscienze.* Turin: Boringhieri.

Vitale, Francesco. 2018. *Biodeconstruction: Jacques Derrida and the Life Sciences.* Albany: SUNY Press.

Wolfe, Cary. 2010. *What Is Posthumanism?* Minneapolis: University of Minnesota Press.

———. 2012. *Before the Law: Humans and Other Animals in a Biopolitical Frame.* Chicago: University of Chicago Press.

Zuolo, Federico. 2017. Equality, Its Basis and Moral Status: Challenging the Principle of Equal Consideration of Interests. *International Journal of Philosophical Studies* 25 (2): 170–188.

———. 2019. Misadventures of Sentience: Animals and the Basis of Equality. *Animals* 9: 1044.

CHAPTER 8

For the Critique of Political Anthropocentrism: Italian Marxism and the Animal Question

Marco Maurizi

1 INTRODUCTION

The present chapter focuses on the reception of the concept of "animality" in the Italian Marxist tradition. In order to limit the scope of the chapter, the expression "Marxist tradition" is here understood in a very strict sense. As Paul Piccone correctly argues, "Italian Marxism" is "neither a clear-cut doctrine nor a well-defined body of ideas" (Piccone 1983: vii). Therefore, I will mainly refer to those intellectuals politically engaged in the Italian workers' movement, especially focusing on the leading figures of the PCI (Italian Communist Party) and some of its main critics "from the left."

My thesis is that the relevance of "animality" in the Italian debate changed according to three different and partially independent lines of development. The importance of the concept is related to the key role played by nature in Marx's and Engels' materialist analysis of human society. Consequently, in the first part I will analyze the concept of animality

M. Maurizi (✉)
University of Tor Vergata, Rome, Italy

© The Author(s) 2020
F. Cimatti, C. Salzani (eds.), *Animality in Contemporary Italian Philosophy*, The Palgrave Macmillan Animal Ethics Series,
https://doi.org/10.1007/978-3-030-47507-9_8

in the context of so-called orthodox Marxism (Labriola, Bordiga, Gramsci, Togliatti). This section includes a brief reference to subsequent Hegelians and historicists (Luporini, Badaloni), as well as champions of anti-Hegelianism (Della Volpe, Colletti), literary critics (Cases), and writers (Calvino, Pasolini).

In the second part, I will see how ecology and ethology helped to discard both the mechanistic scheme introduced by Descartes and the idea of an unlimited exploitation of nature. It is important to underline that the definition of animality is not a merely theoretical question, since it also implies practical consequences: changes in our understanding of animality often go hand in hand with a different attitude toward nature and non-human animals. It is in such context that Peter Singer's *Animal Liberation* marks a crucial paradigm shift. Although its reception in Italy was at first unremarkable, the introduction of animal subjectivity in the ethical debate posed a philosophical problem that Italian Marxism could not ignore in the long run. I will thus briefly focus on the critique of Marxist anthropocentrism from the point of view of materialism (Timpanaro, Paccino) and animal rights (Bartolommei, Veca, Turchetto).

The last part of the chapter shows how Post-Workerism and Critical Theory have challenged the human/non-human binary opposition and, consequently, deconstructed and redefined the very notion of humanity. I will briefly show the original path taken by Italian Marxism in the last decade, especially in the works of Post-Workerists (Negri, Filippi) and scholars of the Frankfurt School (Bellan, Benini, Maurizi).

2   Animality and Materialism

2.1   Marx's and Engels' Dialectical Theory of Animality

The question of animality plays a significant role in Marx's and Engels' writings. A proto-Darwinian stress on the continuity between animals and humans is typical of Marxist *materialism*, since materialism considers the human being entirely part of nature and rejects the idea that our "soul" or our "reason" could make us somehow *transcend* nature. Marx and Engels neither established an absolute difference between humans and non-humans (something that would sever us from the animal kingdom), nor did they preach some sort of absolute identity between them. From one side, we are animals, not spiritual beings; from the other, our society cannot be described according to mere ethological laws. Thus, although nature is governed by necessity, humans fight for their social freedom:

> Communism, as fully developed naturalism, equals humanism, and as fully developed humanism equals naturalism; it is the genuine resolution of the conflict between the human being and nature, and between the human being and man, the true resolution of the conflict between existence and being, between objectification and self-affirmation, between freedom and necessity, between individual and species. It is the solution of the riddle of history and knows itself to be the solution. (Marx 2007: 102)

As we will see, such "dialectical" approach to the human/non-human opposition can be interpreted in various, contradictory ways. The question is whether Marxism is a purely humanist project of emancipation, which places "Man" at the center of its consideration—thinking Communism as a form of total dominion over nature—or it can reject anthropocentrism and accept the demands of environmentalism and animal liberationism.

Ecologists and animal rights theorists have criticized Marxism for its "Promethean" humanism (Redclift 1987: 48) and its "speciesist" (Benton 1993) attitude toward animals. In Engels, one can read apparently contradictory stances on the relation between humanity and animality. On the one hand, for example, Engels maintains that in Socialism, "for the first time, man, in a certain sense, is finally marked off from the rest of the animal kingdom, and emerges from mere animal conditions of existence into really human ones" (Engels 1973: 226). On the other, Engels explicitly denounced our violent dominion upon nature and foresaw the *revenge* that nature was going to take on us, foreshadowing the idea of a forthcoming harmony between society and nature: the more we learn to know nature, "the more will men not only feel but also know their oneness with nature, and the more impossible will become the senseless and unnatural idea of a contrast between mind and matter, the human being and nature, soul and body, such as arose after the decline of classical antiquity in Europe and obtained its highest elaboration in Christianity" (Engels 1962: 451–453). Lines that have been often quoted by Marxist environmentalists and eco-socialists as a necessary corrective to Marx's and Engels' explicit contempt toward the "romantic" naturalism and the animal advocacy of their times. Engels has been often harsh with vegetarianism and antivivisectionism—whom he considered to be examples of "bourgeois" socialism—writing sarcastically of "opponents of inoculation, supporters of abstemiousness, vegetarians, anti-vivisectionists, nature-healers" (Engels 1972: 455). It is important to underline that the protection of animals in England was often characterized by bourgeois disdain for the "savageness" of the lower classes (May 2013: 7). It is no wonder that the fathers of "scientific socialism" were skeptical about such elitist appeals to non-human nature.

2.2   Animality in Italian "Orthodox" Marxism

It is not hard to find a similar suspicion toward animality in Italian Marxism. On the *Avanti!*, the official organ of the Italian Socialist Party, "animality" represents an inferior, violent, and brutal condition, something that history must deny in order to build an authentic "human" society. The term "animal" rarely signifies more than a description of human abuse and degradation. For example, commenting on the Boer War, the spread of nationalism and the justification of violence are described as the re-emergence of "repressed animality" (Tedeschi 1899: 1). Anticipating the ironic piece in which Paul Lafargue (1975) invited workers to ask for "the rights of horses," another article complained that those responsible for killing a horse had been promptly arrested and punished, while those who had caused the death of a soldier were still at large ("certainly the life of a soldier is worth less than that of a horse ...," Unknown Author 1897: 3). Paolo Orano—a revolutionary Socialist who subsequently followed Mussolini and joined Fascism—wrote a vibrant response to the famous Italian criminologist Cesare Lombroso. In line with his positivism, Lombroso was engaged in a reductionist and "ethological" study of the human being that Orano attacked with an argument built around a typically Engelsian scheme:

> We who no longer admit the creator god, the existence of the soul, the providence and, therefore, the purpose of Life for a heavenly predisposition; we know enough to say that men are not animals; we know, indeed, even better that between humanity and animality a fact has been constituted, an antithetical principle, a negation which means a transformation [...]. The first sacrifice of the wild impulsion is the effect of the first arrest of bestiality. It is the animality that holds back. There is an interruption; there is a nervous brain disorder in nature. [...] The whole of humanity—for these thousand and ten times a thousand hundred years of "labor" that have diverted it from animality—has largely created a nature of its own, shaped by the

dynamics of artifice and effort. The beasts teach us that there is a system of strict and specific natural laws; they cannot teach us to do as they do, to be beasts. (Orano 1906: 3)

After World War I, Bruno Fortichiari, who later was among the founders of the Italian Communist Party, published *Lettere a te che leggi* (*Letters to You Who Read*), a text written in a simple and accessible language as an introduction to Socialism. In the very first lines we read: "My friend, you are not a socialist. [...] Are you indifferent? Being indifferent means being a domesticated animal" (Fortichiari 1920: 3–4).

Such attitude is apparent in nearly all the writings of outstanding Italian Marxists of the Second International. Antonio Labriola, the most important and original intellectual of the Italian Socialist Party, for instance, epitomizes very clearly the problems faced by Marxism in theorizing human animality. Marx's method, he argues, surely implies the necessity of "*naturalizing* history" (1966: 114), since "man is without doubt an animal, and he is linked by connections of descent and affinity to other animals. He has no privileges of origin or of elementary structure, and his organism is merely one particular case of general physiology" (115). Yet, Labriola underlines, "we are not here in the domain of physics, chemistry or biology; we are only searching for the explicit conditions of human association in so far as it is no longer simply animal" (98). Labriola speaks of a "hiatus," a "break" in nature's continuum, "thanks to which human life is found detached from animal life to rise, in the sequel, to an ever higher level" which "separates human life by a great interval from animal life" (115). Having witnessed the negative influence of "vulgar materialism," positivism, and "social Darwinism" on European culture in the late nineteenth century, Labriola condemns any reduction of humans to animality. "Darwinism, political and social, has, like an epidemic, for many years invaded the mind of more than one thinker" (114), he writes, explicitly rejecting any identification of Materialism and Naturalism: "Our doctrine must not be confounded with Darwinism" (120). Properly human relations, according to Labriola, "are not the result of the crystallization of customs under the immediate action of the animal struggle for existence" (117). Labriola distinguishes between biological "struggle" and social "antagonism": while the first works according to the laws of Darwinian evolution, the latter is "the principal cause of progress" (152), a distinction that will later surface in Gramsci's use of the term "contradiction."

Under this respect, the opposition between humanism and naturalism could be mirrored in the contrast between Gramsci and Bordiga, the most influential founding members of the Italian Communist Party. Although none of them was willing to fully embrace one of the poles of the humanism/naturalism opposition, they had quite diverging theoretical inclinations. Gramsci's concern about physicalist reductionism was symmetrical to Bordiga's refusal of humanist romanticism and subjectivism.

It is true that scholars have often underlined Gramsci's naturalist interests, speaking of "the depth of his love of nature, specifically of plants and animals" (Germino 1990: 17), and of his "enthusiasm [...] for animal pedagogy" (Rosengarten 2014: 81). His letters are full of detailed, passionate descriptions of animal behavior. Remarkably, he even wrote he wanted to "to take a dip in animality to draw new vigor from it" (1960: 3). Yet, one of his most recurrent polemics was directed against a vulgar-materialist "naturalization" of history. In his view, history was, in the last instance, to be seen in opposition to nature's "fatalistic" (2000: 57) laws. In rejecting idealism and spiritualism, scientific Socialism must not embrace the mechanist look of physical materialism, with its determinist, "animal" description of human behavior. Gramsci, thus, explicitly distinguishes his position from Bukharin's monism, not less than Croce's dualism, since they both "promote a a-historical and a-critical conception of science" (Antonini 2014). History, on the contrary, is the *locus* of meaning production, a specific "human" *and* "materialist" activity: "without humanity what would the reality of universe mean? The whole of science is bound to needs, to life, to the activity of the humanity" (1995: 292). It is not strange, then, that Gramsci's use of the term "animality" has often a negative connotation, his personal love of nature and animals notwithstanding: "Industrialism is a continual victory over man's animality" (1992: 235).

Opposite, but not essentially different, is Amadeo Bordiga's position. The difference seems at first glance to rely on an epistemological conflict, since Gramsci and Bordiga have contrasting opinions about what is to be understood by "scientific" Socialism. As a matter of fact, Bordiga's interest in mathematics and physics makes him less suspicious than Gramsci about the affinity between Marxism and hard sciences, especially for their discard of "individuality" and their idea that knowledge advances through epistemological revolutions which "destroy" the past (2004: 56ff). Bordiga's polemics is primarily directed against historicism, a philosophy which he considers still heavily afflicted by spiritualism and idealism. The point, according to Bordiga, is to reject the idea that humans are carriers of some individual "power" which detach them from animality: what is character-istic of humanity is their ability to consciously shape their destiny, some-thing humans do through technology as a "species" (2004: 73). At the same time, contrary to bourgeois rationalism (Enlightenment and Positivism), Bordiga believes that progress does not happen in a linear way but through crises, leaps, and adjustments. Nature is intrinsically revolu-tionary action: science is just Nature knowing itself (2004: 109). Praising "the strength and intelligence of the human animal" (1978: 123), Bordiga believes that the human being is the conscious part of such revolutionary process, his entire history testifying of its efforts to break free from its animal conditioning: "human history, since we learned to overturn animal praxis and began to plan our future—although for now still in a very lim-ited way—it is nothing but a move towards a goal" (2004: 80). Thus, human animality is always to be described in opposition to non-human animality as the source of a collective, organizational force:

> what distinguishes the human species is not knowledge or thought or some particle of divine light, but the ability to produce not only objects of con-sumption but also objects devoted to subsequent acts of production [...]. This primordial need to organize the production of tools is linked—and this characterizes the human species—with that of subjecting the reproductive process to some kind of discipline and rules, overcoming the accidental nature of the sexual relation with forms that are much more complex than those presented by the animal world. (2014: 6)

Echoing Engels' theory on the role played by labor in the "making of Man," Bordiga emphasizes the collective activity as a conscious and social process, something which exceeds the individual (we will find a similar enthusiasm for animal social behavior in Negri's use of the concept of "swarm intelligence"). The main difference with Gramsci's stress on "human activity" is that Bordiga believes that such "practical materialism" actually leads to the dissolution of the antithesis between "spiritualism" and "materialism" (Bordiga 2004: 71); he also praises disruptive "intu-ition," violent action, more than Reason and "academic science" as means to seize power (87). It is also interesting to note that while Bordiga does not entirely give up the negative use of "animality" as a metaphor for the blind necessity and struggle for existence under capitalism, he introduces a different scientific view on the animal world: Kropotkin's concept of "mutual aid" helps him to get rid of any possible relapse into social

Darwinism. With the forthcoming proletarian revolution, Bordiga writes, "the animalistic concept of competition—struggle for life—is diminishing, while the principle of mutual aid is emerging" (1996: 185).

The imprisonment of Gramsci by the Fascist Regime and Bordiga's political decline left the field open to Palmiro Togliatti, who quickly established his intellectual hegemony on the Party. While Bordiga suffered increasing marginalization due to his opposition to the official line of the Comintern, Togliatti—a pragmatic and "centrist" politician—soon became an important leader during the difficult times of Stalinism. Intellectually closer to Gramsci, Togliatti shared his historicist overview. In Croce's idealism, Togliatti wrote, "it is denied that there exists, as it actually exists, a system of material forces which is the texture upon which all the rest is woven: animality, and humanity and society, life and death, the events of man, his will and his thought, his society, etc. etc.—here the unity of the real, i.e. of the world, is lost" (2014: 1606). At the same time, his capitulation to Stalinism made him a typical representative of the "industrialist" ideology, according to which nature could only be the object of human domination ("a struggle of man against nature, against himself," 2014: 2458). No coincidence that the references to animality remain here marked by the negative meaning that we have already seen to be typical of Marxism: the "animal" is always that which must be overcome by humanity in its effort to self-determination, or the symbol of violence and oppression that the proletariat must defeat and leave behind once and for all: "we believe that man must become master of nature, which is a biblical task, indicated to man by God himself, in *Genesis*" (2014: 1001).

2.3   Developments in Post-War Italian Communism

Although a new group of Communist intellectuals appeared on the scene, the post-war period did not bring significant changes in the theme of animality. This will be evident both in the heirs of the historicist and Hegelian tradition and in the authors who criticized such approach.

Cesare Luporini attempted to elaborate an epistemology of Marxism as "social science" without rejecting neither Hegel nor Gramsci. Though criticizing historicist teleology, Luporini admitted that "human nature" could not be determined in any fixed form (1959: 858). At the same time, he believed that "materialism" was a form of "humanism," since "the production of our species," differently from animality, is essentially grounded on purely human "productive relationships" (1954: 491). From his part, Nicola Badaloni reiterated "the existence and rooting of ideas unconsciously in animal life" (1975a: 4–5) while, at the same time, speaking of Communism as the end of the "animal appearance" of history (1975b: 53). Stressing the need for the proletariat to "dominate the unconscious, the animal in society" (1975b: 99), Badaloni also spoke of the struggle among "contradictory impulses" as a characteristic that makes man an "unstable" being ("unlike the animal, whose instincts satisfy well-defined tasks, man expresses here his illness," 1988: 35). Similarly to Marsilio Ficino and Pico della Mirandola, the rhetorical strategy here consists in raising man through his lack, condemning animals to purity, that is, lack of choice, freedom, and struggle.

As a matter of fact, both Luporini and Badaloni exemplify the anthropocentrism and idealism of Italian Marxism later denounced by Lucio Colletti. Colletti is probably the most important theorist of anti-Hegelian Marxism. A pupil of Galvano Della Volpe, he began a systematic revision of Marxism that led him within a few years to conservative political positions. Della Volpe understood the relationship between man and nature outside the Hegelian dialectic, of which he was a firm opponent. Using the Crocian concept of "distinction" rather than that of "opposition," Della Volpe maintains that in man the unity of nature and spirit is realized in a non-dialectical, non-Hegelian form. Nature is the sphere of the "particular," while history and spirit are the sphere of the "universal," "general." In man these two poles come together as a unity of distincts, rather than as a unity of opposites (1973: 336–37). However, it is interesting that the outcome of this de-Hegelianization (or de-historicization) of the human-nature relationship ends up sanctioning the domination of the former over the latter. Recognizing in man a "natural cause" that acts on other natural causes—and yet a natural cause that is "conscious" of its own action—Della Volpe speaks of the "anthropologization of nature" (1973: 373) as an accomplishment of Communism. In Communism, Della Volpe argues, humanity becomes both cause and effect of itself, since nature—which is always embodied in his social system—does not exist for him as something "external."

Colletti's general view of the human-animal question hardly diverges from the general line of orthodox Marxism. By analyzing the way Enlightenment has historically used the animal world as a model of human society, for instance, Colletti emphasizes both the risks and the merits of such an attempt. And here, not surprisingly, animality still emerges as a symptom of a violence not sufficiently tamed by human society: "capitalism is a river which can carve out its own channel. Mandeville and Smith are therefore in a sense right to celebrate it. On the other hand, if it is 'civilization', it is still true that capitalism is so through inequality and oppression; and that it is therefore a progress which does not humanize man but exacerbates his predatory animal nature" (1972: 216). In his criticism of Hegel, Colletti proceeds partly in the direction of Della Volpe; contrary to Della Volpe, though, Colletti considers history as the sphere of "unrepeatability," creation, and singularity: characteristics that are denied by any "animalization" of human society. According to Colletti, while blurring the distinction between man and animal leads to abstract generalizations and therefore to a false universal, their dichotomization inevitably produces the impossibility of submitting history to the laws of science—which cannot but be universal (1972: 34–35). The most original part of Colletti's reflection, however, concerns his attempt to free Marxism from the influence of Hegel. Here, the fragile Marxian-Engelsian construction of the man-animal dialectic is attacked and dismantled as "ideological." The very concept of "dialectic"—"dialectical" thinking being, according to Hegel, what distinguishes man from animals (1973: 217)—is refuted as mystical and anti-scientific. At the same time, the idea that the human being is a "generic being" (*Gattungswesen*) and that its "true" nature is essentially nothing but the relationship of humanity to itself is unmasked as an idealistic inheritance (233ff). Colletti traces the historical genesis of this idea from the Renaissance image of Man: a being suspended between Earth and Heaven, therefore a "negative" being, constantly open to "becoming," a being that is never reifiable, that can never be reduced to "thing" except when he loses his "truly" human essence (1973: 238ff). It is interesting to note how on September 20, 1959, the date of the launch of the Soviet Lunik, the event had been celebrated on *L'Unità*—the official press organ of the PCI—with an excerpt by Engels and quotes by Pico della Mirandola, Tommaso Campanella, and Leonardo da Vinci.

The entire Italian debate on animality hardly challenges such narrow and abstract theoretical framework. Animals always turn up as the evolutionary, historical, and social "other" of human society; they are a sort of mirror where humanity contemplates itself in a distorted form. Since Italian Marxists do not seem to care about the animal world as such, animality is never the real focus of their discourse: over and over animality gets described or theorized in a very stereotyped way. Even less does such discourse show any preoccupation about animal exploitation, suffering, and death. It is not for chance that hints at a different interpretation of animality often come from the field of literature, where our ontological ambivalence is addressed through the animal. It is thus in the poetic imagination that the Italian Marxist tradition of the 1950s and 1960s manages to force the limits of its longstanding humanism.

Although heavily influenced by the orthodox Hegelianism of Lukács, for example, Cesare Cases casts a rapid, tragic look on animality. In a powerful note to his 1974 review of *La Storia* (*History*) by Elsa Morante, Cases writes that "a truly human world implies respect for the animal," adding: "it seems that even Marx [...] praised Schopenhauer for having expressed among 'the fundamental ideas of his ethics [...] simply the duty, which comes from the essential unity of every organic nature, not to hurt neither men nor animals'" (1987: 115–16).

In some authors of the Marxist tradition there seems to be the need to reflect on the contradictions of civilization, to question the myth of growth and progress at all costs. The protagonist of Italo Calvino's novel *Il Barone Rampante* (*The Baron in the Trees*), for instance, takes to trees, at the threshold between civilization and animality. Calvino's position becomes even more suggestive in his theoretical interventions, where the author engages in a deep reflection on animals and comes to question the primacy of the human. In controversy with his friend Elio Vittorini, Calvino defends not only the continuity between the animal and the human but even goes so far as to undermine the anthropocentrism and humanism of the Marxist culture of his time, arguing that the future may belong to "machines" and "beasts" (1995b: 164). The power of the human apparatus by no means implies, as Togliatti argued, the right to exercise a master power over nature; rather, it implies greater "responsibility" toward other creatures (1995a: 72). And it is here that he imagines the forthcoming dissolution of man, its creative power being replaced by a post-human synthesis. The intersection between cybernetics and animality, on which Calvino carefully reflects, seems to push him in the direction of the techno-scientific, quasi-Bordighist wing of the humanism/naturalism dilemma. Although it must be noted that Calvino's accents become more apocalyptic and go beyond the rhetoric of the "proletarian revolution": "I am ready to accept a world made of machines, animals and plants only, without men: computers, birds, reptiles, magnets, amphibs, fishes, lepidoptera, manometers, with machines that keep alive and expand the information (the 'culture') that men have clumsily tried to secrete" (2000: 1104–5).

As in the case of Bordiga's collective intelligence, we will see this theme re-emerge in Negri's post-workerism.

The tragic reflection of Pier Paolo Pasolini could be set in opposition to Calvino's techno-scientific apotheosis. The purpose of the Bolognese poet is to describe the vital experience of man even in the bleak scenario of advanced industrial civilization. He does this by exploding the Renaissance heritage criticized by Colletti. Pico's Man, set in the middle of the universe, stretched between the beast and the angel, between instinct and the contemplation of God, becomes a *coincidentia oppositorum*:

> The world is more sacred
> where it is more animal:
> without betraying
> its poetry, its original force,
> it is up to us to exhaust its mystery
> in human good and evil.

Pasolini's work proceeds through an open, visionary, hallucinatory synthesis (1993: 208). We could say that human animality emerges from the clash between these two different and opposite tendencies, between the telluric powers of the unconscious, of the body, of sexuality, and the industriousness of labor. In this case it is the blasphemy (*bestemmia*)—the offense against God which works through a juxtaposition of the Divine Name and the name of a beast—which illuminates the demonic power of the human condition.

The theme of human animality is also treated—in all its destructive-creative power—in Pasolini's scandalous works, such as his *Trilogies* of life and death. It is starting from this undecidable ambiguity with respect to the meaning of animality that Pasolini's judgment on the Italian modernization process becomes elusive and at times problematic from a Marxist point of view. Contrary to Calvino, in fact, Pasolini has strong doubts about the liberating power of techno-science, indulging in a "romantic" revaluation of peasant life. For example, in an article published on the *Corriere della Sera*, the "disappearance of fireflies" reveals the void around which power is organized and works, emptying human experience of meaning and beauty in order to fill it with its own nothing (Pasolini 1975: 1). The famous statement that closes the article ("I would give the entire Montedison, even though it be a multinational company, for a firefly") is a declaration of intent, a celebration of the liberating beauty of animality against the horror of civilization. It is also the signal that a different way of relating to non-human animals is necessary and reveals Pasolini's attention to the animal question. In 1969 Pasolini wrote a letter to actress Anna Magnani:

> I have read these days that trains full of slaughter animals, horses or oxen, use to stand still for three or four days at certain border stations. The beasts, packed inside the sealed wagons, do not eat or drink: and often they die. Men also died in freight trains to Germany, and their bodies remained for days among the bodies of the living, on the excrement. In my imagination the two images—that of the carriages full of men and that of the carriages full of animals—are monstrously confused. [...] Yet precisely because I remember, with an anguish that has remained the same, the martyrdoms suffered then by men, I have pity for the martyrdom inflicted on these suffering animals. (Pasolini 1992: 742)

There is no more evident proof of the superiority of the poetic imagination on philosophical reflection than these Pasolinian lines, written fifteen years before Agamben's reflection on Auschwitz and "bare life" and almost forty years before their appropriation by defenders of animal rights (Filippi 2010).

3   ECOLOGY, ETHOLOGY, AND ANTISPECIESISM

By the end of the 1960s, a vast array of discoveries and new theories in the field of science (cybernetics, DNA), linguistics, and philosophy (the rise of structuralism) lead Italian Marxism to face its own cultural delay. New elements, thus, began to slowly shift its perception of animality. On the one hand, Italian Marxists reiterated the need to "better understand the relationship between the essence of man and his animality through the chemical-biological sciences and cybernetics" (Berlinguer 1972: 3); on the other, attempts were made to keep the new discoveries within the frame of "dialectical materialism."

Almost in the same period, the spread of environmentalism forced Marxist thinkers to discuss the relationship between civilization and nature, human and animal, also as a *practical* problem. The conflict between Man and his "Other" became an urgent issue not only thanks to the concern of youth culture (the "hippy" movement) but also by that of members of the establishment (the documents published by the "Circolo di Roma" in 1968), not to mention the energy crisis of 1973. However, even though Gramscianism and Hegelianism began to be regarded with suspicion by some Italian Marxists, Engels still played a major role in the reception of the "green" philosophy.

While Colletti was criticizing Engels for his anti-scientific use of Hegel, Sebastiano Timpanaro and Dario Paccino took up the author of *Anti-Dühring* for his materialist interest in science, as well as for his emphasis on the continuity between humanity and animality, between history and nature. In his influential work *On Materialism*, Timpanaro combines Engels' thought with eighteenth-century materialism. A scholar of the poet Giacomo Leopardi, Timpanaro sees in the contempt against "sensist" materialism an "idealistic" trait of Marxism:

> Materialism entails also the recognition of man's animality (superseded only in part by his species-specific sociality); it is also the radical negation of anthropocentrism and providentialism of any kind, and it is absolute atheism. Thus it represents a *prise de position* with regard to man's place in the world, with regard to the present and future "balance of power" between man and nature, and with regard to man's needs and his drive for happiness. (1975: 249)

Timpanaro contrasts such "idealist Marxism" to Leopardi's cosmic and tragic vision: nature embraces all living beings and binds them to a destiny of despair. At the same time, and precisely for this reason, nature itself drives them to mutual solidarity, triggering our attempts to reduce suffering, to fight the evils produced by human selfishness (82).

What Enlightenment materialism—unjustly condemned as "vulgar"—can teach Marxism is the importance of pleasure and pain as necessary stimuli of human behavior. Human freedom, thus, does not need the rejection of animality: "The something more that man possesses in relation to animals is a greater capacity to foresee and order means in relation to an end and a greater understanding in the determination of the end, but it is not a greater measure of 'free will' in choosing between various ends" (105). Only a disembodied being, an idealistic spirit—or, that is the same, an Omnipotent Party—can claim to represent man and disregard "vulgar" issues such as individual sickness, suffering, and death. It is precisely this somatic layer, this animal background of man that Marxism can never forget if it does not want to betray its "materialism." Freedom has more to do with the pursuit of happiness, than with the independence from natural necessity: "Of course, both Colletti and Havemann are correct in expressing their dissatisfaction with the Spinozist and Hegelian formulation of freedom as 'consciousness of necessity' which Engels adopts in *Anti-Dühring*. But why is this formulation unsatisfactory? Not because of its anti-voluntarism, but because of its anti-hedonism" (106).

Consequently, the biological study of man cannot be ignored as irrelevant. Nature is not erased by culture; animality is not sublimated by humanity: "human history has specific characteristics which distinguish it from the evolution of the animal species (and of man himself *qua* animal), Marxism must not be dissolved into Darwinism. On this question there can be no argument. But Korsch and the other Western Marxists were not satisfied with this; they wanted to reduce nature to a mere object of human praxis" (241).

The "theory of needs" elaborated by Timpanaro, however, finds itself in the same difficulties experienced by previous Marxism: in order to be materialistic, it must emphasize the animality at the heart of those needs, but, at the same time, it cannot let science define the tasks of socialism. Not differently from Gramsci, Timpanaro must then criticize ethology every time it claims to explain human behavior in mechanistic terms, that is, every time it undermines the possibility of political struggle. Therefore, if nature cannot be reduced to an object of human praxis (which would make man a being that acts on nature "from the outside"), neither is it possible to totally erase human activity and make it the result of mechanical forces, ignoring the role played by "social relations of production, the division of society into classes, the class struggle and the new rhythm which the historical development of mankind has taken on since then and which has replaced that of biological evolution"; when they do so, ethology and ecology become "reactionary ideologies" (15). A polemic which will later re-emerge in the debate on sociobiology (Manghi 1982: 23ff).

Similar positions are expressed by Dario Paccino in his important book of 1972 *L'imbroglio ecologico* (*The Ecological Fraud*). In this text, Paccino underlines the relevance of Engels in denouncing ecology as an ideology of the ruling classes. Taking up Engels' and Marx's project of "natural history," Paccino reaffirms the unity between the human and the animal, seeing in the ability of tool making, in the modification of the environment and of ourselves, the specificity of man (23–24). However, Paccino also reiterates that the "cultural" layer produced by the introduction of tools in nature cannot be understood as a unitary phenomenon: there is always struggle; there is always negation and contradiction, since human relationships are also expressions of class dynamics (32–33). Hence, there can be no defense of non-human nature if environmentalism does not become part of the political agenda of the proletariat.

While Italian Marxism was still struggling with dialectical materialism and was just starting to deal with the theme of animality in a more scientific and circumstantial way, the world philosophical community was shaken by Peter Singer's *Animal Liberation* (1975). Initially translated only in a limited edition by the Italian Anti-Vivisection League, Singer's work did not have a huge circulation in the Italian debate in the first years after its publication. Surely his utilitarian and analytical approach did not help in making it fashionable among Italian Marxists. Not to mention his defense of vegetarianism as an individual mean for changing the world: assumptions that could only sound naive and petty-bourgeois to the Marxist culture of the time. As a matter of fact, the Italian Communist press began to mention Singer mainly because of the scandalous consequences of his moral theory, especially the use of human embryos for scientific experimentation (Terragni 1988), a hypothesis that provoked a worried response from Giovanni Berlinguer: "I suspect that someone (few, fortunately) has more respect and love for animals than for men" (Berlinguer 1988: 2). No coincidence that the reception of the Australian philosopher dates back to the end of the 1980s, when the hegemony of classical Marxism in the Left had been definitely questioned. This meant, however, that Singer's philosophy implanted itself in an alien terrain. The first important study on the problem of animal ethics published on *Critica Marxista*—one of the most prestigious journals of the PCI—was a sympathetic analysis by Sergio Bartolommei (1986). Bartolommei was however more interested in the non-anthropocentric foundation of an ecological ethic (a subject that he will systematically explore in 1989) and to distinguish the defense of the environment from the romantic and technophobic strands of Deep Ecology. A few years later the Marxist epistemologist Salvatore Veca (1990) wrote a supportive preface to the Italian translation of *The Case for Animal Rights* by Tom Regan. Even though the limits of Singer's thinking—such as the need to solve the problem of animal exploitation at a systemic level—had been sometimes noted (Passi 1991: 27), no serious attempt was made to find a real synthesis between Marxism and the critique of speciesism. Rather, the eclipse of Socialism slowly led such theoretical encounter to lose all reference to the thought of Marx and Engels. The debate about animal rights that was going to follow in the early 1990s with no exception assumed the liberal-individualistic perspective of Singer and Regan and therefore had nothing more to do with Marxism (Maurizi 2011). It was only at the beginning of the 2000s that some intellectuals who had not renounced Marx's categories began to open up to the animal question, welcoming the theme of animality in a perspective of political liberation for the working class. This is the case of Maria Turchetto, an Althusser scholar, whose interventions underline the need to rethink our relationship with other animals and to question our anthropocentric and speciesist *Weltanschauung* starting from the recent acquisitions of ethology and eco-feminism (2009, 2010a, b, 2013, 2016).

4 Animality and Post-Workerism

It is necessary now to give an account of one of the most successful "heretical" tendencies of Italian Marxism, that is, its workerist, post-workerist, post-structuralist, and post-modern component. This is a very vast cultural area that has crossed the animal question in various ways, above all in the last twenty years. We are obviously talking of Antonio Negri and Giorgio Agamben and of the theoretical paradigm that today goes under the name of "biopolitics." Though such philosophies have culturally gravitated around the Communist Left, it could only be very roughly ascribed to "Italian Marxism" (especially in the case of Agamben). This is the reason why I will focus here on Negri, who explicitly placed himself in continuity with Marx's thought.

Italian workerism develops along the Panzieri-Tronti-Negri line and finds in the author of *Empire* a new theory of animality. This will happen, however, subverting Panzieri's and Tronti's initial approach which had much in common with Gramsci's stress on the centrality of human activity. Panzieri's critique of "determinism," though, is not direct against science, but rather against the idealistic-historicist tendency to see the triumph of socialism secured by an objective historical necessity (1976: 51ff). According to Panzieri, the spread of this kind of determinism had both theoretical and political reasons: on the one hand, it derived from Engels' formulation of the "laws of dialectics"; on the other, it served as a justification for Stalinist and reformist bureaucratism. Marxism, though, is not a universal philosophy of history, but the science of capitalist society. Its object is therefore not "humanity" but the "working class," which Panzieri places at the center of his analysis: it is in the body and in the conscience of the working class that the contradiction between capital and labor lives (1976: 3ff). The apparently "subjectivist" aspect of this position seems to make Workerism a theory little interested in animality: as a matter of fact, what Workerists and Post-Workerists called "General Intellect" cannot but belong to our species (Virno 2003, 2010, 2013; Virno and Hardt 2006). Engels' reference to the unity of science gets therefore ridiculed by Panzieri: "The science of dialectics, applicable to the physical sciences as to the social sciences, is evidently a negation of sociology as a specific science, and instead of this it recreates a metaphysics, which is both the metaphysics of the labor movement and the metaphysics of the tadpole and the frog" (Panzieri 1976: 89).

In Italian *Operaismo*, however, the centrality of the factory is only apparent: right from the start Panzieri maintains that capitalism is organized as a system *around* the factory. Therefore, the factory is the "ideal" center that determines the structure of the surrounding world. Since the Worker is only the empirical mask of a relationship that concerns *all* social phenomena, the workerist discourse *is* a discourse on Man, after all. In Mario Tronti we see how this shift does not lead to a substantial revision of the traditional Marxist prejudice against animality. From one side, he invites us to re-evaluate "that shaggy and deep, anti-humanistic vein, which has been so lightly given to the reactionary culture. The 'animalists', as Haydn put it: against the god-man of Pico, outside the nature-man of Montaigne, lies the man-beast of Machiavelli and Guicciardini" (2017: 249). From the other, he does not work on this "animalist" hypothesis but formulates his anti-humanism in a rather classical way, that is, according to the theory (already criticized by Colletti) which sees man as an indeterminate being. In his late reflections on the "death of the subject" and on the "end of man," in fact, Tronti states that "the human difference must be re-evaluated with respect to its reduction and homologation to the categories of the modern bourgeois universe, bourgeois and *citoyen*, *homo oeconomicus* and subject of power, animal the first, machine the second, nature and history, freedom and law. All that is *humanum* must be *alienum*" (433). The human—neither animal nor machine—finds its essence only in the subversive "difference." Taking up the thesis of Scheler, Gehlen, and Cassirer (440) the human is seen by Tronti as the indeterminate, what resists assimilation, what defies the system. It is *animal symbolicum* and *Mängelwesen* (a "being that lacks something"). Unlike the animal that "lives immersed in concrete reality, in a *hic et nunc*, in a contingent way of being," the human being says "no" to its immediate reality (444). Where this does not fully succeed, man is reduced to an *animal democraticum* (606), a "bourgeois animal spirit" (445) against which Tronti asserts Nietzsche's conception of man as a "mad animal."

Inspired by Deleuze's "becoming-animal," Antonio Negri abandoned workerism's anthropocentrism and proposed an interesting reversal. His anti-humanist perspective opens up to the reconsideration of animality in the process of re-appropriation of the working class of its creative vitality. Negri proposes a project of radical "Spinozist" immanence: once Man is seen inside the natural order, any sign of transcendence begins to waver. We must therefore break down "the barriers we pose among the human, the animal, and the machine. If we are to conceive Man as separate from nature, then Man does not exist. This recognition is precisely the death of Man" (2000: 218). Della Volpe's and Colletti's opposition between singularity and general, history and nature, collapses. This time it is nature that is conceived as "becoming," as a stream in which difference reigns, rather than uniformity and repetition: "the plasticity and mutability of nature really refer[s] to the common—and indeed nature is just another word for the common" (171). Since "there are no fixed and necessary boundaries between the human and the animal, the human and the machine, the male and the female, and so forth," nature itself becomes "an artificial terrain open to ever new mutations, mixtures, and hybridizations" (215). In this scenario, the humanist and anthropocentric project comes to an end. Taking up the legend of St. Francis, pitching "the joy of being" against "the misery of power" (413), Negri and Hardt describe the purpose of Communism as a new form of "humble happiness." By shifting the subject of the revolution from the factory worker to the social worker, from the material work to the immaterial one, Negri can resume Bordiga's identification of techno-scientific development with the growth of collective consciousness and the process of human self-transformation (see his analysis of "swarm intelligence," 2004: 91ff). However, even in this case, revolution does not accept technological progress as such but sees it as part of the political struggle against the dominion of life, questioning its ideological neutrality. On the one hand, Negri can even talk sympathetically of animism ("Amerindians conceive animals and other nonhumans as 'persons'," 2009: 123); on the other hand, he sees in a "politicized" technological development a revolutionary factor, the active transformation of man in its struggle for liberation: "Man is the animal [...] that is changing its own species. [...] Humanity transforms itself, its history, and nature" (2004: 196). In a recent book (2016), Massimo Filippi interviewed

Michael Hardt, trying to make these connections explicit, making room for non-human animals in the idea of a Communist revolution.

5   ANIMALITY IN ITALIAN ADORNOISM

Italian scholars of Adorno have recently underlined the importance of the "animal" in Critical Theory (see Bellan [2006], Benini [2012, 2014], and Maurizi [2011, 2018]). According to Adorno and Horkheimer, civilization implies what Marx called the "appropriation" (*Aneignung*) of outer and inner Nature, that is, the repression of animality which traces an important break-point in the process of estrangement of consciousness from the material world. This is the fundamental structure of all class societies, the hidden core of both material and cultural progress, a structure that cannot be overcome unless its basis—the domination of nature—is also transformed. Thus, a free, classless society cannot be imagined without the liberation of nature, since violence on animals and violence on humans share a common root. This root is our animal self-hatred, that is, the fact that *we despise our animal-being* (Bellan 2006: 158; Maurizi 2011: 173ff).

Thus, the limit of orthodox Marxism was its passive and static notion of animality (Maurizi 2011: 100ff). The question whether animals are part of human society or not cannot be answered in purely theoretical terms. Animals express a living contradiction in our society. What we "are" and how we "consider" and "treat" the other animals is part of our political struggle for self-determination. To discuss, then, the problem of "animal subjectivity" means to re-frame the problem of "sensibility." The problems of animality and sensibility become important for Marxism because they help us develop a consequent historical and dialectical materialism. Even better: a form of solidaristic materialism (Maurizi 2011: 173ff; Benini 2012: 197ff). Contrary to "vulgar" materialism, which understands reality as an inanimate interplay of matter and void, solidaristic materialism is a kind of "relational ontology," in which reality is understood in a dynamic, historical way. Matter is not a static "thing"; it rather should be described in terms of reciprocal "relations" evolving in time.

Two points are important, here. First, human relationships are *conflictual*, and such conflicts shape not only the material production of society but also the whole development of Civilization. It is through social organization that human consciousness becomes self-conscious, posing itself as something other than nature. The elevation and improvement of the human "spirit" is only possible through privilege and dispensation from material labor. Secondly, the "Self" is embedded in societies from which individuals entirely depend. Self-idealization is mediated by material and symbolic structures: our escape from animality becomes the *conditio* sine qua non of our own social existence. Nobody can avoid such *self-domestication* (Maurizi 2011: 43ff).

Human dominion over nature is not the proof of our "unnaturalness"; the cliché of the human as a "mad animal" is just the counterpart of the religious belief, according to which the human being is the "King of the Earth." Defining human beings as "unnatural" beings would simply restore an idea of transcendence. On the contrary, we must recognize that what human society does to nature is a consequence of the illusory opposition between spirit and nature: the inner struggle of human culture is the perpetuation of a *natural intestine war*. Our "second nature," "the human world [...] is not the realm of freedom" (Benini 2012: 53): all human culture is mediated brutality. What we do as *homines sapientes* is to reproduce at a higher level the violence of nature, its relational and oppositive structures. The very concept of civilization implies the idea of going *beyond nature* (Maurizi 2011: 173ff). It is precisely this idea that ought to be articulated in a different, materialistic way.

Thus, any attempt to get out of the present catastrophe must ground its efforts on a simple, materialist notion: that the human being is, after all, still an animal (Bellan 2006; Benini 2012, 2014). At the same time, only by superseding the dialectic of civilization does the human being become *properly human* (Engels). It is only by recalling its own naturalness that the human spirit can solve its inner contradiction, its furious opposition to the rest of the natural world, its urge for power and control. Such move discloses the historical process as a *self-referential natural process*: history is nature's attempt to go beyond nature. Nature is intrinsically divided and seeks a way out of its own condition of general violence: it aims at a state of universal reconciliation that only human society—violent as it may be—could guarantee (Maurizi 2011, 2018). In its inner drive toward universality, in its attempt to recompose the pain of a fragmented nature, human reason proves to be something more than just the logic of cold extermination. As a socially liberated force, reason is an instrument of universal justice, solidarity, and peace. The hope of the oppressed. Even those who are oppressed by nature.

In a way, "it is not just a matter of recovering humanity towards animals, but more radically of developing a form of animality towards humans and nature" (Benini 2014: 15). Here we finally have a different *dialectical* theory of animality. Its decisive, paradoxical thesis can be defined as follows: until we consider ourselves different, we are nothing but animals; yet, it is only by considering us nothing but animals that we can make the difference in nature. This means that our alienation from nature will torment us, until we theoretically accept our own animality; at the same time, by practically putting an end to animal exploitation, humanity could really emerge from nature as a different kind of animal (Maurizi 2011).

WORKS CITED

Antonini, Francesca. 2014. Science, History and Ideology in Gramsci's *Prison Notebooks*. *HOST – Journal of History of Science and Technology*, 9. http://johost.eu/vol9_spring_2014/vol9_4.htm. Accessed 19 Nov 2019.

Badaloni, Nicola. 1975a. *Il marxismo di Gramsci: dal mito alla ricomposizione politica*. Turin: Einaudi.

———. 1975b. *Marxismo e storicismo*. Milan: Feltrinelli.

———. 1988. *Il problema dell'immanenza nella filosofia politica di Antonio Gramsci*. Venice: Arsenale.

Bartolommei, Sergio. 1986. Tre studi sui diritti degli animali. *Critica marxista: rivista bimestrale* 4: 151–168.

———. 1989. *Etica e ambiente: il rapporto uomo-natura nella filosofia morale contemporanea di lingua inglese*. Milano: Guerini.

Bellan, Alessandro. 2006. *Trasformazioni della dialettica: studi su Theodor W. Adorno e la teoria critica*. Padova: Il Poligrafo.

Benini, Erika. 2012. *Mimesi e corporeità. Saggio su Adorno*. Rome: Stamen.

———. 2014. *Ein gutes Tier gewesen zu sein*. Riflessioni a partire da Adorno. *Animal Studies* III (9), 7–16.

Benton, Ted. 1993. *Natural Relations: Ecology, Animal Rights and Social Justice*. New York: Verso.

Berlinguer, Giovanni. 1972. Un'esigenza di materialismo. *Unità*, December 19, 3.

———. 1988. Scienza e coraggio di dire no. *Unità*, June 22, 2.

Bordiga, Amadeo. 1978. *Drammi gialli e sinistri della moderna decadenza sociale*. Milan: Iskra.

———. 1996. *Scritti 1911–1926*. Vol. 1. Graphos: Genova.

———. 2004. *Per una teoria rivoluzionaria della conoscenza. Cinque testi inediti*. Turin: N+1.

———. 2014. *The Factors of Race and Nation in Marxist Theory*. International Communist Party. http://libcom.org/files/Amadeo%20Bordiga-%20The%20factors%20of%20race%20and%20nation%20in%20Marxist%20theory.pdf. Accessed 19 Nov 2019.

Calvino, Italo. 1995a. *Album*. Milan: Mondadori.
———. 1995b. *Saggi*. Vol. 1. Milan: Mondadori.
———. 2000. *Lettere 1940–1985*. Milan: Mondadori.
Cases, Cesare. 1987. *Patrie lettere*. Turin: Einaudi.
Colletti, Lucio. 1972. *From Rousseau to Lenin. Studies in Ideology and Society*. Trans. John Merrington and Judith White. London/New York: New Left Books – Monthly Review Press.
———. 1973. *Marxism and Hegel*. Trans. Lawrence Garner. London: New Left Books.
DellaVolpe, Galviano. 1973. *Opere*. Vol. 3. Rome: Editori Riuniti.
Engels, Friederich. 1962. *Dialektik der Natur*. In Karl Marx and Friedrich Engels, *Werke* (MEW), Bd. 20. Berlin: Dietz.
———. 1972. *Zur Geschichte des Urchristentums*, MEW, 22.
———. 1973. *Die Entwicklung des Sozialismus von der Utopie zur Wissenschaft*, MEW, Bd. 19.
Filippi, Massimo. 2010. *Ai confini dell'umano: gli animali e la morte*. Verona: Ombre corte.
Filippi, Massimo, Michael Hardt, and Marco Maurizi. 2016. *Altre specie di politica*. Milano: Mimesis.
Fortichiari, Bruno. 1920. *Lettere a te che leggi: umilissimo opuscolo per gli umili*. Milano: Avanti!.
Germino, Dante. 1990. *Antonio Gramsci: Architect of a New Politics*. Baton Rouge: Louisiana State University Press.
Gramsci, Antonio. 1960. *Opere*. Vol. 10. Torino: Einaudi.
———. 1992. *Prison Notebooks*, vol. 1. Trans. Joseph A. Buttigieg and Antonio Callari. New York: Columbia University Press.
———. 1995. *Further Selections from the Prison Notebooks*. Trans. Derek Boothman. Minneapolis: University of Minnesota Press.
———. 2000. *The Antonio Gramsci Reader: Selected Writings 1916–1935*. Trans. and ed. David Forgacs. New York: NYU Press.
Labriola, Antonio. 1966. *Essays on the Materialistic Conception of History*. Trans. Charles H. Kerr. New York: Monthly Review Press.
Lafargue, Paul. 1975. *The Right To Be Lazy and Other Studies*. Trans. Charles H. Kerr. London: New Park Publication.
Luporini, Cesare. 1954. Marxismo e sociologia. *Rinascita* XI (7): 491–494.
———. 1959. La scienza e il destino umano. *Rinascita* XVI (12): 857–860.
Manghi, Sergio. 1982. La sociobiologia e la critica 'marxista'. *Quaderni Piacentini*, Nuova Serie/7, 23–50.
Marx, Karl. 2007. *Economic and Philosophic Manuscripts of 1844*. Trans. Martin Milligan. Mineola: Dover Books.
Maurizi, Marco. 2011. *Al di là della natura. Gli animali, il capitale e la libertà*. Aprilia: Novalogos.

————. 2018. *Quanto lucente la tua inesistenza. L'Ottobre, il '68 e il socialismo che viene*. Milano: Jaca Book.

May, Allyson N. 2013. *The Fox-Hunting Controversy, 1781–2004: Class and Cruelty*. Farnham: Ashgate.

Negri, Antonio, and Michael Hardt. 2000. *Empire*. Cambridge, MA: Harvard University Press.

————. 2004. *Multitude*. New York: The Penguin Books.

————. 2009. *Commonwealth*. Cambridge, MA: The Belknap Press of Harvard University Press.

Orano, Paolo. 1906. Discorrendo di bestie e d'uomini. L'animalità può farci scuola? *Avanti!* January 7, 3.

Paccino, Dario. 1972. *L'imbroglio ecologico*. Turin: Enaudi.

Panzieri, Raniero. 1976. *Lotte operaie nello sviluppo capitalistico*. Turin: Einaudi.

Pasolini, Pier Paolo. 1975."Il vuoto del potere" ovvero "l' articolo delle lucciole." *Corriere della Sera*, February 1.

————. 1992. *I dialoghi*. Rome: Editori Riuniti.

————. 1993. *Bestemmia. Tutte le poesie*. Vol. I. Milan: Garzanti.

Passi, Mario. 1991. La Terra salvata dagli animali? *L'Unità*, April 11, 27.

Piccone, Paul. 1983. *Italian Marxism*. Berkeley: University of California Press.

Redclift, Micheal. 1987. *Sustainable Development: Exploring the Contradictions*. London: Methuen.

Rosengarten, Frank. 2014. *The Revolutionary Marxism of Antonio Gramsci*. Leiden/Boston: Brill.

Tedeschi, M.1899. Il nuovo delitto del capitalismo. *Avanti!* October 28, 1.

Terragni, Fabio. 1988. Feti umani come cavie? *Unità*, June 18, 22.

Timpanaro, Sebastiano. 1975. *On Materialism*. Trans. Lawrence Garner. London: New Left Book.

Togliatti, Palmiro. 2014. *La politica nel pensiero e nell'azione*. Milan: Bompiani.

Tronti, Mario. 2017. *Il demone della politica. Antologia di scritti (1958–2015)*. Bologna: Il Mulino.

Turchetto, Maria. 2009. Editoriale. *L'Ateo* 2 (62).

————. 2010a. Editoriale. *L'Ateo* 3 (69).

————. 2010b. Risposta a tre lettere. *L'Ateo* 4 (70).

————. 2013. Le donne, gli animali, la natura e i loro nemici. Tre letture eco-femministe. *L'Ateo* 5 (90): 12–14.

————. 2016. Pensare con l'animale. Dominique Lestel e il paradigma 'biocostruzionista' in etologia. In *Per gli animali è sempre Treblinka*, ed. Monica Gazzola and Maria Turchetto, 253–258. Milan: Mimesis.

Unknown Author. 1897. Crudeltà militari. *Avanti!* January 4, 3.

Veca, Salvatore. 1990. Premessa. In *I diritti degli animali*, ed. Tom Regan. Trans. Rodolfo Rini. Milan: Garzanti.

Virno, Paolo. 2003. *Quando il verbo si fa carne*. Turin: Bollati Boringhieri.

————. 2010. *E così via, all'infinito. Logica e antropologia.* Turin: Bollati Boringhieri.

————. 2013. *Saggio sulla negazione. Per un'antropologia linguistica.* Turin: Bollati Boringhieri.

Virno, Paolo, and Michael Hardt, eds. 2006. *Radical Thought in Italy: A Potential Politics.* Minneapolis/London: University of Minnesota Press.

CHAPTER 9

Experiencing Oneself in One's Constitutive Relation: Unfolding Italian Sexual Difference

Federica Giardini

In the Western philosophical Canon, politics has always been conceived as the specific attribute of human beings; even more, politics is the space where humans are recognized as such. In this perspective, "animal" appears as a notion that is both relational, coupled with "human," and historical—its meaning changes according to the different historical periods and social organizations. In another and yet connected perspective, this same philosophical Canon views "woman" as being isotopic to "animal," that is to say, the notion is assigned to a similar position and function in respect to the political subject. This chapter intends to develop the premises of the approach of difference feminism, as it arose in Italy in the 1980s. Taking as point of departure three different readings of Clarice Lispector's 1964 novel, *The Passion According to G.H.*—on the encounter between a woman and a cockroach—those by Luisa Muraro, Adriana Cavarero, and Rosi Braidotti, it will outline the feminist approach of sexual difference as the opportunity to outline an ontological and epistemological shift.

F. Giardini (✉)
University Roma Tre, Rome, Italy

F. Cimatti, C. Salzani (eds.), *Animality in Contemporary Italian Philosophy*, The Palgrave Macmillan Animal Ethics Series,
https://doi.org/10.1007/978-3-030-47507-9_9

1   *THE PASSION ACCORDING TO G.H.* ACCORDING
     TO MURARO, CAVARERO, AND BRAIDOTTI

A woman, entering her maid's room that she generally does not visit, loses and finds herself by means of an encounter with a cockroach; *The Passion According to G.H.* (Lispector 2012) is the literary expression of the itinerary of a human and feminine loss through the experience of non-human life. What loss is at stake? What succession articulates it? Is it the account of a death, of a rebirth, of a failure, or of a promise? Through the different questions they pose and the uses they make of Lispector's text, Luisa Muraro, Adriana Cavarero, and Rosi Braidotti, who all and each have contributed to the first development of the sexual difference approach, answer differently and, also because of the different years of publication of their readings, identify the distinct internal articulations of this approach.

1.1   *A Failure, an Opportunity*

Luisa Muraro's "Comment to *the Passion According to G.H.*" appeared in 1987. After only a few pages, Muraro's considerations are presented as a dialogue with the statements of another woman; in fact, both the reference to a previous dialogue and the approach to the novel itself testify to the existence of the two different feminist spaces that developed the Italian sexual difference thought.

On the one hand, the Milan Women's Bookstore Collective, founder of the journal *Via dogana. Rivista di pratica politica* (*Via dogana. Journal of Political Practice*) and publisher of *Le madri di tutte noi. Catalogo giallo* (*The Mothers of Us All. The Yellow Catalogue*, 1982)—a collection of conversations and commentaries on some major writers, such as Jane Austen, Gertrude Stein, Elsa Morante, and Virginia Woolf. Reading literature is part of a peculiar political program, that is the discovery and articulation of "political practices" that are initiated at the Bookstore but are fed and regenerated by a multitude of women's groups, feminist archives and documentation centers, feminist informal institutions, and networks of professionals in teaching, in law, in academic research, in public administration, and so on. Conceiving women's liberation according to the sexual difference approach entails some innovative starting points: liberation is a process implying autonomy, that is, it cannot be directed to predetermined goals—becoming equal to man, for example—but it rather has the primary duty to constitute an alternative symbolic realm, autonomous, independent in respect to the hegemonic patriarchal symbolic order. The politics of the symbolic (Cigarini 1995) has thus to do with a collective and relational research for new significations that can account for and liberate human feminine experience; a research that can only be more than individual, indeed the very possibility of saying "I" is the outcome of a constitutive relation—this central assumption is made in the consciousness-raising groups (Lonzi 1978) and in the feminist anti-Oedipal account of feminine subjectivation (Irigaray 1985; Muraro 2018).

On the other hand, the philosophical questions that Muraro addresses to Lispector are the clue of the developments taking place in another major site of sexual difference thought in Italy: the feminist philosophical community of Diotima, established at the University of Verona in 1986. As we read in the first of the collective volumes published by Diotima, the standing point of the research is "the (feminine) part separating from totality, and not for the purpose of defending a particular interest, but for the sake of justice and truth" (Diotima 1987: 20). More than an academic group, Diotima works for a free signification of séxual difference on the women's side, taking into consideration some human and canonical issues, such as justice, truth, and being, and at the same time creating new connections and articulations, from the positional experience of what we may call the subaltern position or, in other terms, transforming absence from the philosophical canon into an opportunity to create new questions.

Therefore, Muraro's reading is nurtured by the collective research for a free signification of sexual difference as a space of discovery of a sexed experience of truth. In order to finally encounter the cockroach, G.H. accomplishes and gives literary expression to an itinerary of disidentification, of the loss of what Muraro believes is represented by the "third leg" in Lispector's text (Muraro 1987: 74), that is, the possibility of identification that a woman receives from the existing social and symbolic order. The third leg is the vivid literary figuration of both the subjective sense of stability and reassurance, emanating from the answer to the question "who am I?" as it is provided by the social and discursive order, and the definite impossibility of running, that is of discovering and saying the truth in touch with one's own experience. Unlike the Cartesian foundation, which assumes thinking and instituting the subject as coextensive, in the feminist sexual difference approach, experiencing truth develops from a move of disidentification. The *Comment* clarifies that this move has nothing to do with the emancipation from the empirical ego toward the rational subject of knowledge, for "passion is more than thought" (Muraro

1987: 71); passion here stands for the embodied subject of experience, where "body" is taken in its affective and yet expressive capability—it is an assumption relating to Muraro's research on the hysteric position (1993) and on the mystics (2003).

Thus, the feminist signification of sexual difference creates this first subversion of the philosophical Canon—experiencing truth takes place on the side of both disidentification and passion—and opens up to another subversion concerning the human faculty of language and the related failure in mastering a systematic sense of reality. Comparing Lispector's novel and Wittgenstein's *Tractatus*—a philosophical gesture inspired by the practice of the *Yellow Catalogue* that combines irreverence toward the philosophical Canon and the creation of a feminist genealogy as the conditions for a symbolic production of women's liberation—Muraro underscores how, whereas for the first the failure in mastering the sense of reality as a whole is trauma, "for a woman [who] knows being a woman with the simultaneous insignificance of the same being and knowledge [...] a reality that doesn't signify is more than nothing, is less than nothing" (Muraro 1987: 77); moreover, this experience results in passion and in the knowledge about the independence of reality, in the discovery that "world independed on me" (Muraro 1987: 77; Lispector 2012: 144). Finally, the cockroach is encountered, an encounter between beings, where the human has lost her privileges and yet is experiencing a common condition: "I had only thought of it as female, since things crushed at the waist are female" (Lispector 2012: 72).

Being a woman has nothing to do with the attribution of fixed and distinctive characteristics; rather, it expresses the possibilities of a position, that is marked bodily by subalternity, in opening up a path to liberation. Sexual difference appears thus as a "historical act of interpretation and its simultaneous transcendence" (Muraro 1987: 73); the historically and socially situated expression of a sexed position can tell some truth to others, as long as it is exposed to the failure in signifying with already given significations; truth is the passional experience of the failure of sense. Throughout the feminist difference approach, the loss of the egological mastery and the belonging to a sexed body become the ontological opportunity to recombine transcendence with new questions, what exceeds reality in respect to language and the experience of truth.

1.2   Countering the Canon

Adriana Cavarero also comments on Lispector (Cavarero 1995) when taking part in the philosophical community of Diotima. As a further sign of this collaboration we can mention the fact that the original version of *In Spite of Plato* appeared in the series "Il pensiero della differenza sessuale" (The thought of sexual difference) that Muraro edited for the publisher Editori Riuniti. Despite this proximity, Cavarero's approach is specific in a way that soon resulted in her leaving the group. The title itself is indicative of her approach: *In Spite of Plato* expresses the choice for a direct relation, even though a critical and polemical one, with the authors of the Canon. In order to further detail Cavarero's option, it is useful to recall that the political practice of reading, and thus instituting a feminist genealogy for a different symbolic order, consists in some specific assumptions: reading a writer—that is, giving value to the symbolic expression of another woman—has to take into account the different readings and understanding of each woman participating, and one's singular position is the result of this multifold exchange—as was the practice of the Milan Women's Bookstore and of the philosophical community of Diotima. Furthermore, the main objective is not to counter the authors of the Canon, but rather to compose (or to accept the decomposition of) significations in order to outline a symbolic realm for the liberation process; it is less a hermeneutical operation and more the research of words for a new experiential order. On the contrary, *In Spite of Plato* has an admittedly hermeneutical intention and addresses the philosophical Canon directly—which is endowed with a specific and transhistorical epistemology (universalism and dualism between nature and history, body and rationality)—for the sake of its inner subversion. The volume thus presents a reinterpretation of some feminine characters—Penelope, the Thracian servant, Demeter, and Diotima—who appear in Ancient Greek philosophy and are treated as clues of what has been canceled and appropriated by the Western patriarchal culture (Cavarero 1995: 6). In fact, in her contribution to the first collective volume of Diotima (Cavarero 1987), the sexed position is assumed by Cavarero as a position of estrangement, that is, intending "woman" as already and always inscribed in the hegemonic discursive order, whereby her space of discursive action consists in distancing herself from that order that does not provide for her participation.

The reading of *The Passion According to G.H.* is therefore a counter-reading; we find it in the chapter dedicated to Diotima (Cavarero 1995:

91–120), where the author denounces the double move made in the philosophical Canon that views human embodiment through the two concepts of birth and death and subordinates the former to the latter: "death becomes the measure of birth" (Cavarero 1995: 99). This logical and temporal inversion establishes the subalternity of the feminine sexed body and the knowledge that she develops: Diotima's speech on love. The related and now negative equivalence between maternity and animality, which in the pre-Platonic Dionysus is still the manifestation of the "primal flow of life where birth and death are but rhythm and cadence" (Cavarero 1995: 113), leads Cavarero to Lispector (Cavarero 1995: 114–120). The itinerary of depersonalization appears similar to the one introduced by Muraro, but the hermeneutical intent aims to counter the conceptual opposition between birth and death, reality and nothingness, and singularity and universalism. The encounter with the cockroach, as Cavarero underscores, is not only a visual but also an oral experience—G.H. tastes the matter pouring from the cockroach—and for a feminine body it reveals "the sexed maternal root that links every 'I' to the impersonal life itself, every single being to his or her beginning" (Cavarero 1995: 117).

The centrality of birth as the motive for an ontological reflection shows the increasing importance that the Arendtian concept of natality has in Cavarero's research, though birth and death are here still conceived as the condition of a common belonging, that of human and non-human beings, to the flow of life. The theoretical move is kindred to the postmodern diagnosis of Western epistemology: what the Greek Canon considers as an ontological deprivation of the feminine body is subverted and becomes a positive attribute, the departure point that allows to redefine humanity.

1.3   A Materialist Difference

Metamorphoses, the volume in which Rosi Braidotti expands on Lispector's novel (160–67), was published in 2002. Not many years have passed since Muraro and Cavarero's readings; nevertheless the text enacts a discontinuity with regard to the position from which the novel is read, the choice of interlocutors, and its philosophical uses. Braidotti came into contact with the approach of sexual difference in France, through Luce Irigaray—also a reference for the Milan Bookstore and Diotima at an early stage: her work is quoted both in Muraro's and in Cavarero's commentary to Lispector. Being a native Italian speaker and having extensively studied the approach of sexual difference, Braidotti's exchange with the Italian philosophers seems to be an opportunity at hand. Even though Braidotti spells out the reference to Cavarero, the exchange does not actually last very long, mainly for reasons that concern the feminist philosophical practice and its outcomes.

A first clue of this appears in the title dedicated to *The Passion According to G.H.*: "Clarice Lispector as the anti-Kafka." By now the position from which Braidotti is reading the text relies on a symbolic space; it no longer tackles difficulties that border on the unutterable and on estrangement; rather, all along her position is constituted by a dialogue with canonical authors, although belonging to a minor canon, as well as with feminist authors such as Hélène Cixous, Julia Kristeva, Luce Irigaray, and Cavarero herself. It is a relevant change in respect to the practice of reading: feminist interlocutors are fully textual authors and interlocution is not limited to the authors consecrated by the Canon. The encounter with the animal has already found the words to be expressed, in a plurality of voices. In another respect, however, an asymmetry remains: sexual difference appears as an unsolved question and as a polemical tension; Braidotti reads Lispector with a sense of divergence from the interpretation that Deleuze and Guattari gave of Kafka's *Metamorphosis* (1986); indeed, the authors "do not take in account the variable of sexual difference" (Braidotti 2002: 163). With this remark Braidotti is in no way identifying some distinct feminine and masculine issues about corporeality; on the contrary, we can assume that sexuation consists of multiple compositions among different bodies, human and non-human ones, whose experiential, semiotic, and discursive expressions Deleuze and Guattari overlook in giving their account.

In Braidotti's view, the encounter with the cockroach becomes an experiential site concerning what "we can neither assimilate nor expel" (Braidotti 2002: 149), no satisfactory and reassuring projection can be made, as with domestic animals, and at the same time we cannot rely on the possibility of gaining a safe and purified ground (Cixous 1986). The composition between the cockroach's body and that of G.H. reveals a possible state of a feminine sexed experience. Lispector writes after having endured an abortion (Braidotti 2002: 163) and the itinerary is altogether a dynamic of desubjectivation and of abjection, the experience of losing the social identifications and of experiencing the non-human organicity we are made of; moreover, by referring to Kristeva's *Powers of Horror* (1982) this experience is also described in its affective dimension, a negative and inglorious affect: the generative capacity of the female sexed body appears thus as the site of the decomposition of subjective boundaries—that Braidotti describes in an intersectional approach as boundaries instituted by gender and class (2002: 161–62)—as well as the site of direct experience of organic life (*zoé*). The itinerary traced by G.H. is thus of positive deconstruction, where the encounter with the void "contrary to the Sartrian nothingness, is a site of interconnectedness and mutual interdependence" (2002: 165). While Muraro and Cavarero conclude for an exceeding reality, Braidotti is anticipating her materialist and posthuman approach, when she summarizes the outcome of the itinerary as the discovery that "the world is not human" (Lispector 2012: 56).

2   Unfolding Difference, Politics, and the Animal

Some decades later—in the midst of a radical crisis that has been surging since 2008, becoming an economic, social, political, and environmental crisis—we can take advantage of these proposals in order to develop some of the political questions that emerge from the conceptual couple "human-animal." In Western traditions, politics has always been conceived of as the specific attribute of human beings. We might even say—thinking of the status of being/not being a citizen—that politics is the place where humans are recognized as such. The fact that this conception has always been in place does not mean, however, that it has always been thought of in the same way, positioning "animal" in the same place in respect to "human"; the philosophical Canon does not rely on a univocal logic of devaluation and exclusion. Rather, we may say that "human" and "animal" in the Western political realm are relational terms, historically changing terms, as they change their mutual meanings depending on historical periods and social organizations. In this respect, "animal" does not appear as a topic, but as a relation by which "human" can position and conceive itself. The conceptual couple human-animal is thus a political operator, dynamically organizing and articulating the field of politics, in its subjects, rules, and representations of the borders that identify the human.

If politics and human tend to identify an equivalent space, in another and specular perspective woman and animal tend to share the other part of the conceptual couple. Now, what happens if we look at this couple from the experiential position of a feminine sexed body? In other words, what kind of significations must we articulate in order to elaborate an itinerary of liberation from a subaltern experience? In the previous section of this chapter some departure points have been indicated. First of all, subjectivity, the very possibility of saying "I," is not originally given; rather it is the outcome of relation. This relation is embodied, and it organizes the whole perceptual field, the body schema, and the symbolic orientation. In fact, assuming a sexed embodied relation is something different from considering relations among corporeal beings, for sexuation entails the restless tension between social significance and insignificance; the biological, affective, and iconic agency of the body—as was the case of the hysterical body—the collective embodied memories; the *habitus*, working as practical intelligibility schemas; and the institution of alternative genealogies so as to liberate a new order of experiences. In the perspective of an embodied—that is, sexed—constitutive relation, knowledge is not a matter of interaction between two already constituted consciousnesses; it is the experiential site of the interplay of different relational forms—the alter-ego-mirror, the analogical-empathic, and the strangeness-estranging relation—that can appear intertwined and contribute to the dynamic constitution of an experiential position (Giardini 2004). In fact, cognitive and conscious perception is neither the only nor the principal access to knowledge, as failure in signification can be the sign of an experience that includes non-human reality, which goes beyond the exclusively human faculty of language. If failure is the loss of significance as well as of previously established borders, we can assume that crisis, in its corporeal implications, is a loss in orientation, that it is the loss of the embodied memories that are working as practical intelligibility schemas. Therefore, this failure can also be cultivated as the demand for new and different genealogies, in order to access other means for conceptualization. In this respect, the Canon should be considered as neither a unique and coherent order of exclusion nor a memory already assigning the preeminent model to masculinity; rather, it should be considered in the light of the conceptual needs of the present.

2.1 Canonical Zoopolitics

In Greek philosophy, the relation politics-human/animal appears as being characterized by two main aspects: because the real objects of knowledge are human beings—with a shift from the previous philosophers, the *physioi*—politics too has to be defined as a properly human domain. We may think of the Aristotelian statement "he who is unable to live in society, or who has no need because he is sufficient for himself, must be either a beast or a god" (*Politics* I 2, 1253a), where the community of the *polis* is precisely what politics is about; and so this is the scene that on a theoretical level achieves the equivalence between human and politics. In this first Aristotelian argument we can see that the beast (*therion*) stands not only for non-human but also for an un-political being; it marks the border of what must be conceived as properly human (alter ego relation—I am what you are not).

But Aristotle presents a second aspect of the relation politics-human/animal. Human is the "political species" (*zoon politikon*)—among others: bees, ants, and cranes naturally live together, as living together constitutes the first level of political being. Human beings, however, not only communicate with each other—this is specific to bees too—they can also formulate judgments about what is pleasant and unpleasant, useful and useless. The human species is *logon echon*; it has *logos*: in Aristotle's view this means that human beings are characterized by a deliberative capacity. This relation between human and animal is more articulated than the negative relation—I am what you are not; it is rather a gradualist or continuist relation, a matter of degrees, which nonetheless designs the real members of a political community. With respect to the deliberative capacity, slaves are non-human as they do not possess it; (male) children are becoming human in that they will possess it; women belong to the *oikos*, the domestic sphere, because by nature (essence) they possess it but only in an incomplete way. This gradualist relation is what ensures both the identification of the due members of the political community, Greek male adults, and the functional and therefore graduated participation of the others.

In modern political theory, we may think about the wolves in Hobbes as another example of the human-animal relation. Human beings are "wolves to each other"—incidentally, Hobbes seems to neglect the fact that wolves are political animals in the Aristotelian sense as they live in packs, though he does this on purpose—as long as they do not enter the political Contract. In this case the relation is more specifically a mirroring relation: the other, the animal, is the inverted image—like that reflected in a mirror—of the human; it is constructed in an inverted and concealed continuity with the characters of what the philosopher wants the human to be. In fact, there is an interesting use of the mirror inversion in Hobbes' contract theory: even though he presents the state of nature as the origin, in fact the logical beginning of the theory is sovereignty as absolute power; and the fact that the un-political human has to be represented as a natural beast is functional to this. Furthermore, "beast" has a radically different meaning from the Aristotelian one; it does not stand for isolation or lack of deliberative capacity, but it is the name for a radical and compulsive tendency to destruction. In order to appreciate the extensiveness of the animal-human dynamic relation in the Western political and philosophical Canon, we could mention La Boétie's description of the same beastly beings, animated by a love for liberty they defend with their teeth and claws, as opposed to humans who adhere to voluntary servitude (La Boétie 1997 [1576]).

2.2 Alterations of the Canon

Renaissance political theory offers another form of relational couple human-animal, with the Centaur that Machiavelli views as the figuration of the political subject. Major commentaries on this Machiavellian use of animality focus on the statement that "there are two modes of fighting: one in accordance with the laws, the other with force. The first is proper to man; the second to Beast " (Machiavelli 2005: 60). By merely taking into consideration this statement, Machiavelli appears to be reiterating a somehow generic tradition that puts man on the side of civilization and animal on the side of brutality. However, a more comprehensive view can be put forward in respect to the historical social and cultural context of Machiavelli's thought on politics. In *The Prince* (1513) he warns about the fact that "having a half-beast and half-man as a teacher, a prince must know how to employ the nature of the one and the other" (Machiavelli 2005: 60); that is to say, the political subject participates in both humanity and animality and this does not imply any hierarchy between the two; this is indeed confirmed by the primary importance that Machiavelli gives to passions in politics. Moreover, the Centaur as a *mythical* figure acts as an interruption of historical time; it indicates a symbolic space where human cannot represent itself according to the previous traditional borders between what was and was not "political." In fact, we know that Machiavelli is the author who introduces a rupture in tradition, as he conceives politics as a realm that functions according to specific and distinct rules, while the Christian ethical virtues become useless and ineffective for politics. I would here suggest that Machiavelli is positioning the relation human-animal in an analogical way; it is a proportional dynamic that traces a complex web of common elements, where human and non-human cannot be distinguished as such (Giardini 2004: 127–138). In support of this different approach to the Canon, we may think of the centrality that the

Platonic *Timaeus*—the dialogue presenting the doctrine of the mathematical connections between human and non-human realities—gained in Renaissance culture; we can also look with different eyes to da Vinci's Vitruvian Man ("The proportions of the human body according to Vitruvius"), as the map of the human body in analogy with the map of the universe (Cassirer 1963), an epistemological approach that concerns alchemy, the political guide composed by Marsilio Ficino in 1529 (Ficino 2002), as well as the allegoric style in arts and politics (Sternberger 2001: 157, on Machiavelli's Centaur). In this perspective, Renaissance appears as the epoch of a creative disorder in respect to the previous borders of politics and of a new conception of politics in its connections with more than human principles; moreover, it is significant that in those times sovereignty was also discussed as a female capacity, with an effort to represent it being made in Italy (Andreini 2004 [1622]), also becoming the field of both apocalyptic scandal (Knox 1985 [1558]) and of symbolic and political invention in England (Yates 1975; see also Schmitt's reference to Elisabeth I as the inventor of the Atlantic model in sovereignty, Schmitt 1997: 20).

2.3   Concerning a Failure in Being Fully Human

The sexed difference feminist approach entails the subversion of positional attributes—the view that the position of a woman must be considered on the basis of characteristics, possibilities, and aims—pre-assigned by the hegemonic symbolic order. In this respect, the canonical equivalence between woman and animal, as constituting the negative side of the rational subject, is radically reconsidered: being not fully human has to be conceived as a relational dynamic, historically variable, providing for a multiplicity of diversified outcomes.

In the feminist theoretical production of the twentieth century, the isotopy between woman and animal has been discussed with different approaches and objectives. Authors such as Evelyn Fox Keller (1983) have assumed and claimed the proximity between woman and nature. This new starting point in sexuating scientific knowledge has represented a radical change in epistemology; at the same time, it implies that sexuating knowledge develops through the inversion between the valuable and the devaluated poles of the sexed relation (male as objectivity and female as empathy); in so doing, an experiential feature of the constitutive relation is lost, the benefit of the relation of estrangement, that is both a guarantee against assimilation and a limit to overwhelming subjectivation. Other authors, such as Beauvoir (2011) and Judith Butler (e.g., 1993), have harshly countered the isotopy, refusing to consider that woman can be identified with her biology on the basis of a deterministic approach; being a woman or rather becoming a woman is thus reconsidered as a social construction; "woman" is a social and cultural agent, situated at a maximum distance from "nature." Human dissolved into agency is nevertheless limited to experiencing what this same agency is producing; in this new feature, animal is either a social artifact or it does not exist. Another option, related to the "death of the subject," contests the isotopy woman-animal by deconstructing the very same theoretical framework that produces that isotopy: human becomes "woman," becomes "animal"; it *monsterifies* itself. However, this approach does not consider that, if putting the hegemonic Subject to death has an effect of liberation, this does not work for subaltern subjects seeking expression and subjectivation (Spivak 1996: 288). In respect to animal, even if it blurs the borders identifying humans, this approach risks reiterating an assimilative relation, for it assumes that human can be or can stand for the animal (Deleuze and Guattari 1986). In this respect, the three commentaries on Lispector's novel overlook that the itinerary of desubjectivation is nonetheless a linguistic and symbolic itinerary: Lispector is writing about it.

2.4 The Constitutive Relation as a Dynamic Topology

Starting once again from the assumption that the relation human-animal is a relation constituting the human itself, in historically different outcomes; therefore it cannot be conceived according to a univocal epistemological principle—and sexual difference works the same way, for "the meaning of being a woman/man is like a four dimensions battle field, the fourth dimension being the historical creation of the meaning of sexual difference" (Muraro 1995: 114)—we have thus to account for the plurality of positions emerging from the constitutive relation: the same possibility of defining the human, of saying "I," of assuming a sexed position, is always the result of an embodied relational dynamic.

Position is thus an inner dynamic constituted by the interaction with others; it is neither immediate nor evident; it cannot assume the function of a superior principle organizing the field of experience, as this is already a relational outcome. The constitutive relation operates through the vicissitudes—they all contribute to her outlining and consistency—of similarity, partial or total opposition, inversion, as well as the different degrees of proximity and distance. The crisis, that is the loss of borders and orientation, the loss of previous established frames of experience, appears as the opportunity to recombine the figures of the constitutive relation. Constitutive relation is thus more than just questioning subject and subjectivity; it has to do with the experiential field and its historical and political organization. The space defined by the dynamic of the constitutive relation, which is no more organized by the centrality and priority of an egological and self-evident subject, can therefore take advantage of the non-Euclidean, topological conception of space.

"Which 'subject' has taken an interest in the anamorphoses produced by the conjunction of such curvatures? What impossible reflected images, maddening reflections, parodic transformations took place at each of their articulations?" (Irigaray 1985: 144). A woman outlines the possibility of a sexed position, starting not from the center of a unique subject, but rather from a dual principle, from the dual centers of the elliptical space of topology. In fact, while Euclidean geometrical figures are constituted by fixed and visible perimeters separating the inside and the outside, the void and the plenty, and the openness and the closure, topological figures are constituted by tactile continuity and its permutations. The Klein bottle, for instance, reveals a superficies with a peculiar structure playing in an elliptical space; its "surprising" characteristic is that even if it is a "closed superficies and its radius of curvature is zero, its content is infinite" (Ströker 1965: 298). Moreover, the Mobius strip and the torus are transforming inversions from the fixed spatial organization of the specular relation to the tactile and sensitive continuity of a dynamic of reversibility and reveal the interdependent necessity of void and movement (Giardini 2004: 103–113).

2.5   Rethinking Politics in a Topological Space

In the perspective of the constitutive relation as dynamics and topology, the isotopy of both woman and animal becomes the opportunity to reconsider the political realm in an innovative way; in the topological space, "animal" and "human" appear in a plurality of positions where the inside and outside, proximity and distance, and similarity and alterity/strangeness are not fixed properties but dynamic outcomes. There are two major experiences we can grasp thanks to this theoretical reconsideration of the experiential field, which go beyond the discussion on a supposed and generic relation between human and animal, and the eventual positioning of "woman" in this relation.

In the Western urban context of the twenty-first century how does this relational experience work? If we do not want to metaphorize the animal, this experience is part of a literally assimilative relation: animal appears mainly as part of human nutrition. The distinction between the inside and outside, the oppositive identification of we human and they animal, and the unequal attribution of mastery and availability are ineffective here. On the contrary, if we assume the topological dynamics of the constitutive relation, a plurivocity appears and articulates the political questions and relational moves we are confronted with: estranging the relation of assimilation will consist in recovering a distance, an externality between human and animals, making the latter a stranger again, who has to be seen, experienced again in its autonomy and independence from human needs and functions. The plural forms of the embodied constitutive relation offer us further perspectives, in respect to the option of humanizing animals, and therefore raising an ethical question, or devaluating them as an organic resource for human reproduction.

A second experience we are all exposed to is global epidemics such as mad cow disease or avian flu. The characteristics of these epidemics—being altogether the outcome of human action on the animal body; becoming a threat to humans because of some animal agency; not respecting the internal and external national borders; raising public alert and consequent public policies addressing citizens as living bodies—subvert the meaning of politics as the properly human domain. Rather, they configure it as a transitional space (Giardini 2013) and demand a conception of politics that is in line with an embodied relational and topological dynamic. This approach brings about a difference in respect to thinking in terms of biopolitics—which still, as Braidotti reminds us, has to do with qualified, that is human, *bios* (Braidotti 2002: 132)—or in terms of a subversion of the modern hierarchy between the subject and the object, thinking from the side of the object, or of an hyper-object that we fail to experience because it is beyond our perceptive capacities, beyond subjectivity (Morton 2013). Politics reconceived according to the topology of the constitutive relation is rather an epistemological and political program, which calls for an experiential description of the dynamics one's position is constituted by and consists of, giving a plurality of accounts about the mutual positions, identifications, and distances, as well as for the detection of the blocks—the outcomes of power—in the movement of the relation, should it be intoxication, famine, massive exploitation, or global threats.

WORKS CITED

Andreini, Giovanni Battista. 2004 (1622). *La centaura*. Genova: Il Nuovo Melangolo.

Beauvoir, Simone. 2011. *The Second Sex*. Trans. Constance Borde and Sheila Malovany-Chevallier. New York: Vintage Books.

Braidotti, Rosi. 2002. *Metamorphoses: Towards a Materialist Theory of Becoming*. Oxford: Blackwell.

Butler, Judith. 1993. *Bodies That Matter: On the Discursive Limits of Sex*. New York: Routledge.

Cassirer, Ernst. 1963. *The Individual and the Cosmos in Renaissance Philosophy*. Trans. Mario Domandi. Oxford: Blackwell.

Cavarero, Adriana. 1987. Per una teoria della differenza sessuale. In *Il pensiero della differenza sessuale*, Diotima, 41–79. Naples: Liguori.

———.1995. *In Spite of Plato: A Feminist Rewriting of Ancient Philosophy*. Trans. Serena Anderlini-D'Onofrio and Aine O'Healy. New York: Routledge.

Cigarini, Lia. 1995. *La politica del desiderio*. Parma: Pratiche.

Cixous, Hélène. 1986. L'approche de Clarice Lispector. In *Entre l'écriture*, 115–138. Paris: des femmes.

Deleuze, Gilles, and Félix Guattari. 1986. *Kafka: Toward a Minor Literature*. Trans. Dana Polan. Minneapolis: University of Minnesota Press.

Diotima. 1987. *Il pensiero della differenza sessuale*. Naples: Liguori.

Ficino, Marsilio. 2002 (1529). *Three Books on Life*. Trans. Carol V. Kaske and John R. Clark. Tempe: The Renaissance Society of America.

Fox Keller, and Evelyn. 1983. *A Feeling for the Organism: The Life and Work of Barbara McClintock*. San Francisco: W.H. Freeman.

Giardini, Federica. 2004. *Relazioni. Differenza sessuale e fenomenologia*. Rome: Luca Sossella.

———. 2013. Cosmopolitiche. Ripensare la politica a partire dal *cosmos*. *Babel* 1: 147–163.

Irigaray, Luce. 1985. *Speculum of the Other Woman*. Trans. Gillian G. Gill. Ithaca: Cornell University Press.

Knox, John. 1985. *The Political Writings of John Knox: The First Blast of the Trumpet against the Monstrous Regiment of Women and Other Selected Works*, ed. Marvin A. Breslow. Washington, DC: Folger Books.

Kristeva, Julia. 1982. *Powers of Horror: An Essay on Abjection*. Trans. Leon S. Roudiez. New York: Columbia University Press.

LaBoétie, Etienne. 1997 (1576). *The Discourse of Voluntary Servitude*. Trans. Harry Kurz. Montréal/New York/London: Black Rose.

Libreria delle donne di Milano. 1982. *Le madri di tutte noi (Catalogo giallo)*. Milan: Libreria delle donne di Milano.

Lispector, Clarice. 2012. *The Passion According to G.H.* Trans. Idra Novey. New York: New Directions Publishing.

Lonzi, Carla. 1978. *Taci, anzi parla. Diario di una femminista*. Milan: Scritti di Rivolta femminile.

Machiavelli, Niccolò. 2005 (1513). *The Prince*. Trans. Peter Bondanella. Oxford: Oxford University Press.

Morton, Thimothy. 2013. *Hyperobjects: Philosophy and Ecology After the End of the World*. Minneapolis: University of Minnesota Press.

Muraro, Luisa. 1987. Commento alla *Passione secondo G.H. DWF* (5–6): 65–78.

———. 1993. *La posizione dell'isterica e la necessità della mediazione*. Palermo: Udi.

———. 1995. Oltre l'uguaglianza. Identità umana e differenza sessuale. In *Oltre l'uguaglianza*, ed. Diotima, 114. Naples: Liguori.

———. 2003. *Il Dio delle donne*. Milan: Mondadori.

———. 2018. *The Symbolic Order of the Mother*. Trans. Francesca Novello. New York: SUNY Press.

Schmitt, Carl. 1997. *Land and Sea: A World-Historical Meditation*. Trans. Simona Draghici. Washington, DC: Plutarch Press.

Spivak, Gayatry Chakravorty. 1996. Subaltern Talk: Interview with the Editors. In *The Spivak Reader*, ed. Donna Landry and Gerald Maclean, 287–308. New York/London: Routledge.

Sternberger, Dolf. 2001. *Le tre radici della politica*. Trans. Rosamaria Scognamiglio. Bologna: Il Mulino.

Ströker, Elisabeth. 1965. *Philosophische Untersuchungen zum Raum*. Frankfurt a.M.: Klostermann.

Yates, FrancesA. 1975. *Astraea: The Imperial Theme in the Sixteenth Century*. Boston: Routledge and Kegan Paul.

CHAPTER 10

Paolo De Benedetti: For an Animal Theology

Alma Massaro

An inquiry into the status of Italian animal theology needs to be focused on Paolo De Benedetti, a major figure of contemporary theology, Judaism, and biblical exegesis in Italy. Starting from a consideration of the Shoah, De Benedetti elaborates an anti-metaphysical analysis of the problem of evil which leads him to reject traditional Western theology. In dialogue with ancient rabbinic sources and modern European writers, such as Dietrich Bonhoeffer and Fyodor Dostoevsky, he came to recognize how the quandary of evil affects not only humans but animals and plants as well. In this sense, for De Benedetti theodicy becomes a preferential way to reflect on the mysterious and generally ignored relationship that connects God to non-human creation. The present chapter deals with the life and thought of Paolo De Benedetti in order to outline the origins and borders of the animal theology he elaborated, which is today foundational for those who want to examine the relationship between the Judeo-Cristian tradition and animals.

A. Massaro (✉)
Genova, Italy

© The Author(s) 2020
F. Cimatti, C. Salzani (eds.), *Animality in Contemporary Italian Philosophy*, The Palgrave Macmillan Animal Ethics Series,
https://doi.org/10.1007/978-3-030-47507-9_10

1   A Short Biography of Paolo De Benedetti

Paolo De Benedetti was born in Asti on December 23, 1927, to Teresa Alieri, a practicing Catholic, and Ettore, a Jewish doctor. This double origin played a fundamental role in the development of his philosophical and theological thought. As he explains, it was precisely his father who let his wife educate their two kids, Paolo and Maria, in the Catholic faith:

> Our father, Ettore, was a physician, the head physician of "special clinical medicine" at the Hospital of the city of Asti. He was a Jew, but having received a secular education he allowed our mother to instruct us religiously – provided that she would not delegate it to any church but would teach us herself. For this reason I had the privilege not of going to catechism classes but rather of hearing the New Testament from our mother's voice. She came from a Catholic family [...]. This is how, little by little, I became a Christian. (Caramore 2008: xi)

At the University of Turin, where he studied philosophy, De Benedetti became better acquainted with Jewish culture, of which he would later become one of the most authoritative experts in Italy. With some classmates he began studying ancient Semitic languages such as Hebrew, Aramaic, Babylonian, and Syriac, assisted by the then President of the Court of the city of Asti, Giuseppe Invrea. It was actually precisely this linguistic interest that allowed De Benedetti to encounter, first, the Hebrew language and then its culture. From this moment on, in fact, his twofold Jew-Christian identity started to take shape explicitly. With the delicate irony that distinguished his thought, De Benedetti defined his condition as "marrano," composed, as he explained, "of the simultaneous presence of Jewish mental categories and allegiances together with some Christian beliefs, in a precarious but irrevocable combination" (De Benedetti 1992: 5–6). A strenuous supporter of the right to coexistence of different religions and cultures, when asked about his faith, he replied with delightful sarcasm: "Christian on Sunday, and Jew on Saturday" (qtd. in Altamore 2016: 183).

As Assistant Professor of Near Eastern Languages at the Catholic University of Milan, in 1952 he started his publishing career by joining the publishing house Bompiani. Here he would become the managing editorial director and enter into contact with the most eminent intellectuals of that effervescent period. Valentino Bompiani described De Benedetti joining his company with these eloquent words:

> Celestino Capasso brought to the *Dizionario delle opere e dei personaggi* (*Dictionary of Works and Characters*) another friend: Paolo De Benedetti. "He is a saint," he explained to me [...]. I did not really understand what relationship there could be between sanctity and philology. I understood it later, in his involvement and in his detachment, in his research for the unique word, or verb, and in his vision of culture as an event of universal character. De Benedetti, or P.D.B., as we have always called him, is a Christian with a Jewish family tree who maintained fraternal relationships with Christianity and filial relationships with Judaism. After the death of Capasso, he took charge of the *Dizionario*, which he has carried to a conclusion. (qtd. in Giacomoni 2006: 63)

During these years of intense work, his interests gradually focused on Judaism (Giuliani 2018). He introduced the thought of Dietrich Bonhoeffer, the Lutheran pastor who brought Jewish-Christian dialogue to the foreground, into Italy. De Benedetti translated some of Bonhoeffer's works, such as *Letters and Papers from Prison* and *Ethics*. He did the same several years later with the works of some Jewish authors until then unknown in Italy, such as Shmuel Y. Agnon, Nelly Sachs, Martin Buber, Elie Wiesel, and Jakob Petuchowski (Giuliani 2018). After leaving Bompiani, De Benedetti moved to Garzanti publishing house and began collaborations with other important Italian publishing houses, such as Morcelliana, Marietti, and UTET (Unione Tipografico-Editrice Torinese).

His professionalism notwithstanding, De Benedetti's childlike gaze and rare wit set him apart. He was able to look at the world without preconceptions and go beyond its immediate semblance. Gianandrea Piccioli's words of description are pleasing but illuminating:

> PDB was also extraordinary in choosing the oddest collaborators: existences on the fringes, solitary figures or eccentric characters – however all precious for integrated revisions, translations from arduous languages, expertise in arduous sectors of knowledge. PDB had the nose for these kinds of people. I never understood if it was them who found him or it was him who looked for them. Possibly they converged: just as a magnet attracts iron, so PDB attracted everything that lied outside of the norm or outside the banal. Anyway, these are some sketches of memories: a theater expert who got by as a stilt walker in the streets of the city; two spies (I never had proof that they really were spies, but I still have the moral certainty: one was a former priest who lived in Berlin during the Wall period and knew everything one needed to know about the publishing industry and German *intelligentsia*); a philology lecturer who supported his family by selling rugs – which he also brought to the publishing house – sent to him by his wife's family, who

belonged to some nomadic tribe from the Libyan Sahara; a translator from an Asiatic language – I do not even remember which one – who lived in the countryside surrounding Vicenza and came to the publishing house always with new, shady female students, either Circassians or Kyrgyz, really young, really pretty, whom he claimed he was helping in their *cursus studiorum* at evening school. (Piccioli 2017)

When during the 1960s the Second Vatican Council (1962–1965) allowed the use of vernacular languages at mass, an official CEI (Italian Episcopal Conference) Italian version of the Bible became necessary. As a basis for this new edition, an already-existing Italian translation was chosen, one published by UTET (Unione Tipografico-Editrice Torinese). De Benedetti was brought into the project by his friend Cardinal Carlo Maria Martini in order to suggest possible corrections and for copyright matters. A clear indication of De Benedetti's spirit are his memories of that period about Cardinal Martini's broken-down car and his strange way of driving in the city of Rome. As Silvia Giacomoni narrates:

When the CEI Bible draft was ready [...] PDB was sent to Rome. By morning he was passing by the Pontifical Biblical Institute where, with his extremely beat-up Fiat 600, Father Martini drove him to CEI offices – where they would compare the two versions in light of copyrights. Remembering this episode makes always PDB in an extraordinary good mood. (Giacomoni 2006: 63)

At the same time as these activities, De Benedetti was called, as the most eminent expert in Hebrew history and culture, to hold the chair of Judaism at the *Theological Faculty of Northern Italy* of Milan, as well as the chair of the Old Testament at the Centers for Religious Studies in Urbino and Trento. In spite of his numerous professional achievements, he remained a deeply humble person, avoiding celebrity, always ready to smile at his own weakness, which he did not hide either from himself or from others. He was not concerned to show off or preserve a dignified image of himself. Once retired, he spent his time writing books, giving conferences, and taking part in public and private debates as well as radio programs, without disdaining parish halls. He died in Asti, where he lived with his sister Maria, at the threshold of his eighty-ninth birthday, on December 11, 2016. An eclectic figure, Jewish "marrano" with different facets: "editor, biblical scholar, philologist, theologian, public speaker, poet, spiritual director of a Sardinian monastery, and keeper of the synagogue of Asti" (Bertoletti 2017: 13). Maybe the most fitting words to describe his earthly life were found by the Italian journalist Piergiorgio Cattani (2006): "God on man's lips."

2   BROAD OUTLINES OF DE BENEDETTI'S THOUGHT

The rich Debenedettian bibliography, composed not only of theological and exegetical studies but also of poems and limericks, has been recently published in a volume edited by Agnese Cini Tassinario (2016). Thanks to the efforts of the publishing house Morcelliana, it will soon be available in an archive devoted to De Benedetti at the Catholic University of Brescia. Those interested in examining De Benedetti's entire legacy can access these sources. In the following pages, I will focus instead only on those areas of research particularly relevant for the development of his animal theology.

First of all, of paramount importance is the *anti-metaphysical perspective* from which De Benedetti approaches the study of the Holy Scriptures, and his rejection of theology understood as a systematic discourse on God. In place of a rigid thought based on human logic, he preferred the form of the *midrash*—the traditional rabbis' commentaries—as well as Talmudic discussion. He favored forms of thinking much closer to orality than to writing and governed by the logic of comment and irony, two aspects of foremost importance to understand his "theological" work. For De Benedetti, in fact, the overflow of God's word with respect to human discourse excludes every attempt to reduce it to a logical and consequential discourse. Doubt is thus intended as a fundamental dimension of human life, and this is because human beings are not allowed to fully understand the truths about God.

Referring to Talmudic language, De Benedetti himself adopted some linguistic devices, precautionary expressions directed to reiterate the hidden richness of God's Word: "as it were" (*kevayakhol*, a Midrashic expression used in connection with daring expressions about God), "I do not know" (*enijodea*), "maybe, perhaps" (*ulaj*), "another meaning is" (*davaracher*), and "unresolved" (*teku*). These are small devices devoted both to the avoidance of falling into dogmatization and fundamentalism and to "trying to maintain in the scale of the relative the thinking and saying of God and on God" (Morganti 2005: 127) by keeping an attitude of constant questioning of the Word. But if, on the one hand, the richness of the revealed Word induces De Benedetti to take a precautionary attitude, on the other it encourages him to recognize the way the truth here contained offers itself to a multitude of times and persons (Rossi 2013). It is actually in this direction that the idea at the roots of the concept of "the seventy-first sense" moves. In his work, *Ciò che tarda avverrà* (*What is Long in Coming Will Arrive*, 1992), he opens his reflection to the senses included in the Scriptures. Here De Benedetti emphasizes the priority of interpretation, that is, of research, with respect to the answer. By recalling the old opinion that attributed seventy meanings to each word contained in the Torah, he suggests commenting on holy texts by bringing to light all those meanings already present, even if hidden, inside revelation (De Benedetti 1992: 7–21).

However, De Benedetti raises the bar a little further, and to these seventy meanings, he adds a seventy-first, that is, the one belonging to each single person who is called to listen to the Word—*Shemaʿ: listen*. The peculiar way each person listens to and interprets the Word will be unique and unrepeatable as much as will be the meaning—the seventy-first—that they will give it (Giuliani 2018). In so doing, what comes to light is the role attributed to each person, the original contribution that each one can—and should—make to the discovery of unknown aspects of the truth. To the continuous and inexhaustible offering of Scriptures corresponds, therefore, the duty of each single individual to discover a piece of truth, in an infinite process.

As he explains: "The infinite interpretation (that I have symbolically identified in the seventy-first sense) is, at the same time, a guarantee against every hermeneutical authoritarianism and an anticipation of future life" (De Benedetti 1992: 6). And it is actually by recognizing the freedom of thought and research of each single individual that De Benedetti is able to offer an interpretation of the category of "neighbor" capable of including not only humans but, together with them, all that is endowed with life, therefore also animals and plants:

> it is necessary to enlarge the concept of "neighbor"; my "neighbor" is the whole creation, and it is all the more neighborly the more two elements are developed: sensibility – in all its degrees up to consciousness – and the dependence on me, for better or for worse. (De Benedetti 2007: 19)

In a dialogue with authors such as Bonhoeffer and Dostoevsky, De Benedetti rejects the traditional anthropocentric way of thinking that "not only has prevented a reading of the Bible attentive to the signs of respect towards, and solidarity with, animals, but has also fostered that frequent prejudice that considers the caring for animals as a disregard for humans:

as if love towards creatures was a blanket too small to cover them all (De Benedetti 1999: 38–39).

From these premises De Benedetti begins an uninterrupted inquiry into the mystery of evil, the *mysterium iniquitatis*. In his book *Quale Dio?* (*Which God?*, 1996), he explores the question of evil in the history of the world and the role bestowed to God in the face of it. As he observes, Auschwitz seems to summarize all the aporias about God. But all the answers theodicy has offered to them went up with the smoke of the camps. *The Bible*—and specifically Job's life—appears to the author as the archetypical question for all those questioning the relationship between justice and mercy, guilt and punishment (De Benedetti 1996: 20–21; Bertoletti 2013: 26–27).

This is how De Benedetti embraces the cry of Ivan, one of the Karamazov brothers: in his monologue, Ivan—Dostoevsky's mouthpiece—desperately reflects on the inconsistency of the kinds of answers given to the problem of evil in the face of the suffering of children:

> if everyone must suffer, in order to buy eternal harmony with their suffering, pray tell me what have children got to do with it? It's quite incomprehensible why they should have to suffer, and why they should buy harmony with their suffering. [...] And if the suffering of children goes to make up the sum of suffering needed to buy truth, then I assert beforehand that the whole of truth is not worth such a price [...]. Besides, they have put too high a price on harmony; we can't afford to pay so much for admission. And therefore I hasten to return my ticket [...]. It's not that I don't accept God, [...] I just most respectfully return him the ticket. (Dostoevsky 1992: 207–8)

De Benedetti recognizes, with the Russian writer, the insufficiency of traditional theodicy in the face of the problem of innocent suffering. That is why he goes back to the Holy Scriptures, the Word, and, following up the invitation to listen—the *Shema'*—he gives life to his proper seventy-first meaning, that is, his proper contribution to the listening community. His most original contribution is his *theology of God's debt* (cf. Bertoletti 2013), where the absolute qualities of the Aristotelian God are overshadowed in order to make room for an unpopular attribute of God: His fragility (De Benedetti 1996: 25–26). As De Benedetti makes clear, fragility is a divine quality metaphorically present in the Holy Scriptures, which fundamentally emerges from the experience of the Cross. This quality, he explains, does not invalidate God's greatness but rather offers a new sense of the divine, giving a name to the divine closeness to the living community. By overturning the traditional ontological argument for God's existence, it follows that "God cannot not be, not because He is a being than which no greater being can be conceived, but because His messianic debt is such that he must exist" (Bertoletti 2013: 31–32).

After life is therefore a "necessity" for God, because there He will have the chance to explain all those things that in this life He keeps unexplained—first and foremost the suffering of the innocent (De Benedetti 1996: 23, 64). It is actually the possibility of this fragility that allows De Benedetti to speak of God even after Auschwitz. As he affirms, after the Shoah:

> the existence of unjust suffering "saves God" only if there is a time or a place [...] where He will explain Himself and will explain to us [...]. God, as our allied, owes us an explanation: for this reason, we believe in Him and in the afterlife. But for the moment we do not want to hear other discourses [...]. We do not want because God asked us for too big a proof: there are no useful words until He will speak His words. (De Benedetti 1992: 122–123)

The extreme outcome of this argument is that God Himself is waiting for the eschatological future. De Benedetti finds inside the Holy Scriptures the image of a God who walks, loves, suffers, debates, and even changes with his people. A picture of God really far away from the idea of the Aristotelian unmoved mover and that bestows on humans a responsibility not only toward fellow creatures but also actually toward God:

> That God is fragile precisely because he is Love, is the only metaphor or myth that saves Him from the siege of evil and guilt: but then we are responsible towards Him (as He is towards us, because we too are fragile). To understand what it means to be responsible for God we need to recall the fact of being the image and likeness that links humans to God: being responsible for God means being responsible for his image, saving it in us and in everything that has life (that is, His "breath"), like a flame from the terrible blows sent to extinguish it by evil, suffering, injustice, death, and oblivion. (De Benedetti 2003: 75–76)

3 ANIMAL THEOLOGY

De Benedetti's close intimacy with animals—in his books he often speaks about dogs, cats, donkeys, and other creatures he shared his life with—leads him to direct his attention toward non-human creatures whose lives, as he observes, would appear insignificant only to a distracted gaze. Thanks to a careful consideration of their being, he arrives at an entire acknowledgment of the value of their worldly, sensible, and emotional experiences. For this reason, through a style "tending to force the horizons of meaning in order to grab new nuances of reality" (Cattani 2005: 18), animal suffering can be taken into serious consideration. The problem of theodicy is now pushed beyond human borders: "Among the mysteries of the world – have we ever thought about this? – there is that of animal suffering" (De Benedetti 1999: 44). De Benedetti makes clear that, just as with the children Ivan Karamazov was speaking about, animals are innocent beings subjected to pain and death. But, unlike children, he adds, in animals not even the seed of sin is present. This mismatch between (past) guilt and (present) sufferance pushes him to recognize animal—and plant—suffering as the biggest mystery theodicy has to face (De Benedetti 2007: 12). As he writes: "The true Job, the suffering righteous, is the animal, and, somehow, also the plant" (De Benedetti 1999: 23).

At this point De Benedetti undertakes a reading of Scriptures which pays attention to the role here bestowed on animals. Through a close analysis of several biblical passages, what emerges is both the dignity conferred to every living being—human and non-human alike—and God's deep care for the whole creation. As De Benedetti makes clear, the final redemption is no longer only an exclusive human affair, but is now located within a universal horizon. The theology he is speaking about is, therefore, a new way of understanding the relationship between God and humans, and God and the world. This new theology is in fact able to promote a deeper respect for all living beings. De Benedetti names it "animal theology" because it transcends the mere acknowledgment of the value of animal life and rather leads to a reflection on life as a whole. As he explains, animal theology is an unexplored way to "find a more 'just' measure of the relationship of the human creature between the world we inhabit and the infinite worlds we do not know" (De Benedetti 2012a: 191).

As has already been noted, it is difficult to summarize De Benedetti's anti-systematic thought. I will here try to illustrate three cornerstones of his animal theology, always keeping in mind the incompleteness of any attempt to classify his work. The first principle is an essential divine attribute quite often forgotten: God's *compassion* toward all creation. On the basis of the Hebraic doctrine of *Tzimtzùm* (*contraction, withdrawal*) that describes God in the very act of contracting himself in order to make space for creation (Di Filippo 2007), De Benedetti describes creation as a divine *need*: "If I may say so, God created the universe and life because he 'needed' a 'you,' a reality to listen and to turn to. This is the reason of life that stands at the very core of the creation narrative. As it emerges from the Bible, among God's needs I would emphasize the creatures' 'gaze' that indeed corresponds to the divine need of a 'you.' God is love and love requires eyes, ears, and heart. Love is not only a question but also an answer. This is why animal creation (and plant creation) is the first need of God against loneliness" (De Benedetti 2015: 9). He finds a confirmation of this divine need in the miscellaneous narration of God's love toward all living beings present in the *Holy Scriptures*.

The second principle is the *twofold role* attributed to the human being, guilty for dragging the whole creation to unhappiness and death but, at the same time, called by God to be its co-redeemer (De Benedetti 1999: 53–54). Not surprisingly, a constant of De Benedetti's thought is indeed the dynamic reconsideration of what it means to be "in the image and likeness" (of the Lord): from his point of view, this sentence does no longer describe a definitive condition of the human creature before the sin but, more compellingly, a task to carry out, or, better put, the first and ultimate aim of each human being. As he put it in an interview with Valentina Gelmi and Massimo Giuliani: "The similarity with God is an end point, a goal to achieve. The human being must be the God of Creation; not, however, a God in the sense of an owner, but as a caring and responsible governor" (2005: 15). It follows that if God is concerned with animals' lives, then human beings cannot avoid caring for them; and if God is compassionate, then humans need to be likewise: therein lies the meaning of the expression "in the image and likeness."

It is indeed in order to describe the relationship humans should maintain with the rest of creation that De Benedetti proposes an analysis of the verbs present in the two narrations of creation contained in the first two chapters of *The Book of Genesis*: *'avad* (to cultivate), *shamar* (to care for), *kavash* (to subdue), and *radà* (to dominate). These are, as he explains:

> four verbs influenced by a biblical anthropology that we have to keep in mind if we do not want to distort the meaning of the relationship between

humans and the world – as has been done for a long time and is still done today. Human beings [...] have their foundation and goal in being the image of God. It must be said that the "image" and "likeness" of Genesis 1:26–27 must be fulfilled also in the way the human being carries out the works expressed by the aforementioned verbs. "To cultivate" and "care for" are the tasks reserved to the mythical condition in Eden by the symbolic narration about sinless humans. We could say they are pacific verbs, that convey an idea of the happy human being as a worker accomplishing a work, which is a cultivation (the verb *'avad* indicates both work as well as service and cultivation, that is, to serve God). That human beings in Eden have a responsibility: to work, to cultivate, to care for the beautiful creation where they are and to which they belong means being God's representatives and developing His creation [...]. Because the earth belongs to God and human beings have not only to care for it [...] like the servant of the parable who buried the talent (Matthew 25:24–25). But they must "work" it, "cultivate" it, in order to extend beyond sin, as it were, the first chapter of the Book of Genesis. The divine self-mirroring of Genesis 1:26 mentioned above refers to the human being that "does," that operates, and not to the static one. We can now begin to understand how the verbs "subdue" and "dominate" should be understood. (De Benedetti 1999: 16–18)

It is up to human beings to carry on God's work, cultivating and caring but also subduing and dominating, just as God does. In the context of a world fallen because of sin, Adam's guilt becomes the human duty to bring the whole of creation to salvation (De Benedetti 2007: 50). In order to understand the verbs "to subdue" and "to dominate," De Benedetti makes clear how the world discloses itself not as an object of discovery, nor of conquest, but of revelation through the Word (De Benedetti 1992: 137–139). The relationship between humans and the world is therefore mediated through God's word. Furthermore, as he explains, human action on creation has its premises in the biblical idea of creation as "neighbor," a concept intrinsic to the Edenic horizon of *Genesis 1–2* (1992: 105–6). Therefore, "the male and the female are authorized, or better encouraged through a divine blessing, to 'subjugate' and 'dominate' the creation. A blessing that works as a mediator between human and the world, and marks the believing relationship with the creation as different from the secular one" (1992: 137–39).

The third and last cornerstone of De Benedetti's animal theology—which follows from the first two—is the doctrine of the *complete redemption* of everything that had life. Moving from the Jewish and Cristian eschatological vision, found primarily in the *Book of Isaiah* and in the *Epistle to the Romans*, De Benedetti infers the *definitive and total victory* of God over death. But a victory, in order to be such, requires that the whole of creation, that is, every form of life, be saved. And this must hold even for what appears to our eyes as the most insignificant thing, even for a "simple" blade of grass. As De Benedetti explains:

> In fact, even if theology has gravely neglected this aspect, it is necessary to affirm with full faith the resurrection of everything that had life, animal and plant. Should this not happen, one should admit that death is more powerful than God, that death wins eternally over life. (De Benedetti 2013: 19–20)

And this because:

> If everything that the Creator has endowed with life, with existence, and has lost it in death would not receive it again, one should conclude that death is more powerful than God, because it wins over existence. This has been a real danger for God. For this reason, he wanted to choose his own death in order to announce the resurrection. If not only a blade of grass, but even a small pebble did not receive existence again, Paul's question would be false: "Oh death where is your sting?" (De Benedetti 2009: 36–37)

In the same vein, even with regard to resurrection, De Benedetti refers to the "needs" of God. However, on this occasion the need would not derive from a divine debt but from God's superiority over everything that is opposed to the life he has created. Moreover, he finds a confirmation of the universal dimension of salvation in the experience of awaiting that, as he observes, unites all living beings in a more or less conscious way:

> One of the images of the messianic awaiting that most astonishes me – please do not be scandalized – is that of a fly bumping against the window glass. Truly this is the symbol of a salvation that does not arrive. And precisely moving from this minimal and common experience, I would say that awaiting is the mood that unites all living beings: not only humans, not only animals, but also plants, with their sprouts stretched towards light. It is an awaiting, let's say a hope, that finds its realization sometimes in life, sometimes in death, and that makes the human being the helpless messiah to whom animals look. (De Benedetti 2013: 19)

Therefore, at the very core of this thought lies not only suffering but also the hope for a transformation, for the eschatological Saturday yet unknown but for which the whole creation yearns: "Paradise is a destination for everything that breathes, for the fly against the window glass as for the mystic, for the flower as for the dove" (De Benedetti 2013: 21–22).

What emerges from these reflections is how animal theology cares to give due consideration to God's love for the whole creation. By recognizing the animal—together with the plant—as the "neighbor" the Scriptures speak about, animal theology shows the way human beings might follow in order to place themselves "in a bigger logic of the living." Only in this way they will be able to reinterpret their relationship with God. Looking at life as an experience endowed with a universal meaning, they could put an end to the environmental deterioration carried out in modern society. As De Benedetti explains:

> Animal theology is able to grasp in animals an intrinsic fragility which becomes more evident the more their lives get closer to human lives [...]. The gaze of the suffering animal – as well as that of suffering children, dying human beings, helpless victims – clearly "shows" in which direction God's gaze is oriented. Looking at animals with attention, responsibility and, finally, with love is not, therefore, a simple sentimental act but an act that could help us humans to get rid of our disastrous pride [...]. From here a new "theology" should begin, a new way of thinking the relationship between God and humans and between God and the world. (De Benedetti 2012a: 190–91)

From the above it follows that animals and plants are also neighbors. The Jewish tradition teaches the inviolability of life. Life "is a value even outside the human, and even outside the animal too [...]. As the famous Rabbi Nachman of Breslov (1772–1811) affirmed, '*Cutting* down a tree before its time is like killing a *soul*'" (De Benedetti 1999: 35). Not surprisingly De Benedetti affirms: "God's gaze is over all living beings. We should not forget that among living beings there are also plants, the 'vegetal creatures': pain, hope, and prayer sprout out even from the fronds, the roots, and the blade of grass" (De Benedetti, qtd. in Bertoletti 2013: 47).

However, De Benedetti takes a further step. It is not only the human, not only animal, not only the vegetal realms that are proper subjects of God's love but the whole of creation. Everything that exists is a "creature," since it received its being from the Creator: as De Benedetti affirms:

"even the blade of grass and the rock exist because they are in God's thought" (2009: 27). It is God that bestows value on every being (endowed or not with biological life). The world is, therefore, the place where humans have to uphold their being in the *image and likeness*. Human responsibility engages the whole planet. As Bertoletti observes: "The theology of God's debt [...] becomes firstly [...] animal theology, and in the end recognizes itself as a theology of creature, because each suffering living being that suffers the pain of its own death is unredeemed" (2013: 47).

4   CONCLUSIONS

By going beyond traditional theology, primarily focused on the human being, Paolo De Benedetti aims to return to the Creator and his creation the space traditionally denied to them. In so doing, he suggests a serious downscaling of the role generally attributed to the human being—who is no longer understood as the center of the universe but as a *creature among creatures*. Between humans and animals, he says, there exists a communion of origin and destiny that, if wholly understood, could affect the everyday life of a human. In this sense, all biblical precepts about animals are of foremost importance, not because they are theological statements but because they are everyday life rules meant to promote wisdom. De Benedetti argues that theology today needs to direct its gaze toward those minor, marginal, and forgotten creatures: toward all non-human beings. In so doing theology can open its doors to a deeper inquiry into God and the world. Animal theology is therefore described as a theology against forgetfulness, capable to bring light to the darkness of Western thought that has hindered the memory of millions of human and non-human lives:

> Surely theology should deal with numerous subjects. If, as it has been said, it has forgotten women, the same has done with our "younger brothers," animals, and with our "older brothers," the Jews. For centuries both older and younger brothers have been victim of theological silence, or worst, victim of a bad Christian theology. Somehow, they have had the same unjust and undeserved destiny. I think no Jew would be hurt if their destiny (hopefully now overcome) is compared to that of animals, but it is really true: human anthropocentrism – and, within it, Christian anthropocentrism – caused both those today called older brothers and those called younger brothers to suffer the same treatment. This is already a proof that caring for animals is not a luxury of the spirit. (De Benedetti 1999: 15–16)

Caring for animals is today a compelling duty for us humans, who need now to abandon metaphysical thought in order to recover a theology of "little things." But what animals is De Benedetti speaking about? As seen above, human responsibility toward creation is total: "my 'neighbor' is the whole creation, and," he adds, "it is all the more neighborly the more two elements are developed: sensibility – in all its degrees up to consciousness – and the dependence on me, for better or for worse" (De Benedetti 2007: 19). In this sense De Benedetti elaborates the concept of "dialogical identity": as Cattani glosses, "the more I can come in contact with an animal endowed with superior identity, the more my responsibility increases; and vice versa, it decreases with those animals in which the species prevails over the individual, as it could be for the ant or the microbe. Dialogical identity could be summarized by two elements: the gaze and the word. The animal I look at and who looks at me, with whom I speak and who, in its own way, speaks to me, completely belongs to my neighbor" (2005: 18).

At his death in 2016 Paolo De Benedetti bequeathed to us a number of inspiring works on animal theology. It is worth noting the eclectic editorial criterion that he followed in writing them: the ideal book, he says, is the one that, should it by accident fall on your feet, would not hurt them. The reader approaching his legacy, therefore, should expect short works, full of ancient knowledge infused with a delicate irony. As noted above, his sister Maria and some of his affectionate pupils are trying to make his legacy available as a public archive. A legacy made of books, lecture notes, writings, poems, and limericks, all imbued with a call to each human being to take the responsibility of their role inside the history of the world:

Thtat is how human history begins: in the beginning was the garden. And in a garden history will end, because "paradise" means nothing more than "garden" […]. We could say that all misfortunes, from Adam until us, come from the bad care that caused us to be expelled from the garden. But homesickness remains and colors, as we have already seen, the Jewish and Christian eschatological views […]. In our gardens one can meet God without dying: in grass and flowers, in plants and in the color of the sky, in water and birds, in butterflies and in children, and – why not – also in the gardener who "cultivates" and "cares." The garden is not only a landscape, it is also the living creature where memory and awaiting are mirrored. (De Benedetti 2017: 9–10)

Acknowledgments I would like to express my gratitude to Dr. Maria De Benedetti, for her unconditional support and enthusiasm along the way. I would also like to thank the Italian editors MC Editrice and Morcelliana for the permission to republish and translate Paolo De Benedetti's poems.

ADDENDUM

Besides theological and philosophical works, De Benedetti's legacy includes several poems and limericks. A consistent number of them are devoted to those animals he shared his life with. This kind of poetry is helpful to understand the broadness and depth of his thought, and I propose here some examples as a way of conclusion.

A Bobi
Bobi, che su nel cielo
muovi la coda a Dio,
essere amato e amare
è stata la tua sorte
in vita come in morte.
Ora, ti prego, insegnaci
a varcar quella porta
mentre si fa più corta
la nostra attesa; e un filo
di luce dal tuo pelo
ci guidi a ritrovarti
nel prato di asfodelo.
(De Benedetti and Bianchi 2013: 17)

To Bobi
Oh Bobi, who up in the sky
wags your tale for God,
to be loved and to love
has been your destiny
in life as in death.
Now please teach us
to cross that door
while our waiting
shortens; and a glimmer
of light from your hair
may lead us to find
you again in the Asphodel Meadows.

Per Dove sei
Ti avevamo chiamato Dove sei
perché a questo appello tu venivi
e avevi scelto che fosse il tuo nome.
Ma ora non rispondi, e non sappiamo
dove sia la tua anima pensosa
di grande gatto, che scrutava il mondo
con ansia, confidando in pochi affetti
e fuggendo nel sonno. Forse Dio
ti ha detto: Dove sei? Perché voleva
qualche cosa di morbido sul grembo,
fra tanti santi un poco soffocanti.
Ma ti ha rapito a noi, che nel tuo
esserci credevamo che al mondo pur ci
fosse
qualche spiraglio ancor di paradiso,
e che il creato fosse "molto buono".
La morte è amara, ma più amaro assai
è vedere morire. Forse questo
potevi risparmiarcelo, Signore,
ancora un po'. Ora conserva l'anima
piccolina del nostro Dove sei
per quando arriveremo, e se tu puoi
consolala. Ma forse tu non puoi,
perché la morte è troppo anche per te.
(2012b: 39)

For "Where are you"
We called you "Where are you"
because you answered to this call
and you chose this as your name.
But now you do not answer,
and we do not know where is your pensive soul,
big cat, who peered at the word
with bated breath
just trusting few loved ones
and fleeing into sleep.
Maybe God said to you "Where are you?"
because he wanted something soft on his lap,
among many slightly suffocating saints.
But he took you from us,
who in your being there
found some glimmer of paradise in the world
and believed creation was "really good."
Death is bitter but even more bitter is
To see someone dying. Maybe this you could
have spared us, oh Lord,
for a while.
Now please preserve the tiny soul of "Where are you"
until our arrival, and console it, if you can.
But perhaps you cannot,
because death is too much even for you.

A Martino	To Martino
Matino, gatto mio,	Martino, oh my cat,
tu credi ch'io sia Dio,	you think I am the Lord
e mi guardi adorante	and you look at me adoring
come fanno le sante	as do female saints
e gli estatici frati	and the ecstatic friars
scolpiti o pitturati.	carved and painted.
E non sai che i tuoi occhi	And you do not know that your eyes
insegnano anche a me	teach even to me
senza piegare ginocchi	without bending the knee
come pregare il re,	how to pray the king,
un re come te zoppo	a king lamed like you
di un regno ove il dolore	of a land where pain
anche per lui è troppo.	is too much even for him.
Per consolarlo, salga	To console him, may our tenderness
la nostra tenerezza	climb up,
affidata al tuo sguardo	entrusted to your gaze
e a una tua carezza.	and to a caress from you.
(2012b: 13)	

Works Cited

Altamore, Giuseppe. 2016. *Dalla stessa radice. Ebrei e cristiani, un dialogo intra-religioso.* Turin: Lindau.

Bertoletti, Ilario. 2013. *Paolo De Benedetti. Teologia del debito di Dio.* Brescia: Morcelliana.

———. 2017. Paolo De Benedetti. 71 significati di una sigla. *Corriere della Sera – Brescia*, December 9, 13.

Caramore, Gabriella. 2008. Con il 'rabbi' De Benedetti. Tra Karl Barth e i Buddenbrooks. Tuttolibri *La Stampa*, December 20, xi.

Cattani, Piergiorgio. 2005. La dignità degli esseri viventi. *Il Margine* 25 (6): 16–20.

———. 2006. *Dio sulle labbra dell'uomo. Paolo De Benedetti e la domanda incessante.* Trento: Il Margine.

Cini Tassinario, Agnese, ed. 2016. *Fare libri. Panorama completo delle opere di PDB.* Brescia: Morcelliana.

De Benedetti, Paolo. 1992. *Ciò che tarda avverrà.* Magnano: Qiqajon.

———. 1996. *Quale Dio? Una domanda dalla storia.* Brescia: Morcelliana.

———. 1999. *E l'asina disse... L'uomo egli animali secondo la sapienza di Israele.* Magnano: Qiqajon.

———. 2003. Afterword to Paul Ricoeur, *Il male. Una sfida alla filosofia e alla teologia.* Brescia: Morcelliana.

———. 2007. *Teologia degli animali.* Brescia: Morcelliana.

———. 2009. *Il filo d'erba. Verso una teologia della creatura a partire da una novella di Pirandello*. Brescia: Morcelliana.

———. 2012a. Per una teologia degli animali. In *Emotività animali. Ricerche e discipline a confronto*, ed. Matteo Andreozzi, Silvana Castignone, and Alma Massaro, 189–191. Milan: Led.

———. 2012b. *Gatti in cielo*. Brescia: Morcelliana.

———. 2013. Teologia degli animali. In *Collaboratori del creato. La scelta vegetariana nella vita del cristiano*, ed. Guidalberto Bormolini and Luigi Lorenzetti, 17–22. Florence: Lef.

———. 2015. L'ebraismo e gli animali. *Animal Mundi. Le grandi religioni e gli animali*, ed. Alma Massaro, 13: 9–14.

———. 2017. In principio era il giardino. *Il Margine* 27 (1): 9–10.

De Benedetti, Paolo, and Michela Bianchi. 2013. *Animali e noi. Un destino in comune*. Milan: MC Editrice.

Di Filippo, Claudia. 2007. Anche Dio ha i suoi guai... Paolo De Benedetti e Maurizio Abbà a colloquio con la Genesi. *ACME. Annali della Facoltà di Lettere e Filosofia dell'Università degli Studi di Milano* 15 (3): 339–365.

Dostoevsky, Fyodor. 1992. *The Brothers Karamazov*. Trans. Richard Pevear and Larissa Volokhonsky. London: Everyman's Library.

Gelmi, Valentina, and Massimo Giuliani. 2005. La Bibbia e gli animali, 'prossimo dell'uomo.' Intervista a Paolo De Benedetti. *Il Margine* 25 (3): 12–16.

Giacomoni, Silvia. 2006. Quell'ebreo che venne battezzato. *La Repubblica*, April 11, 63.

Giuliani, Massimo. 2018. Paolo De Benedetti: ermeneutica intertestuale, midrash, teologia storica. *Nuovo giornale di filosofia della religione*, June 30. https://www.filosofiadellareligione.it/index.php/2-primo-piano/104-paolo-de-benedetti-ermeneutica-intertestuale-midrash-teologia-storica. Accessed 14 Oct 2019.

Morganti, Martino. 2005. *Mai dire fine. Orizzonti di speranza*. Trapani: Il Pozzo di Giacobbe.

Piccioli, Gianandrea. 2017. Paolo De Benedetti Maestro di editoria. *Le parole e le cose: Letteratura e realtà*, December 13. http://leparoleelecose.it/?p=30354. Accessed 14 Oct 2019.

Rossi, Rodolfo. 2013. Idolatria, conflitto, debito. La teologia di Paolo De Benedetti. *Città&Dintorni* 109: 54–62.

PART III

Fragments of a Contemporary Debate

CHAPTER 11

"*Il faut bien tuer*," or the Calculation of the Abattoir

Massimo Filippi

*It's amazing when you come to think of it how the human spirit seems
to blossom in the shadow of the abattoir!*
—*Lowry (*Under the Volcano. *New York: HarperCollins, 2007: 90)*

1 The Algorithm of the Calculation

1.1. As Derrida argues and as it is by now evident whichever way one looks at it—philosophical, scientific, ontological, or moral, political, or anthropological—the Subject is a "fable" which, nonetheless, is to be taken "seriously," since "it is the serious itself" (Derrida and Nancy 1991: 102). The Subject is the main character of the *fable* which invented it, since it is not—as we are told—a universal invariant, a fact of nature, self-sufficient, and autonomous. It is not created ex nihilo by the performativity of some

Trans. Carlo Salzani and Enrico Monacelli (Previously published in Italian as "'*Il faut bien tuer*' o il calcolo del mattatoio," Chapter 1 of Massimo Filippi, *L'invenzione della specie. Sovvertire la norma, divenire mostri* (Verona: Ombre corte, 2016). The translators thank Dave Mesing for his help and advice.)

M. Filippi (✉)
Vita-Salute San Raffaele University, Milan, Italy

© The Author(s) 2020
F. Cimatti, C. Salzani (eds.), *Animality in Contemporary Italian Philosophy*, The Palgrave Macmillan Animal Ethics Series,
https://doi.org/10.1007/978-3-030-47507-9_11

divinity's word, nor is it born via parthenogenesis from the (decapitated) head of some thinker reflecting (himself) by the fireplace. Nor is it thrown into the world, already given, then and there—human, adult, rational, male, and heterosexual—from God knows where and made by God knows who.

The Subject does not make itself, as Robinson Crusoe thought, strolling in circles on the beach of an uninhabited island, carefully retracing the footprints he himself left, to make sure they are his own and achieve thus some solidity (cf. Derrida 2011: 35).[1] Nor is it the result of a disembodied dialectical incorporation, difficult, perhaps, but with a happy ending nonetheless, as in the dashing adventures of the Self narrated in *The Phenomenology of Spirit*, that sort of anomalous *Bildungsroman*: a *coming-of-age story* narrating the vicissitudes of a Self, which we eventually get to know everything about except the event of its birth. All the stages of the journey which construe the Self are presented to us, and we witness the Self growing, struggling, falling and getting up again, regretting what it had to leave behind at each turning point, feeling remorse and nostalgia, facing the most unlikely twists and turns in order to test its own strength, engaging in various sexual experiences—from the penetration into the other to the pacified onanism in which the other is finally abandoned—until, as it grows old and weary, it rids itself even of the pleasures of art and the beliefs of religion. Despite all this, however, a disquieting obsession haunts this Self: the insight that, in the end, "the *being of Spirit* [that is, the Self itself] *is a bone*" (Hegel 1977: 208).[2]

1.2. The Subject is the result of a "sacrificial structure," of "a noncriminal putting to death" which involves the "ingestion, incorporation, or introjection of the corpse" (Derrida and Nancy 1991: 112)—an operation which is both symbolic and real. It is the result of a *calculation* which, even when it appears at its most perfect state of stability, is still incessantly *haunted* by the return—obsessive and spectral—of the "Animal," of all those (non-humans reduced to things and humans reduced to animals) and of all that (instinct, sensitivity, corporeality, imagination, etc.) which it has incorporated in order to become *incarnate* in the monstrous form of an immortal deity.

The Subject, caught in the economic circle of guilt, responsibility, and retribution, is "an animal [bred] with the prerogative to *promise*," an animal on which one can count because it is always able to account for itself. Therefore, it is "reliable, regular, necessary" and produces "modern men" who "have inherited *millennia of conscience-vivisection and animal-torture*" (Nietzsche 2006: 35, 36, 66; last emphasis added).

That is why the Subject is "the serious itself": it is the One which *adds* itself to itself, after *dividing* from the other to *subtract* it, in order to finally present itself as undivided and thus endlessly *multiply*.[3] The Subject is *the result of a complex reproductive algorithm* which subtracts (excludes) by adding (appropriating) and which multiplies (itself) (privatizes property) by dividing (mutilating and self-mutilating). It is a *ratio* and an (autopsy) report which entails no connections, relationships, or entanglements.

1.3. Even though it cannot even start to conceive itself without the *capital* accumulated by means of the millenary domestication of animals, the Subject is a recent invention arising *from the blood and sweat* spilled over the fields of Europe ravaged by religious wars and plague, over the ruins upon which the modern nation states have built themselves, over the lands hit by the violence of colonialist expansion, over the oiled and shiny wheelworks in the factories of the industrial revolution and the ever more sharp and pervasive instruments of the techno-scientific revolution (cf. Foucault 2002: 375–422). And from the blood and sweat which have cemented the architectures of the palaces of sovereigns and the offices of bureaucrats (cf. Foucault 1978: 133–59). The Subject is the fable which justifies the infinite horror of *this* History (which we still live in) and by which it is justified; it is the calculation that made this history possible and that was made possible by it. It is simultaneously the operation and the result of a calculation that is anything but neutral and natural, but which presents itself *naturally* as neutral. Of a calculation that adds, subtracts, multiplies, and divides humans and animals, which removes animality and excises the other, which forcefully bans any imagining of the monstrously other, of the other-than-human, and that produces, among others and beside It-Self, the Woman, the Homosexual, the Abnormal, the Migrant, the Criminal, and the Animal.

The Subject can claim that it is self-generated because it has disowned the endless series of female bodies from which it was born; it can declare itself self-sufficient because it has removed its dependence on parental care, its vulnerability, its ability to play, and its finitude; it can claim to be rational because it has renounced the unconscious.[4] It claims to be absolute because it self-mutilated and dissected animals and relationships and to be universal insofar as it claims that its own constitutive traits (male, white, adult, heterosexual, etc.) are universals. The Subject is the spectral remnant of an endless labor of dissecting those individuals it has locked up in the various collective singulars—"the Animal" being the broadest and most consolidated of them; in order to bury them in the coffin, they have to vacate at night, so that it can rest after the labors of the day and wake up, stronger than ever, at every new sunrise. Without the mask of the Person and of the universal Good, the Subject is a bloodthirsty monster who reduced itself with its own hands, piece by piece, shred by shred, word by word, to a bleeding stump. It is a diurnal vampire feeding on the blood of the "Animal"—another monstrous entity composed by the remnants of wombs and vaginas, of anuses and sinuous movements, of stupidity and backwardness, of fangs and claws, of delusions and follies, of babblings and vulnerabilities, of darkness and shadows—created by the Subject itself in its operating (operational and operative), mutilating, and destructive fury.

1.4. All too often we have been trained to think that the calculation of the Subject is exclusively a metaphorical abstraction, unable to exceed the limits and (in)stances[5] of thought. In fact, this calculation is also, and above all, the sum of a series of material operations happening again and again, relentlessly, every day. Since, when counting, the one counted no longer counts, calculation is always a *judgement day*. That is why we should abstain from keeping an improbable tally of the pleasures and pains (of the others), which will always be violent, even when performed in good faith. No wonder, then, that cultural traditions as different as the Greek-Western and the Chinese—traditions which François Jullien (2012) considers so "external" to each other that by juxtaposing them we can glimpse the *unthought* of both—describe the essence of the "real" resorting to metaphors exalting, in one case, the spiritualized perfection of the measured, precise, and balanced gesture of the good butcher and, in the other, of his shiny knife.[6]

In the abattoir,[7] the structure where the production of meat is the rule of law, the calculation of the Subject—with its operations of addition, subtraction, multiplication, and division—attains the peak of horror.

2   THE MATHEMATICS OF SACRIFICE

2.1. *Addition* is the basic operation of the abattoir. Animals are killed one after the other to produce the exorbitant numbers of their massacre—so exorbitant that they become imprecise, from 13 to 50 and more billions per year depending on the estimate (which says a lot about the inflationary devaluation of their lives), and not including fish and smaller animals.

Furthermore, the addition of the abattoir is an addition *à la cyborg*, an addition adding pieces of animal bodies to the labor of the workers and the activity of the machines in order to create a monstrous hybrid. The result of this addition is the serial exposition of corpses on the shelves and counters of the supermarkets. The system abattoir is also the center of an even more dreadful addition, the one sequencing the transport systems beforehand, and the disposal, refrigeration, and sales systems afterward, so that every single part of the butchered bodies can yield the maximum profit. All this is described with mathematical precision in Upton Sinclair's *The Jungle*, in which the metastatic proliferation of Packintown (Sinclair 1985: 43–45) reveals itself in the form of a "Cartesian grid" (Pynchon 2006: 10).

The extreme power of the abattoir addition lies however even further upstream, in the patient enumeration of livestock units in the farmers' registers, in the vets' reports, in the firms' books, and, above all, in the scientific codification of all the steps that, summed up, make possible the "aestheticized" functioning of the abattoir. Georgi Gospodinov perfectly understood this point, and in *The Physics of Sorrow* reports, word for word, a section of the "Guidelines for Humane Handling, Transport and Slaughter of Livestock in accordance with the European Convention for the Protection of Animals for Slaughter." Between the lines of which he comments: "That's almost poetry," "What mathematics of death, what geometry of murder," and finally concludes: "Now that's what I call an innocent, hygienic text, as cold and aseptic as the tiles of a slaughter-house – washed sparkling clean once the job is done. No animal would do that" (Gospodinov 2015: 155–57).

No animal would do that, because no animal has ever defined itself as the sum of the "Animal" and of something else—the soul, language, agency, laughter, weeping, the ability to use tools, and so forth—counting. No animal, adding itself to the other, has ever excluded the other, transforming into an other. No animal has ever been able to polish the horror, to make it hygienic and aseptic, by compiling a chilling list of operations that make sense only when added one to the other with cold determination.

Though with a totally different agenda, also Damien Hirst seems aware of the hyperbolic dissecting power of the abattoir addition. What are his cows, sliced in half and placed side by side, if not the exhibition of pieces of the corpse of the "Animal," the result of the calculation of the abattoir? Here the spectator, through the *summary execution* of the addition hidden in the thin gap between the two halves, can enable the artwork to finally appear as a whole. If the "Animal" and the abattoir are abject objects provoking a "stifling nausea," Hirst, just like many other contemporary artists, "turns the sublime into an excremental excess," in a *sublime trash* that cannot cease to pile up, addition after addition, as response to the additional lining up of the "trash itself" in every moment and in every space of our everyday life (Žižek 2000: 39).

2.2. The abattoir *subtracts* life. Not only from animals but also from the people working there. From the perspective of the system that it feeds (the society of simulacra and consumption) and on which it feeds (intensive animal farming), the abattoir performs its subtraction in the form of alienation. In the abattoir, understood in this wider sense, both animals and workers are alienated from the product, from the productive activity, from their own kind, and from nature.[8] Moreover, the abattoir subtracts itself from view because it is consciously hidden outside the city walls and from all possible gazes and because, as essential and inseparable component of the naturalizing ideologies and praxes, it is so much in the foreground that it saturates the view, and falls thereby, by subtraction, into the sphere of invisibility.

Esteban Echeverría's short story, emblematically entitled "The Slaughteryard" (2010), poignantly illustrates the ambiguity of this operation: a subtraction which subtracts itself from view because it is exposed in the most vivid light. The plot of the story is pretty slim: it is Lent of some unspecified year in the 1830s, and Echeverría describes with anesthetizing precision first the activities taking place at the *Matadero de la Convalecencia*—at the time Buenos Aires' abattoir, confined to the district of *El Alto*—then, with more emphasis, the rebellion of a bull soon subdued and killed, and finally the butchers assaulting a young man who manifested his dissent against the oppressive politics of dictator Rosa. Echeverría uses the abattoir to speak of something else, to denounce the atrocities perpetrated by the Argentinian regime: the activities of the abattoir are the activities of that regime, and the fate of the bull is that of the people opposing it. As he writes in the last lines of the story, leaving no doubts about the operation he is performing:

> At that period, the cutthroats of the slaughteryard were the apostles who by rod and fist spread the gospel of the rosy federation, and it is not hard to imagine the sort of federation that would spring from these butchers' heads and knives. In accordance with the cant invented by the Restorer, patron of

their brotherhood, they dubbed "barbarous Unitarian" anyone who was not a barbarian, a butcher, a cutthroat, or a thief; anyone who was decent or whose heart was in the right place; every illustrious patriot or friend of enlightenment and freedom. From the events related above [the attack on the young dissident], it can clearly be seen that the hotbed of the Federation was in the slaughteryard. (Echeverría 2010: 32)

Despite the fact that we are clearly in an abattoir, the place disappears with a clever sleight of hand. This short story makes as evident as ever the result of the subtraction of the abattoir, what Carol Adams calls the absent referent:

Through butchering, animals become absent referents. Animals in name and body are made absent *as animals* for meat to exist. Animals' lives precede and enable the existence of meat. If animals are alive they cannot be meat. Thus a dead body replaces the live animal. Without animals there would be no meat eating, yet they are absent from the act of eating meat because they have been transformed into food. (Adams 2010: 66)

This also happens in the *still lifes* painted by the Dutch masters: the butchered meat is before our very eyes, but it is made completely invisible by the harmony and complexity of the composition and, even more deceitfully, by the signs of the transience of all life (the necrophagous insects and the scrolls with the memento mori) that conceal the massacre by hiding it behind the *naturalness of dying*.

2.3. As emphasized by both J.M. Coetzee and Derrida, what happens in the abattoir is a carnage through *multiplication*. The abattoir is "an enterprise without end, self-regenerating, bringing rabbits, rats, poultry, livestock ceaselessly into the world for the purpose of killing them" (Coetzee 1999: 21). In the abattoir,

the annihilation of certain species is indeed in process, but it is occurring through the organization and exploitation of an artificial, infernal, virtually interminable survival, in conditions that previous generations would have judged monstrous, outside of every presumed norm of a life proper to animals that are thus exterminated by means of their continued existence or even their overpopulation. (Derrida 2008: 26)

In these passages is clearly outlined the calculation of multiplication which the abattoir unceasingly performs: in the abattoir what is implemented is a carnage through multiplication and, with unprecedented monstrosity, the multiplication of an "artificial survival" resting on the endless manipulation of the *life for which we live*, the impersonal mechanisms of the life underlying the living of individuals and which enables *the life that we live.*[9]

Obviously—how could it be otherwise?—the reproductive calculation of multiplication primarily exploits the anatomy of reproduction, in particular that of female animals. Without the complete control over *reproduction*—rigorously denied in order to completely appropriate it in the form of the most violent removal—the *reproductive power of the abattoir* would simply be impossible.[10] That is why Elias Canetti can evoke in a few lines the full horror of the abattoir he visited as a young student in the company of his natural history teacher. The latter, faced by Canetti's "emerging sensitivity on all questions of eating and getting eaten," counterpoises the idea that "whatever happened in nature was not subject to our moral judgements" (Canetti 1989: 228). But this fairy-tale narrative suddenly collapses in the face of the irruption of the more-obscene-than-the-obscene: "We came to a ewe, who had just been slaughtered and lay there open before us. In its water bag, a lamb was floating, tiny, scarcely an inch long, its head and feet were perfectly recognizable, but everything about it looked transparent" (Canetti 1989: 229). The act of bodily reproduction, induced but interrupted in favor of the capitalist reproduction of money—which ostensibly reckons as more profitable the immediate killing of the mother compared to waiting for the birth of the lamb—leads the young Canetti, still mindful of the experience of the "war, which had just ended," to issue "quietly," "in a sort of trance," his irrevocable sentence: "Murder" (Canetti 1989: 229).

The multiplication performed by the abattoir does not stop even in this extreme point; it goes beyond the multiplication of the animal reproductive survival which is indispensable to multiply the productive accumulation of commodities. In order to keep operating, it needs in fact to indefinitely multiply the story of speciesism's justificationist ideology, and it needs to perpetuate, by continuously repeating them, the long-term mechanisms called, at the level of "species," *culture* and, at the individual level, *education*. Hence the necessity, for Canetti's mother, to remove the "cruelty of the fattening process" of geese in the obsessive and self-satisfied celebration of "how good such fat geese tasted" (Canetti 1989: 226). Hence the urge of the natural history teacher to carefully prepare the visit to the abattoir:

> During several lessons beforehand, he often talked about it, always explaining over and over again that they didn't let the animals suffer, they made sure – in contrast to earlier days – that the animals die a quick, painless death. He went so far as to use the word "humane" in this context. (Canetti 1989: 228)

And finally, his very attitude during the visit:

> He explained each device as though it had been thought up for the sake of the animals. His words imposed themselves as a protective layer between me and everything I viewed, so that I couldn't clearly describe it myself. When I think back upon it today, I felt that he acted like a priest trying to talk a person out of believing in death. It was the only time that his words seemed unctuous to me, though serving to shield me against my horror. (Canetti 1989: 228–29)

The intensive exploitation of survival in order to produce "meat" necessarily transforms animals into living dead, specters, and *revenants* who, as such, return, multiplying in the mind of those who witnessed their killing: Chaïm Soutine cannot stop painting butchered oxen to free himself from the scream he was not able to utter and which he keeps feeling in his throat since, as a child, he saw "the village butcher beheading a bird and draining off the blood" (Szittya 1955: 107–8); the female *alter ego* of Patrick Modiano in one of the short stories of the collection *Des inconnues* (*The Strangers*, 1999: 107–155) is haunted in her dreams, at the end of every night, by the rhythmic sound of the hooves of the horses walking down rue Brancion toward the abattoir; the main character of *These Are the Names* by Tommy Wieringa (2015) is haunted throughout his whole life by the painful, unforgettable, and unprocessable memory of the pigs he saw behind a fence, ready to be hung up by the hind legs to bleed to death; Canetti himself is haunted by the maternal geese with which he ends up identifying: "the implacable thumb of a maid stuffing more and more corn mush into the beak of a bird […] became a terrifying image in my dreams, in which I myself had turned into a goose and was getting stuffed and stuffed, until I woke up screaming" (Canetti 1989: 226).

It was perhaps Francis Bacon that more than others let himself be haunted by the specters of the abattoir. Looking at the exhibition of butchered animal corpses, he exclaimed: "Well, of course, we are meat, we are potential carcasses. If I go into a butcher's shop I always think it's surprising that I wasn't there instead of the animal" (Sylvester 1999: 46). And he did not hesitate to identify with animals captured in a photograph as they were pushed toward the abattoir (Sylvester 1999: 23). That is why his paintings, even when portraying other *subjects*, just obsessively repeat the obscure abyss, the gloomy vortex, and the irreparable rift of "butchered meat" (cf. Deleuze 2003: 20–26). Unbearable darkness multiplied by the mirrors he used to paint and that appeared as enigmatic presence in his canvases: the mirrors reflect his deformed portraits, his own butcherable flesh, they multiply and reproduce, they multiply because they are able to reproduce.

As it is well known, for Lacan (2006) the mirror stage constitutes the decisive phase in the development of identity, a stage that comes just before the establishing of the symbolic order, the discourse of the Father, the logocentric discourse, in which "man" recognizes and translates his own phallic fullness into superiority over "woman," who in turn is reflected as lacking and passivity. It is however also well known that Luce Irigaray identifies in another mirror, in another "specula(riza)tion," the possibility of recognizing, where patriarchal *reflection* only saw an absence ("the opaque and silent matrix"), a full and complex sexuality ("fires and mirrors are beginning to radiate, sapping the evidence of reason at its base"), analogous to what Kristeva called the semiotic order of the mother, the *disorderly order* of "touching, caressing, parting the lips and vulva" (Irigaray 1985: 144, 29).[11] A disorderly order not only preceding the paternal order of law, but that never ceases to secretly accompany it, destabilizing and making it, at least occasionally, *impotent*. It is this disorder that haunts Canetti's dreams and Bacon's paintings. It is this disorder that represents the *paradoxical hope* to disarrange, with the incalculability of politics, the order of police's classifications which put everyone in her place until it fills up all slots of the taxonomic and taxidermic system.[12] It is again this disorder that will be able to achieve a "free identification, which can only realize itself *beyond* man with all that is alive and, consequently, suffers" (Lévi-Strauss 1978: 40), capable of making the wheelworks of History run idle in order to suspend it in the miracle of being here: "And the sliced up pieces of meat turn back into an animal, beef turns back into cow. The intestines slip back into its belly, the steaks plaster its haunches. [...] And it's as if the butchers' knives have become thick sewing needles, while they themselves are the tailors, dressing it in the skin" (Gospodinov 2015: 160).

2.4. "The cubicle at the end of the locker room / is called the jerk-off snug / pinned to three walls pictures of women / with hairless vaginas /

to the fourth the poster of a cow / revealing in different colors its delicious cuts" (Ferrari 2004: 3). Thus begins *Macello* (slaughterhouse), a
poetry collection by Ivano Ferrari, who worked for many years in Mantua's
abattoir.

These few verses illustrate the last operation of the abattoir, the most
terrible and definitive: *division*. This division is above all material dismem-
berment of bodies, the final result of the whole calculation of the abattoir,
presenting an uncontrollable penchant for infiltrating every space: it shows
itself already before we cross the threshold of the abattoir—in the locker
room—in the alienation of the workers reduced, and self-reduced, to sexual cogs of the machinery. And it continues to dissect, metastatically, what
lies beyond that threshold, infecting society as a whole and transforming
it into an immense reserve of meat to be turned into commodity, through
the reproductive juxtaposition of butcherable human flesh—typically
women, reduced to their sexual organ which is completely exposed to a
gaze more calculating than lustful—and what has always been *naturally* fit
for this end. This shows us, if proof were still needed, the almost incalculable power of the calculation of the abattoir: the order of the *factors* can
be changed without this modifying the *product*. And at the center of the
dissecting explosion, there is again the onanist Subject: *the jerk-off snug* is
the place where the Subject can rise in complete solitude, after severing
itself from any other form of life and relationship.

Since it is the deepest operation of the abattoir, its most proper factor,
which allows to isolate from the rest what is "proper to man," division is
everywhere and is unrepresentable. That is why in this case it is impossible
to trace out emblematic examples of this operation, although one can
rightly say that division is still recognizable in all passages of the calculation of the abattoir. Division is the result of the aseptic procedures dili-
gently transcribed by Gospodinov, of which the still lifes of the Dutch
masters are in turn the result. It is once again division that guarantees the
sovereign detachment allowing Echeverría to transform the rotten (subject) matter of his story into a metaphor and Bacon to dive into the cuts
of the wounds opened by it. It is division that allows Hirst to put back
together the pieces of butchered animals in the abstract wholeness of an
artwork—just another "piece," immediately available for new consumption—and Canetti to see the horror of the a priori—the transcendental—
sovereignly separating the mothers from the cubs.

Nobody—not even Ferrari—can tell us what the division of the abattoir
is. Nobody can, because this division involves neither archives nor

witnesses (cf. Agamben 2002). Those who plunged to the bottom of the abattoir's dark heart cannot come back to tell us their experience, if not oneirically, in our nightmares, almost secretly, popping up, here and there, intermittently, as runaway specters hunted by everybody. Moreover, unlike what happened in the case of other animalized groups, the division of the abattoir produces no archives. The being-without-name, taken to the highest degree of abjection, burns without glowing, producing not even the dim light of the bureaucrat filling in his register, or the bright light of the sovereign who makes the law coincide with the state of exception, or even the sinister light of the infamous men brought on stage, for an instant, by the gaze of power. The indivisible compactness of the division of the abattoir means absence of witnesses and absence of archives. It is absence of place, the zero degree of place, void fully void, where the real can reveal itself in all its voiding excess of presence, in all its unconceivable monstrosity, in all its bloody horror.

Perhaps only the pictures and footage "stolen" by animal activists bring somehow to the *scene* the ob-*scene* excess of presence of this division of bodies, what should remain off-*scene*, what, despite all, is still alive although it is already dead[13] and, in the suspension of time that each image performs, is always on the verge of dying. These shots and footage, where the living and the dead reach the point in which they are to the utmost indiscernible, stage the absence—of the victims and of witnesses—the obscenity of the putting off-stage the outraged bodies and of the hidden stage where they were produced. This is the most haunting point (beside the disfigured extras these images portray): the eminently spectral character of these images, exponentially emphasizing the painful spectrality they exhibit in the mortal clash between bodies and gaze.[14] The most disturbing thing is the empty scene they overexpose, the absence—more than ever present—against which they stand out, the scotoma of the gaze, their "own" lacking the mark left by the irrevocable disappearing of bodies, even if (or precisely because) put in front of our very eyes. Flashes of "inscrutability itself," in which "one has the impression [...] of something stirring in them, as if one caught small sighs of despair," spaces "between which the living and the dead can move back and forth as they like" and where "we who are still alive are unreal in the eyes of the dead" (Sebald 2001: 258, 261).

The division of the abattoir is the dark of night in which "all cows are black" (Hegel 1977: 9)—and, one could add, in which all vulvas are hairless.[15] It is the zero—and without the zero the whole system of the calculation would be impossible—showing the failure of what was promised by the spirit, which now increases the darkness it once wanted to dispel. *In that darkness* the spirit shows itself for what it is: not a firefly but, as already mentioned, a bone. Perhaps the very bone that, at the end of the parabola of dialectics, Günter Anders (1979: 86) digs up from the remnants of Auschwitz; the bone painted by Hans Holbein in *The Ambassadors* and that our gaze cannot catch because it is hidden by the distorted visibility caused by its being suspended in the foreground; the bone epitomizing the "great labor of death": "at the edge of the birch wood, the earth itself is constantly causing the traces of the mass massacres to resurface. The washing of the rains, in particular, has brought countless splinters and fragments of bone back up to the surface" (Didi-Huberman 2017: 106).

3   The Remnant of the Good

3.1. The fable of the Subject is a serious fable, with no happy ending. It is a fable, like many others we were told as children—from *Hansel and Gretel* to *Little Red Riding Hood* and *Hop-o'-My-Thumb*—hiding, under the glossy and twinkly surface of the story, horror and blood, mutilations and loss, massacres and destruction. It is a fable told for the greater Good (the universal Good) by those who, by calculating, accumulate goods.

In all its many variations, the fable of the abattoir always tells us that *in order to live it is inevitable to kill* and that, if need be, *one must kill well*—without suffering, in the good, clean, and just way, after a life it was worth living. In order to take leave of this fable, we must face the calculating abyss upon which towers, concealing it, the bloody and bleeding horror. We must transform *il faut bien tuer* into *il faut le bien tuer* (cf. Derrida and Nancy 1991: 115); we should "kill" the Good in order to incarnate illogical goodness—illogical because incalculable; we should kill the very Good that spiritualizes private property (enclosures) and the privateness of properties (taxonomies). We should incorporate the "Animal" into class struggle or better *incarnate (animal) flesh within class struggle*. Class struggle should no longer be seen merely as a material and political concern revolving around "Man" and aiming at a more equitable redistribution of goods and means of production but also and above all as a symbolic and impolitical event, capable of messing up the ranks and lines of *classifications*. "Killing" the Good would mark the last universal gesture of the Good itself, which, reaching the end of its historical function, would dissolve in the indistinction of love[16]: exuberant proliferation of joyful powers, festive production of worlds, desiring ontological machine. Love brings on stage the monsters of class society in order to break the bank(s) and the counters of calculation; it marks the passage from appropriating inclusion, which is also and always excluding appropriation, to what Deleuze defined "disjunctive inclusion" (Deleuze 1990: 169–76). This is the infinite play taking place between a non-occupied place and an occupant without a place, a play that can discern the senseless sensuality of sense, which we could call—way beyond the acceptations the term, derived from *jouissance* (enjoyment), has taken in Lacanian literature—*jouis-sense*, enjoyment-in-sense (enjoy-meant) (cf. Žižek 2008: 43), or perhaps, even better, *jouis-sans*. An incalculable play, unproductive and excessive, that brings to a halt and makes inoperative the heterosexual, reproductive, and self-reproducing *couples* of binary and hierarchizing classifications. A play well aware that "inside and outside never cover the entire space," that "there is always the excess of a third space which gets lost in the division into outside and inside" (Žižek 2010: 259).

3.2. If the complexity of the calculation of the Subject and of the abattoir—the banality of Good and of the goods—aims at excluding those who are not, have not been, do not intend to be, or will never be a Subject, and if this exclusion is symbolically and materially an incorporation, then the Other hides in the very heart of the Same, the outside is always already inside. This is the irredeemable remnant of any calculation, the beginning of existence (exit from oneself) and of subversion (exit from the Self). This indisputable permanence is the space—minute, fragile, and yet incoercible and unclassifiable—of love play: "The sum of the remainders constitutes, par excellence, the territory of planetary mixing" (Clément 2004: 7).

3.3. The love play of inoperativity is the insuppressible *waste/remnant* that the calculation of the Subject/Good/abattoir cannot bring, in spite of everything, to a halt. It is what remains, what can drive the humans forward and beyond themselves, to that outside whose proper place is the most intimate recess of their inside and to that inside dwelling in every intimacy of the outside. Only here can start to materialize what cannot be ingested, incorporated, introjected: the corpse of the living that therefore we (already) are. In this place can materialize the paradox that "only an element which is thoroughly 'out of place' (an excremental object, a piece [sic] of 'trash' or leftover) can sustain the void of an empty place" (Žižek 2000: 27). This is what delivers us to our political task—which is no longer a duty but rather *the pleasure to exist and to let exist*: discard (from) the Subject and the calculation, the Good and the goods, properties and property, the private and the privations, categories and enclosures, the cadaster and the stacks of excrements rejected by History to the political task of "creat[ing] history with the very detritus of history" (Rémy de Gourmont qtd. in Benjamin 2002: 543), of saying "no," of refusing by affirming, of becoming waste.

3.4. Usually, to begin to play and to get involved in play one does not wonder "What is it?" but exposes oneself to the other, asking "Who are you?"[17] This exposure allows us to take leave—without removing it from memory—of the horror preceding the horizon of play and to let us "get involved in memorable game" (Schmitt 2012: 129). A memorable game we call life, "life [that] might be understood as precisely that which exceeds any account we may try to give of it" (Butler 2005: 43).

NOTES

1. Cf. Also Alice Munro's short story "Boys and Girls" (1998: 111–127), in which the "favorite book in the world" of the main character's father, a fox farmer and horse butcher, is precisely *Robinson Crusoe*, and in which, among others, are narrated the killing of an untamed mare and the "taming" of the main female character through the progressive construction of her gender.

2. Cf. also p. 210, where Hegel, with a gesture simultaneously sexist and hetero-centered, points out the coupling of "the organ of its highest fulfilment, the organ of generation, with the organ of urination." No wonder, then, that the "activity of dissolution is the power and work of the *Understanding*, the most astonishing and mightiest of powers, or rather the absolute power" (Hegel 1977: 18). And no wonder that also the "rehabilitation" of Hegelian thought carried out by Žižek through infusions of Lacanianism involves the understanding of the Hegelian system not as "oral economy" which devours and swallows, but as "defecation" which, in order to reach completion, releases, abandons, and expels the object previously appropriated/ingested, until the "self-relating gesture of sublating itself" (Žižek 2008: xv).

3. Cf. Esposito (2015), which partially describes the complex calculation—subtraction and division—through which Western political theology derives the One from the Two, without however taking into any account the non-human, and Agamben (2004, especially 29–38), for the chiasmatic workings of the anthropological machine and its system of inclusion/exclusion.

4. It is interesting to note how easily even the unconscious can, perhaps unconsciously, vanish in the calculation of the Subject: "The important

thing, for us, is that we are seeking here [in the unconscious] – before any formation of the subject, of a subject who thinks, who situates himself in it – the level at which there is counting, things are counted, and in this counting he who counts is already included. It is only later that the subject has to recognize himself as such, recognize himself as he who counts" and, even more eloquently: "In order to illustrate this [the subject constituted as secondary with respect to the signifier], I will remind you that the thing may be presented in the simplest possible way by the single stroke. The first signifier is the notch by which it is indicated, for example, that the subject has killed one animal, by means of which he will not become confused in his memory when he has killed ten others. He will not have to remember which is which, and it is by means of this single stroke that he will count them. [...] When this signifier, this *one*, is established – the reckoning is *one* one. It is at the level, not of the one, but of the *one* one, at the level of the reckoning, that the subject has to situate himself as such" (Lacan 1978: 20, 141).

5. Translator's note: we have decided to translate the word "(i)stanze," which is a wordplay which fuses the Italian words for instances and rooms, describing both the occasion in which thought takes place and the spatiality of thought's appearance, as "(in)stances," since the English wordplay mimics closely the meaning of two Italian words. The instances and the stances of thought describe both the happening of thought and the stances which thought spatially assumes in order to happen.

6. For the Greek-Western tradition, cf. Plato's *Phaedrus* (1995: 64 [265 d–e]), where Socrates describes dialectics as the combination of two opposing procedures: the overall view giving shape to the dispersed multiplicity and the ability "to cut up each kind according to its species along its natural joints, and to try not to splinter any part, as a bad butcher might do." For the Chinese tradition, cf. Jullien (2012: 62): "In Chinese thought, what we call the 'real' – thereby reifying it – is considered in terms of breath, flow and respiration [...]. It is here where, in the famous scene from Zhuangzi, slides the knife of butcher Ding. The blade moves 'between the joints' in order to take apart the ox's flesh: encountering no obstacle, no resistance, in this 'in-between' of the interstice, the knife does not wear out and remains always sharp, as if it had just been honed." The material and metaphorical role of the knife in the process of hominization is central, albeit still in an anthropocentric version, also in Sloterdijk's reenactment of the "anthropogene islands" (Sloterdijk 2016: 333).

7. With the term "abattoir," I will designate from now on the institutionalized industrial complex that also includes animal farming.

8. For the "animal translation" of Marx's tetralogy of alienation, cf. Noske (1997: 18–20).

9. The difference between the life that we live (*bìos*) and the life for which we live (*zoē*) is referred to in Agamben (2011: 248–49).

10. Paraphrasing Benjamin (2002), we could say that it is reproduction that hands over the bodies of the living into the sphere of commodity, bringing them close to art.

11. Cf. also Sloterdijk's remarks (2011: 197): "Supplying large parts of populations with mirrors only really began in the nineteenth century, and the process would not have been complete in the First World until the middle of the twentieth. Only in a mirror-saturated culture could people have believed that for each individual, looking into one's own mirror image realized a primal form of self-relation. [...] Even Lacan's tragically presumptuous theorem about the mirror stage's formative significance for the ego function cannot overcome its dependence on the cosmetic or ego-technical household inventory of the nineteenth century."

12. "The essence of the police is to be a partition of the sensible characterized by the absence of a void or a supplement: society consists of groups dedicated to specific modes of action, in places where these occupations are exercised, in modes of being corresponding to these occupations and these places. [...] The essence of politics, then, is to disturb this arrangement by supplementing it with a part of the no-part identified with the community as a whole" (Rancière 2010: 36).

13. Roland Barthes, for example, states: "In Photography I can never deny that *the thing has been there.* There is a superimposition here: of reality and of the past" (1981: 76). Or: "In each of [the images], inescapably, I passed beyond the unreality of the thing represented, I entered crazily into the spectacle, into the image, taking into my arms what is dead, what is going to die, as Nietzsche did when, as Podach tells us, on January 3, 1889, he threw himself in tears on the neck of a beaten horse: gone mad for Pity's sake" (1981: 116–17). Cf. also Susan Sontag: "All photographs are memento mori. To take a photograph is to participate in another person's (or thing's) mortality, vulnerability, mutability. [...] A photograph is both a pseudo-presence and a token of absence" (1973: 16). And Georges Didi-Huberman: "Whenever we are before the image, we are before time" (2003: 31); "Before an image, [...] we have to humbly recognise this fact: that it will probably outlive us, that before it we are the fragile element, the transient element" (2003: 33).

14. "There is no image of the body without the imagination of its opening," writes Didi-Huberman, and concludes: "There is no image of the body without the *opening* – the unfolding until it wounds, until it cuts – *of one's imagination*" (1999: 99, 112).

15. No wonder that George Bataille (2012: 79) can state: "The brothel's nudity calls for the butcher's knife."

16. I take this concept of love from Hardt and Negri (2009: 193–95).
17. These two questions, which mark a radical change in perspective from the calculable to the incalculable, are freely taken up by Adriana Cavarero (cf., e.g., Cavarero and Restaino 2002: 124).

WORKS CITED

Adams, Carol J. 2010. *The Sexual Politics of Meat: A Feminist-Vegetarian Critical Theory*. New York: Continuum.

Agamben, Giorgio. 2002. *Remnants of Auschwitz: The Witness and the Archive*. Trans. Daniel Heller-Roazen. New York: Zone Books.

———. 2004. *The Open: Man and Animal*. Trans. Kevin Attell. Stanford: Stanford University Press.

———. 2011. *The Kingdom and the Glory: For a Theological Genealogy of Economy and Government*. Trans. Lorenzo Chiesa with Matteo Mandarini. Stanford: Stanford University Press.

Anders, Günther. 1979. *Besuch im Hades. 1. Auschwitz und Breslau 1966. 2. Nach "Holocaust" 1979*. Munich: Beck.

Barthes, Roland. 1981. *Camera Lucida: Reflections on Photography*. Trans. Richard Howard. New York: Hill and Wang.

Bataille, Georges. 2012. *My Mother, Madame Edwarda, The Dead Man*. Trans. Austryn Wainhouse. London: Penguin Classic.

Benjamin, Walter. 2002. The Work of Art in the Age of Its Technological Reproducibility. In *Selected Writings, Volume 3 (1935–1938)*, ed. Michael W. Jennings et al., 101–133. Cambridge, MA: The Belknap Press of Harvard University Press.

Butler, Judith. 2005. *Giving an Account of Oneself*. New York: Fordham University Press.

Canetti, Elias. 1989. *The Tongue Set Free*. Trans. Joachim Neugroschel. London: Pan Books.

Cavarero, Adriana, and Franco Restaino. 2002. *Le filosofie femministe*. Milan: Bruno Mondadori.

Clément, Gilles. 2004. *Manifeste du Tiers paysage*. Paris: Sujet Objet. http://www.gillesclement.com/fichiers/_admin_13517_tierspaypublications_92045_manifeste_du_tiers_paysage.pdf. Accessed 23 Feb 2019.

Coetzee, J.M. 1999. *The Lives of Animals*. Princeton: Princeton University Press.

Deleuze, Gilles. 1990. *The Logic of Sense*. Trans. Mark Lester with Charles Stivale. London: The Athlone Press.

———. 2003. *Francis Bacon: The Logic of Sensation*. Trans. Daniel W. Smith. London: Continuum.

Derrida, Jacques. 2008. *The Animal That Therefore I Am*. Trans. David Wills. New York: Fordham University Press.

————. 2011. *The Beast and the Sovereign*, vol. 2. Trans. Geoffrey Bennington. Chicago: The University of Chicago Press.

Derrida, Jacques, and Nancy, Jean-Luc. 1991. 'Eating Well,' or the Calculation of the Subject: An Interview with Jacques Derrida. Trans. Peter Connor, and Avital Ronell. In *Who Comes After the Subject?*, ed. Eduardo Cadava, Peter Connor, and Jean-Luc Nancy, 96–119. London: Routledge.

Didi-Huberman, Georges. 1999. *Ouvrir Vénus. Nudité, rêve, cruauté.* Paris: Gallimard.

————. 2003. Before the Image, Before Time: The Sovereignty of the Anachronism. Trans. Peter Mason. In *Compelling Visuality: The Work of Art in and out of History*, ed. Claire Farago and Robert Zwijnenberg, 31–44. Minneapolis: University of Minnesota Press.

————. 2017. *Bark*. Trans. Samuel E. Martin. Cambridge, MA: The MIT Press.

Echeverría, Esteban. 2010. *The Slaughteryard*. Trans. Norman Thomas Di Giovanni and Susan Ashe. London: HarperCollins.

Esposito, Roberto. 2015. *Two. The Machine of Political Theology and the Place of Thought*. Trans. Zakiya Hanafi. New York: Fordham University Press.

Ferrari, Ivano. 2004. *Macello*. Turin: Einaudi.

Foucault, Michel. 1978. *The History of Sexuality. Volume 1: An Introduction.* Trans. Robert Hurley. New York: Pantheon Books.

————. 2002. *The Order of Things: An Archaeology of the Human Sciences.* London: Routledge.

Gospodinov, Georgi. 2015. *The Physics of Sorrow*. Trans. Angela Rodel. Rochester: Open Letter.

Hardt, Michael, and Antonio Negri. 2009. *Commonwealth*. Cambridge, MA: The Belknap Press of Harvard University Press.

Hegel, G.W.F. 1977. *Phenomenology of Spirit*. Trans. A.V. Miller. Oxford: Oxford University Press.

Irigaray, Luce. 1985. *Speculum of the Other Woman*. Trans. Gillian C. Gill. Ithaca: Cornell University Press.

Jullien, François. 2012. *L'écart et l'entre. Leçon inaugurale de la Chaire sur l'altérité.* Paris: Galilée.

Lacan, Jacques. 1978. *The Seminar, Book XI: The Four Fundamental Concepts of Psychoanalysis*. Trans. Alan Sheridan. New York: W.W. Norton & Company.

————. 2006. The Mirror Stage as Formative of the I Function as Revealed in Psychoanalytic Experience. In *Écrits: The First Complete Edition in English*, trans. Bruce Fink with Héloïse Fink and Russell Grigg, 75–81. New York: W.W. Norton & Company.

Lévi-Strauss, Claude. 1978. *Structural Anthropology*, vol. 2. Trans. Monique Layton. London: Penguin.

Lowry, Malcolm. 2007. *Under the Volcano*. New York: HarperCollins.

Modiano, Patrick. 1999. *Des inconnues*. Paris: Gallimard.

Munro, Alice. 1998. *Dance of the Happy Shades*. New York: Vintage.

Nietzsche, Friedrich. 2006. *On the Genealogy of Morality*. Trans. Carol Diethe. Cambridge: Cambridge University Press.

Noske, Barbara. 1997. *Beyond Boundaries: Humans and Animals*. London: Black Roses.

Plato. 1995. *Phaedrus*. Trans. with introduction and notes, by Alexander Nehamas and Paul Woodruff. Indianapolis: Hackett.

Pynchon, Thomas. 2006. *Against the Day*. London: Penguin.

Rancière, Jacques. 2010. *Dissensus: On Politics and Aesthetics*, ed. and trans. Steven Corcoran. London: Continuum.

Schmitt, Éric-Emmanuel. 2012. *Les Deux Messieurs de Bruxelles*. Paris: Albin Michel.

Sebald, W.G.. 2001. *Austerlitz*. Trans. Anthea Bell. London: Modern Library.

Sinclair, Upton. 1985. *The Jungle*. London: Penguin Classics.

Sloterdijk, Peter. 2011. *Bubbles. Spheres Volume I: Microspherology*. Trans. Wieland Hoban. Los Angeles: Semiotext(e).

———. 2016. *Foams: Spheres Volume III: Plural Spherology*. Trans. Wieland Hoban. Los Angeles: Semiotext(e).

Sontag, Susan. 1973. *On Photography*. New York: Picador.

Sylvester, David. 1999. *The Brutality of Fact. Interviews with Francis Bacon*. New York: Thames and Hudson.

Szittya, Emile. 1955. *Soutine et son temps*. Paris: La Bibliotèque des Arts.

Wieringa, Tommy. 2015. *These Are the Names*. Trans. Sam Garrett. Melbourne/London: Scribe.

Žižek, Slavoj. 2000. *The Fragile Absolute – Or, Why is the Christian Legacy Worth Fighting For?* London: Verso.

———. 2008. *The Sublime Object of Ideology*. London: Verso.

———. 2010. *Living in the End Times*. London: Verso.

CHAPTER 12

Philosophical Ethology and Animal Subjectivity

Roberto Marchesini

1   Introduction

Tradition has bequeathed us a notion of animality that struggles to be coherent with the nature of subjectivity, as we normally recognize it, or rather in the aspects of freedom and expressive protagonism, in being the bearer of intrinsic interests and in the ability to emerge from predetermination through an albeit faint glimmer of self-determination. The difficulties are undoubtedly due to the way in which we interpret the animal dimension, in terms of both the explanatory model used to explain behaviour and the interpretative framework adopted to define categories. Indeed, the paradigm generally employed to describe animal expression denies all forms of protagonism, as it is based on (i) the sequential mechanism, that is, a model that transforms behaviour into a chain reaction of motor units, confined within a rigid track; (ii) innate predetermination, algorithmically organized and thus allowing no possibility for discretion or

Trans. Sarah Ponting

R. Marchesini (✉)
School of Human-Animal Interaction, Bologna, Italy

© The Author(s) 2020
F. Cimatti, C. Salzani (eds.), *Animality in Contemporary Italian Philosophy*, The Palgrave Macmillan Animal Ethics Series,
https://doi.org/10.1007/978-3-030-47507-9_12

creativity; (iii) passive learning, the result of atomic and not strictly experiential associative processes, destined to give rise to automatisms; and (iv) the trigger induced by external stimuli, which activate the ordered cascade of the units.

An even more important factor is the effect of the interpretative framework with which humans have constructed the comparison with the character of animality, setting it as a term of contraposition to their own condition, in order to make the uniqueness of the human dimension emerge through the shifting of the background. The tautological nature and the fallaciousness of this process are easily revealed, since using the category "animal" rather than the species in the dialectical formulation effectively assumes that which it claims to demonstrate. All the ensuing predicative dichotomies—reason versus instinct, culture versus nature, *technē* versus the body—are affected by this fundamental flaw and are simply the consequence of this arbitrary axiom. Animality, strictly speaking, is either the condition inclusive of human beings or has no reason to exist from an ontic standpoint, and consequently its use as a term of comparison has no foundation.

Commencing with the work of Donald Griffin in the 1970s,[1] cognitive ethology has tried to challenge this image of the animal entirely confined within behavioural automatisms, and as a term of contraposition to the human, by introducing the factor of consciousness, as a reflective element capable of endowing even non-human beings with a sort of inner eye able to bring out a sense of self through a glimmer of awareness aimed at feeling, interpreting, desiring, believing…all the way through to the dark realms of self-consciousness. Griffin's work, and that of the authors who followed him in the field of cognitive ethology, was undoubtedly commendable, because it allowed us to cast light on many aspects of the levels of animal intentionality. However, it had, and in my opinion still has, a flaw, in that it links subjectivity and consciousness, thus giving rise to a considerable paradox. Indeed, if we consider the explanatory principle of animal behaviour, which we have seen is based on the automatism model, assigning awareness to this process does not ipso facto produce subjectivity, but awareness of the automatism.

If consciousness is a light that illuminates—to use a metaphor that allows us to understand the very definition of "intentionality" as referring to a content—it is thus evident that the character of subjectivity must precede, or rather underlie as a content, the consciousness that sheds light on it. I do not then wish to deny the different forms of awareness of feeling or reflection that ethology has been able to highlight over the past 40 years with extensive evidence, but to remark on the fact that, when we speak of subjectivity, we are referring to a condition that precedes intentionality itself. Subjectivity appears first and foremost to us as a condition of self-ownership that also manifests itself in the subject's unconscious expression. This self-ownership derives from (1) emotional feeling, that is, the assignment of a meaning in itself to external events, and (2) the desiring condition, that is, being projective towards an elsewhere and proactive in expressive protagonism.

These aspects—more dispositional than reflective, and not necessarily conscious—which we also refer to as the "voice of the heart" or "gut feelings" in humans, define a sort of ownership of ourselves or rather a "being holders of interests that regard us." In defiance of Cartesian thought—as rightly shown by Antonio Damasio[2]—these dispositions constitute the deep roots, able to pump the life-giving sap that nourishes the canopy of our thoughts. It is not enough, for example, to know that you need to be careful in order to cross the road. In other words, semantic-representational knowledge alone is not sufficient to put a prudent behaviour into practice: the activation of the amygdala, that is, a state of alert, is essential to translate knowledge into a state. Clinical research on people with degenerative alterations of these centres has shown the difference between knowledge and a behaviour per se. Another interesting case is represented by the insula, a centre that activates the sense of disgust/aversion but has also been shown to be fundamental in behaviours displaying moral disapprobation and in the inhibition of reprehensible behaviours.

An animal perceives the world, but it is more accurate to say that it "feels it" or rather that it somatizes certain references transforming them into points of reference, that is, values in themselves, because relevant incidents are marked with emotions and thus go from the being events of the world (other than self) to events experienced in the world (other in self). Emotional disposition, on the other hand, is not determined directly by an external event, because it is influenced by the subject's mood, and so the resulting emotion is always an emergency result of the relationship between the subject's mood and what is happening around them. Furthermore, the emotional colouring assigned to the event influences the type of cognitive experience that the individual has, thus directing subsequent cognitive processes, whether they are unconscious or possess a certain degree of intentionality. Hence it is evident that the emotional condition underlies the nature of sentience or rather that transformation of certain events into a feeling—an introjection we might say—and therefore the event can no longer be said to be external but defines a "value of its own," albeit not of a representational type, that is, before the cognitive correlate translates it into a meaning.

Similarly, an animal is never passive or in a state of rest, as if movement were imposed on it from the outside, but is always in a state of projection towards something, whether it be a target, a goal, or a result, and so it exists in an implicitly peripatetic-tensional dimension, even when apparently motionless. This projective propensity is the result of an inner motivational force, namely a drive directed at the outside world that leads it to action. Consequently, we say that an animal is a proactive entity that intervenes and modifies external reality on the basis of specific functional coordinates. This intrinsic state of tension, meaning that it comes from inside, ensures that the subject actively defines the practicability of a context and, in this case too, the type of experience that ensues. If we compare a cat, whose motivation is prevalently predatory, that is, based on the verb "chase," to a human being, whose motivation is, vice versa, prevalently gathering, we discover two different ways of interpreting the practicability of the context (perhaps even the same one) in which they find themselves.

Both emotions and motivations should be considered intrinsic prereflective dispositions that determine the type of intersection that the subject makes in its relations with the outside world. Dispositions are thus selective copulative principles that make the animal interested and connected to everything that surrounds it but in an active manner, that is, as a protagonist, capable of giving a specific colouring to the world on the basis of elective affinities. However, it would be wrong to consider this feeling and desiring of the animal as finite expressions and thus detached from the self-ownership of the individual, as they merely connect the subject in its here-and-now "in a certain way." Indeed, it is precisely because of this, that is, due to the fact that reality is unique in nature, that they are open structures, which can be expressed via actions of modal adjustment. To be clear: the cat is inclined to chase but, through experience, learns different modes of predatory expression in the areas of (i) what to chase, (ii) with which kinaesthetic procedures and (iii) where and when to perform it.

Consequently, modal endowments—those which define the what-how-where-when of behaviour—whether innate or learned, should not be understood as automatisms that move the animal puppet but rather as tools that are used by the subject through canonical, co-optative, generative and creative style structures. In this respect, animal subjectivity is manifested not only through the nature of the intrinsic interest but also by full ownership of the endowments possessed: it is the animal that uses its endowments, not the latter that move it mechanically. Furthermore, this epistemological drift or shift in the explanatory model is much more parsimonious, that is, it responds far more closely to Morgan's canon than to the analytical one based on individual automatisms, because it provides for several operative functions for each schema/tool and does not impose the 1:1 ratio between endowment and function that would make the subject's system of endowments redundant. Indeed, it is far more economical to have a map of a city that allows several itineraries to be made than to have as many endowments as the itineraries completed in the city. Additionally, as reality is experienced through the principle of resemblance and not identity, it is clear that an animal must necessarily always have the capacity to cope with some degree of novelty.

2   Going Beyond the Automatism Model

As I argued in my book *Etologia filosofica* (*Philosophical Ethology*, 2016), subjectivity is the implicit condition of the animal being, an expressive dimension that may also be unconscious but that proceeds from (i) a constructivist conception of behaviour, which we could explain by considering it a work in progress, always open and never predetermined, but equally never imposed and nonetheless always co-factorial to external contributions, and (ii) a propositional view of the individual, or rather from being supported by verbal structures such as to delight, to fear, to want, which join themselves to modal predicates declined according to context, but which likewise render the individual the protagonist. Consequently, subjectivity does not emerge from consciousness, but from disengagement from automatisms themselves or rather from those concepts of trigger (1) and cascade (2) that would make the animal a sort of machine incapable of choosing, evaluating, deciding, programming, planning and adapting its behaviour on the basis of an orientational, interactive and creative protagonism in relation to the opportunities with which it is presented.

The automatism-based model is only apparently more parsimonious than an elaborative model, because it is actually redundant and cumbersome, as anyone can see by reading the descriptive-explanatory formulations of the behaviourist chaining process or the psychohydraulic model. In addition to the rather naive expressions of the early scholars of animal intelligence, the notion of explanatory parsimony in the automatism-based model can be traced principally to two errors: (i) linking subjectivity—and cognitivity more in general—to consciousness and (ii) believing that it is possible to speak of superior and inferior cognitive functions. If we extricate ourselves from this bottleneck, which reveals only the human need to distinguish ourselves from other species and to trivialize the unconscious mind by assigning it to the realm of automatisms and the repressed, almost as if it were a sort of dumping ground of the self, we realize that the analytic automatism-based explanation is anything but parsimonious. Considering animal behaviour to be guided by S-R automatisms—whether configured innately via drives and instinctive predetermined schemata or learned through stimuli and conditioning—creates a profusion of operational titles and a confinement within expressive lines that certainly does not correspond to Occam's razor.

Subjectivity brings us back to a freedom that cannot be attributed to chance, a partiality that cannot be assigned to a deficiency, an ownership that is not disconnection from the outside world. Subjectivity is first and foremost the condition of openness that makes the animal amenable towards the undetermined. We have sought to interpret the animal through the machine model, and I do not deny that this has yielded important knowledge, but it has not allowed us to fully access the heart of animal expressivity, that continuous versatility, that ability to be the originator of one's own behaviour, which inevitably calls for creativity, that is, the ability to invent new solutions, apply old strategies to new problems and modify one's problem-solving recipes on the basis of feedback received. If this is the case, then we can understand the statute of self-ownership that is based on (i) the possession of intrinsic interests, because the world is felt, as well as perceived, desired and encountered, and (ii) the ownership of one's endowments, which can be explained as instruments and not automatisms, instruments at the disposal of the subject who uses them and modifies them according to individual need.

When Stanley Kubrick needed to imagine a versatile interactive computer—the famous HAL 9000—in his memorable *2001: A Space Odyssey*, he necessarily had to animalize artificial intelligence, giving it emotional feeling—anger and resentment when the crew want to exclude it, fear when David gets ready to shut it down—and a desiring projectivity, in other words a need for self-assertion and a plan towards the end of the mission. It is only by virtue of these qualities, that is, by being the bearer of "intrinsic interests," that we derive the impression that we are dealing with otherness. An artificial intelligence may be capable of performing highly complex calculations, often even superior to those of humans, and may even learn new functions and thus go beyond what it is programmed to do, yet still lack the characteristic of otherness precisely because it does not feel nor desire, but merely performs functions.

The animal, on the other hand, even the smallest and simplest, immediately displays its feeling and desire and, in doing so, takes a propositional stance towards the eventiality of the world, or rather it bends reality according to its own specific horizon of practicability. The expression is propositional in that it implies a non-intrinsic content, a reference to something that is offered to it by the outside world. In this respect, subjectivity shows an intentionality that is unconscious or simply precedes all forms of awareness. It is propositional in that it is founded on fear-that or delight-in, where the emotional feeling is always an attitude directed at the outside, which serves as a copula and appeals to the world to assume a configuration. The motivational disposition also replicates this copulative and dialogic character, inasmuch as the projective-proactive tendency that it implies can transform itself from act to action only by virtue of a morphopoietic operation that is the result of the relationship that the subject establishes with the outside world. Consequently, subjectivity is always situated, and it responds to the principle of partiality-infidelity: in other words, an animal is subject in that it never merely repeats an expressive pattern. This tells us that the automatism model is totally inappropriate.

Therefore, behaviour can be considered as a work in progress, which is achieved on the basis of building relations with the outside world; hence, when we speak of intrinsic interests or ownership of own endowments, we are not asserting a self-referential constitution of the subject, but rather a dialogic amenability that transforms the expression in an ongoing process, where the predicates developed are always the result of a co-factorial relationship with everything that surrounds the individual. Rather than a repeated chain reaction of automatisms triggered by the serial input of the sequence, we are looking at a structure where the emotional and motivational dispositions take on the role of propositional clauses, while the endowments, as entities that can be shaped by immersion in experience itself, define the modal here-and-now of the action, which is consequently always correlated and endowed with a certain degree of creativity. Subjectivity, then, concerns a meta-predicative state, which cannot be attributed to consciousness—we do not turn into clocks during our sleep—or even to some *homunculus* lurking in some particular recess of the encephalon, but as an emergency product of synthesis of the entire physiology, not just the neurobiological part but the somatic whole.

Both the constructivist and the propositional principles are based on the protagonism of the animal and on the openness of behaviour, which thus cannot be imposed on the animal by either phylogenetic heritage or the external environment, inasmuch as both the innate and the learned do not take the form of imperative automatisms but of (i) endowments available to the subject, who is thus the owner of them, as instruments that allow multiple solutions to be experimented; (ii) material that can be assembled in different expressive arrangements, which always take on a unique structure in this respect; and (iii) context-oriented "binders," which thus make the subject a relational and hybrid entity. In discussing propositionality, understood as an attitude towards the world capable of introducing contents, I have referred to verb structures, such as fear-that or delight-in, that support the entire expressive structure and equally give a sense of self, perhaps even unconscious, and a protagonism to the individual that causes the emergence of subjectivity, which in turn may display different levels of intentionality, until becoming self-awareness.

However, this is not the point, also because I believe that the discussion of consciousness risks generating more confusion than anything and above all risks acting as a sort of deus ex machina called upon to solve the problem of subjectivity via concealment and circularity rather than resolving the question, because it does not place the focus of reflection on the nature of the behaviour and the explanatory model. If we consider the animal condition as dialogue with the world, sustained by two factors, feeling and desire, we immediately realize that behaviour is always open; in other words, it is in the making, a making that follows the principle of singularity. The individual is subject as the originator of this opening and not as the sole responsible agent of the resulting expression. Consequently, however paradoxical it may seem, the individual's ownership of his or her actions is the result of deferring to multiple levels of influence—the innate, memory, the momentary condition, that which enables or recalls the context—without any of them assuming control of the action itself. Ownership is derived from the opening of the individual to singularity, in other words, of giving rise to an expression that is never repetitive or imposed, but always in some way creative or, indeed, unique and innovative.

Openness does not, however, indicate some kind of neutrality, objectivity or exposition, but is a coordinate of orientation towards the world, as is evident when we speak of perception, but it is equally true for emotional and motivational perspectives. Each expression supported by a disposition is thus an opening, an induction of experience capable of creating behaviourally singular events. If, indeed, the disposition is a copula, the induction of a dialogic event, then all the more reason to consider it as the foundation for creativity, or at any rate expressive adaptation, and certainly not at the basis of blind repetition or bewilderment. We have seen how dispositions compare the state of the body with what the environment makes available, all in a here-and-now that is never repeated, due to that law of singularity of reality that always allows room for novelty. Consequently, each behaviour always has a dual role: (i) that of reformulating or readapting previous schemata, whether the fruit of phylogenetic heritage or the individual's past experiences, and (ii) the creative one of rewriting the behavioural phrase by redefining and modifying its components and the new organizational drafting of the proposition itself to adapt or correlate it to the situation.

We can thus say that openness, which distinguishes the characteristic of protagonism, unpredictability, relationality and creativity of the individual, is accomplished through the verb phrases of the proposition, imputable to (1) the emotionality of the subject, that is, the subject's feeling that, in verb phrases such as fear-that or delight-in, opens itself to contents of specification; (2) the motivational state of the subject, that is, the subject's desiring being, expressed in verb structures such as chase-something, compete-with-someone, which also require contents regarding the four specifying coordinates of what-where-how-when. On the other hand, the resulting expression necessarily requires modal contents that, however, are never repetitions of the past, whether it is innate or experiential, nor are they mere conformations induced by the context. On the contrary, they emerge from the interaction of the past with the singularity of the here-and-now, following the drive for openness towards what is actually present, which characterizes animality in terms of its feeling and desiring.

3   THE ERROR OF ANIMAL AS A TERM OF CONTRAPOSITION

In order to understand the significance of subjectivity, it is also necessary to return to the concept of animality, a dimension that lies above all taxonomic characterizations: indeed, to be defined animal, it does not matter if you are a dolphin or a spider. The term "animal" was used by the humanists and translated into common thought simply in contraposition to the human condition. Indeed, humans are said to differ from animals because of this or that characteristic, turning the designation of animal into an all-inclusive category that does not refer to any particular features, since it is obvious that the different species have different characteristics, but rather to the fact that they belong to a condition that is not comparable to the human one. Therefore, the term "animal" indicates a phenomenal being serving as a comparison, which makes the human appear by opposition and separation, and so it is difficult to recognize ourselves in animality, if not to imply those physiological or somatic characteristics that Descartes assumed in the concept of *res extensa*.

This, however, means that animality is a measurable and calculable dimension, which is not comparable to the entities that, vice versa, elude the metrics of the Cartesian coordinate system capable of translating various phenomena into algebraic formulae. This difference is frequently marked by the use of pairs of opposite qualities, such as reason versus instinct, desire versus drive, love versus copulation and so on. We can thus see that humans locate their existential dimension on a plane that draws on animality as a counter-term or rather as a background whose shifting allows us to emerge, that is, by means of oppositional dichotomies. There is no doubt that human beings have constructed their own identifiability by means of a clear separation from other species. Constructing an insuperable boundary between the human and the animal dimensions also means facilitating recognizability and maintaining the integrity of identity that, on the contrary, would be questioned by any admission of continuity and processes of cross-fertilization. It is actually a process of identification via the practice of *polemos*, which not only regards the emergence of humans but can also be found in other areas of identity construction. The clearest example of this is the subdivision made by the ancient Greeks to define other populations, who were all lumped together under the category of "barbarian."

We are looking at a mechanism of distinction based on three principal processes: (i) the construction of a wall that creates an unbridgeable distance from other populations; (ii) the differences between these populations are voided by the indistinctness of the background created by lumping them all together; and (iii) one's own superiority over otherness is sanctioned by defining the distinguishing process with a missing predicate—in this case the *logos*. This process is even easier to perform with other species due to the perspectival distortion (cognitive bias) that tends to both emphasize the distance between "us" and "them" and standardize the characteristics of otherness that, seen from a perspective of distance, are inevitably compressed to display uniform features. Consequently, we say that animals behave in a certain way, live in complete harmony with nature, fully adhere to their phylogenetic heritage, are unable to do certain things and so on based on the notion that the term for comparison with human beings is not a particular species—as would be correct from a logical point of view—but a generic construct that we call animal.

In general—and here we discover a fundamental aspect of the dialectic artifice—the animal is not characterized on the basis of possessing this or that requirement, but by lacking this or that predicate that the human, on the contrary, displays. A great deal could also be said regarding the issue of "lacking" because, if we do not know the languages of other populations, and perceive them as a confused, as barbar, barbar does not mean that they are actually lacking. Indeed, it is often the case that those who speak of animals do not even bother to become acquainted with the characteristics of the different species, but simply complete the shifting background process, whereby for each predicate they attribute a plus to the human and a minus to the animal. In other words, it is merely a matter of prejudice. It is not my intention here to illustrate the many errors of this approach, but to try to better understand the concept of animality and how it fits into our reasoning regarding the implicitly subjective and inclusive condition of animality.

Each time we try to understand what it means to be human, or rather to fully express our being, that is, those dispositions that make us recognizable as humans, we are inevitably drawn—due in part to cognitive bias and in part to cultural tradition—to seek an opposite side that allows us, by means of background gestalt operations, to make an image of ourselves emerge that is cleansed of all grey, overlapping and common areas and evidence of continuity with the characteristics of other species. A claim that is often heard, and that forms the opening words of numerous arguments about the human being, goes more or less like this: "the human is the only species that…" which is then completed according to the speaker's preference. Once again, we encounter the need to widen the identity range that dissociates us from otherness and at the same time standardizes other species on the basis of the lack of a predicate, which on each occasion is assigned as an element of ontological and not simply ontic divergence, or in other words as something that makes us special and not just specific, belonging to another dimension in respect to animality.

However, all of this has resulted in a refusal to consider humans as a particular expression of animality, no matter how unique (or however one wishes to define it), leading us instead to regard the emergence of humans in terms of opposition, disjunction, negation, submission or elimination of any possible animality. As I said, this process is apparent in the thought of the founding fathers of humanism, in the form of a full-fledged denial of the human being belonging to a natural hierarchy. By declaring the human a Promethean entity, our species is inevitably released from participation in the animal community, to give rise to a different condition. Although this notion was present in a condensed form in the great philosophical and religious traditions that founded Western thought, it becomes particularly important in the modern age, where it has initiated a disjunctive concept of the human. In manifesting themselves as human, and in displaying their qualities to the world, human beings are seen not to disclose their animality, but to abandon it, in particular by declaring their non-citizenship of the Epimethean community, that which is set in a highly specific natural hierarchy.

Let us, then, take a step back in order to better understand this ambiguous and ambivalent concept that we call animality. There is no doubt that "animal" does not mean being a dog or a cat, but refers to something that they have in common, and so we say that both are animals. Consequently, when we speak of animals, we are not referring to one particular species or to specific anatomical or behavioural characteristics, since it is already very risky to identify predicates common to a bee and a chimpanzee. In defining them as animals, we are referring to a dimension that they share and that at the same time distinguishes them from humans, namely animality, which we presume to be above all these predicates, and so we say that dogs and cats are both animals, regardless of how different they are. It is thus obvious that this metapredicative condition (i.e. that goes beyond specific characteristics) shows itself in aspects that are highly generic, but nonetheless distinguishable, albeit through specific prisms of opposition. Reason would demand the search for a lowest common denominator possessed by all animals except humans and vice versa, and, before this even, that humans are capable of performing this descriptive operation. However, this is not the case.

Let us try for a moment to put aside the epistemological reasons that make this dialectic impossible, to discover that, even if it were a plausible operation, it is nonetheless not followed by the humanists, who prefer to construct the two opposing categories in a preconceived manner based on the obsessive need to make the human stand out. The inevitable consequence of this is the definitive banishment of animality from the human *cosmopolis* in the name of another affiliation, which in the ancient Greek mythopoeia takes the form of the dual predicative genealogy effected by the two titans, and in the Jewish tradition is expressed by the concept of *imago Dei*. The various philosophers subsequently established this difference by referring to different predicates—reason, language, technology, morality, politics—but without ever abandoning the use of the human/animal contraposition, which follows the entire path of modern thought, from Descartes to Heidegger. Consequently, humans do not seek to characterize themselves within the concept of animality, but outside of it, and maintain that human development and protagonism are the result of the act of distancing ourselves from the animal condition.

Let us, therefore, return to considering this dimension—animality—to try to understand if it tells us anything about the being it intends to designate and about the possible horizon that it reveals. While some philosophers, for example, Derrida, consider the term a contrived one to be rejected or at least reformulated, this is not the case for naturalists, who retrace it along the undeniably hazy boundary that separates it from the other realms of the *bios*. Animality is thus a matter of semantic contradiction and, in my humble opinion, of philosophical irreducibility, among scientists, who see the human being as one of the various animal forms that populate the planet—moreover placeable within taxonomic groupings and specific degrees of relationship—and the humanists who, on the contrary, consider the animal as that which stands in contraposition to the human. I would thus like to begin with an ontological examination of the term "animal," as I have done more thoroughly in other essays, commencing with the premise that the term nonetheless has its own value, or rather it indicates an existential dimension, but it must not be founded on a process that separates human beings from it, but rather includes them. In other words, I intend to treat the philosophical concept of animality in the same way that science does.

From a physiological point of view, no one doubts the fact that humans are anything but animals, and biology offers a collection of characteristics, for example, heterotrophy, which show us what it means to be animal, regardless of the individual species. The animal condition is not a mystery to the natural sciences and has no need to be extracted through a dichotomy exclusive to humans, which does not make it lacking in descriptive references. Consequently, it should not be so complicated to understand animality in ontological terms either, which are capable of defining the condition of "being animals" from an existential point of view. We should obviously lay down the arms of distinction, oppositional dialectic and the obsessive need for anthropocentric supremacy. All of this should thus help us to better understand the propositional condition that makes us hunger for the world and likewise comprehend the relationship that we have with our phylogenetic heritage, which we normally define with the term "human nature." Indeed, precisely because it is marred by the need for another attribution, the humanist turn risks annihilating human nature, considering it irrelevant or even impedimental in respect to the anthropoiesis, with the danger of making the human lapse into indistinctness and fluidity. But in this case, what does being animals mean from an ontological point of view? How is it possible to speak in philosophical terms of animality in an inclusive manner, that is, by referring to a dimension that is not the result of the background shifting and separation effected by humans? I think this is an important matter, not only to establish greater coherence between the so-called descriptive sciences and philosophical research but also to better understand the human being, because this separation has given rise to several problems in understanding what the conditions of existential completeness are. The idea that the human is a non-animal—the basic concept of humanist philosophy since its earliest formulation by Pico della Mirandola (2016), revived by Martin Heidegger with the "world-poor animal" (2012: 185–96)—renders our emotional and motivational propositionality apparently devoid of a root system.

The contemporary separation from animality that underlies many aporias and equally numerous turns seems a final desperate attempt to reaffirm, and in many respects reformulate, the dichotomy, possibly also due to the bursting of those dividing banks caused by the Darwinian revolution. We thus see a focusing on the distance, which purports to be ontological, separating human from animals, a disjunctive procedure that considers the human a special rather than a specific entity, condemned to a foundation *iuxta propria principia* that does not stand up, as demonstrated by transhumanist ravings on disembodying the human and translating it into an extrabiological dimension. Many contemporary drifts of thought undeniably seem to be acts of immunization against the evolutionist disease, whose anthropo-decentralizing result is evidently feared.

4   REDEFINING ANIMALITY

Reconciling the humanist outlook with Darwinian thought is no easy task, and I would argue that it is impossible, for it can be achieved only through conceptual artifices and semantic twists and turns that are patently strained and highly precarious in terms of their underlying assumptions. Perhaps the most trivial of these attempts to domesticate Darwinian theory is to interpret it as a process aimed at a purpose that, in the case of humans, is expressed in the idea that the evolution of our species constituted a sort of emancipation from animality, censurable in the childish statement "humans are derived from animals." In this case, the relationship with the animal dimension is drawn back into such a remote past that it no longer arouses the same progenitorial embarrassment that it still did in Victorian Britain. Another way of dismantling the explosive force of evolutionism is by considering the animal dimension a minority portion of the human being, and not the characterizing one; a sort of psychological background correlated to the needs and drives that may emerge like an imp during madness, dreams and irrational thought and in the anonymous crowd. In this case, while a dog expresses its animality in its state of being a dog, and likewise for the diverse world of other animals, when it comes to our own species, being human means keeping our animality under control and above all keeping it in the background. Here, animality does not represent the declinable root of the human, as in other animals, but assumes the guise of a wild turn, a descent into bestiality, the sleep of reason, the dark mirror; in short, the loss of those human predicates that, vice versa, continue to be considered its opposite. My proposal, illustrated in particular in my book entitled *Emancipazione dell'animalità* (*Emancipation of Animality*, 2017), is, on the contrary, to consider, the human dimension as a declination of animality. This requires three fundamental operations: (i) the reformulation of the concept of animality, (ii) the definition of the process of declination in animality and (iii) the recalibration of the relationship between innate and learned. These three shifts regard the theme examined here, even though their detailed description is beyond the scope of this essay.

In relation to point (i), beyond the criticism already levelled at the presumption of extracting human and animal via an antithetical dialectic, I will simply say that it is essential to consider animality as a metapredicative and inclusive dimension of the human being. This means that being animals does not signify having this or that adaptive predicate, but adhering to a certain condition that humans and animals share, despite it being expressed and above all experienced in different declinations. Consequently, "animal" cannot be used as a counter-term, because the term of comparison for the human being can only be that of another species and not the generic category of animal. However, each comparison with the individual species does not allow the emergence by opposition of what it means to be animal, because this can, on the contrary, be found by observing what human and dog share, beyond their predicative dimension. On the other hand, since each adaptive feature is merely a version of animality itself, we inevitably reach the conclusion that all the manifestations of the human being are derived from the declinative process of its animality.

We have seen how the animal displays an ontological condition of openness towards the world, experiential peripatesis, and introjection of external references. We can therefore say that this openness, which forms the basis of the animal qualities, is characterized prevalently by three aspects: (1) sentience, as the ability not only to feel that which surrounds the individual but also to transform entities/events into a corporeal condition, or rather to somatize eventualities into a psychological state; (2) projectivity, arising from the desiring condition of the individual, which inevitably leads it to go beyond the state in which it is placed, hence an animal is always oriented towards the outside; and (3) ownership of endowments, in other words, not being enclosed in them, but being able to avail oneself of them, thus making them available and versatile, never repetitive and constantly shapable. We can thus state, in a very succinct and perhaps rather simplified manner, that being animal does not mean possessing specific characteristics—those which distinguish a dog from a cat and are the result of the adaptive process—but being brought into a psychological condition based on feeling and desiring, as well as on being constantly dialogic-constructive in the expression deployed.

Point (ii) is also fundamental for understanding the relationship between animality and subjectivity. Indeed, declination is a way of explaining animality, which is realized in a diachronic manner through processes of introjection of events. Declination therefore occurs not only in the expressive moment of the subject but also in its constitution, commencing with the declinative events that preceded it during phylogenesis. In this respect, we can see that belonging to a certain phylum, class, order, family, genus or species is nothing more than the derivative of a declination event related to what happened at a particular moment of phylogenesis. In order to simplify, we can say, for example, that being a dog or a cat can be considered two different declinations of being carnivores, just as being carnivores or primates are two different declinations of being mammals. Every living being is the result of a large number of declinations that occurred at different times and is thus a diachronic result of declinations of events that have occurred over time. This means that each declination is a reflection of the place-time in which the process occurred, where what had previously been external was subsequently internalized becoming a declining root in the successive shifts.

Animality, therefore, manifests itself not by expressing pure essential qualities but by declining the phylogenetic heritage, which is also the result of previous declinations. It does so through the relationship that the subject initiates with everything that surrounds it, and consequently, in order to understand the resulting predicates, it is vital not to rely on an essentialist interpretation of the individual but on the relational drivers that animality deploys. Declination is a somatic interpretation of what happens to us, just like an oak tree picks up the characteristics of the context, although it does not do so in a neutral or objective way, but through its characteristics of an oak. In this case too, it is evident that, in order to comprehend animality, it is essential to refer to its introjective character, that is, its somatic translation of relationships. Consequently, it is crucial to understand the drivers of this implicitly relational condition of animality, which I have defined as the driving forces of openness, that is, the sentient and desiring character of the animal. In other words, being animal means bringing to the surface or reproposing the events of a story that have taken place over time to decline them in a subsumptive manner in the here-and-now.

When we consider being human in terms of its basic predicates, point (iii), it is therefore essential to refer to both everything that preceded it— that which we define with the term of heritage or innate—and everything that binds it to past experiences from a cultural, parental or experiential point of view. There is a very close relationship between the two terms, which cannot be considered juxtapositive, as though the former were alongside the latter but disconnected, almost like water and oil in the same glass. In actual fact, the innate realizes the learned through two basic factors or principles: (i) the dispositional factor, responsible for the conjugative and declinative openness of the subject in the experience that produces learning, and (ii) the endowment factor, responsible for those processes of reformulation and reconfiguration that, commencing with innate modal elements, gives rise to learned modals. We could therefore say that the innate is realized through the learned and vice versa or rather that there is a directly proportional or dimensional relationship between the innate and the learned. Both the innate and the learned are the fruit of a relationship and, in turn, predispose relationships, so that each consequent expression is always a hybrid fruit.

We cannot, then, consider it in an essentialist way for two basic reasons: (1) because our heritage does not characterize us exclusively, for we have spent the greater part of phylogenetic time, and thus also most of the characterizing declinative processes, in other guises and as progenitors of multiple species, and so the majority of the predicates that we display are not exclusive to humans but shared, and (2) because our animality does not consist in disclosing internal qualities, but in declining them on the basis of the encounters that the subject makes during its existential journey, so it is never possible to base the individual on itself or on internal recognition because the qualities assumed are relational (hybrid) fruits and not emanations. This view of identity as multiple relations established throughout the entire phylogenetic process, which I have described in other essays as a sort of *dialogo ergo sum*, leads us to consider the projectivity and sensibility of the individual and its openness as the two sides of the relational condition that characterizes animality.

NOTES

1. Since the 1970s, Donald Griffin's work has emphasized the importance of considering behaviour in a mentalistic or systemic light rather than viewing the individual patterns as separate mechanisms. This marked the birth of cognitive ethology that, while not rejecting the classic explanation of driving mechanisms, places them within the mental framework as co-factorial elements that the individual can manifest through different levels of intentionality. See Griffin (1994).
2. The innate emotional structures are considered fundamentals of reflective thought, that is, of those processes of evaluation, judgement, planning and decision that underpin rationality. See Damasio (1995, 2000).

WORKS CITED

Damasio, Antonio. 1995. *Descartes' Error: Emotion, Reason, and the Human Brain*. New York: Avon Books.
———. 2000. *The Feeling of What Happens: Body and Emotion in the Making of Consciousness*. London: Vintage.

Griffin, Donald. 1994. *Animal Minds*. Chicago: University of Chicago Press.

Heidegger, Martin. 2012. *The Fundamental Concepts of Metaphysics: World, Finitude, Solitude*. Trans. William McNeill and Nicholas Walker. Bloomington: Indiana University Press.

Marchesini, Roberto. 2016. *Etologia filosofica. Alla ricerca della soggettività animale*. Milan: Mimesis.

———. 2017. *Emancipazione dell'animalità*. Milan: Mimesis.

Pico della Mirandola, Giovanni. 2016. *Oration on the Dignity of Man: A New Translation and Commentary*. Trans. and ed. Francesco Borghesi, Michael Papio, and Massimo Riva. Cambridge: Cambridge University Press.

CHAPTER 13

From Renaissance Ferinity to the Biopolitics of the Animal-Man: Animality as Political Battlefield in the Anthropocene

Laura Bazzicalupo

1 ANIMALITY: GENEALOGY OF THE QUESTION

The great human-non-human dichotomy has been, and still is, a formidable device for the organization of power relations. The political meaning of the device that conceives the human through hierarchical separation from the animal is so significant and remarkable that we can think, in a Nietzschean way, that it is a priority with respect to the descriptive concept of animality, or rather that there is a coalescence of the representational order and the hierarchical order of subordination: it is the Möbius strip of the truth-power nexus. This dichotomy, however, is present in all descriptive terminology, which is always also prescriptive: from the concept of Order to that of Law, which find their synthesis in the descriptive-prescribing ambivalence of Physis-Nature.

The twentieth-century deconstruction of this humanist hierarchy, revealing its power-will matrix functional to domination and economic

L. Bazzicalupo (✉)
University of Salerno, Fisciano, Italy

© The Author(s) 2020
F. Cimatti, C. Salzani (Eds.), *Animality in Contemporary Italian Philosophy*, The Palgrave Macmillan Animal Ethics Series,
https://doi.org/10.1007/978-3-030-47507-9_13

exploitation, remains in an ambivalent relationship with the naturalization and biologization of the human, driven by the paleontological and biological sciences that, from the nineteenth century, have paved the way for an ontology of immanence and continuity of life, as well as highlighted the persistence of animality (in which it is now possible to include the plant world) within the diversification of specializations. What is new is the fact that, even if slowly, this ontology of immanence has acquired a certain credence in common sense and in common morals, although moral and juridical language, and political and economic praxis, persist in using animal metaphors for selective purposes of the living being. The deconstruction lever therefore is the decline of transcendence (both religious and secular) that marked the break between human beings endowed with reason, soul, and language, with animals devoid of all this, and the contextual turn toward biological immanence. This turning point, however, is ambiguous and has not produced the end of the function of domination and exploitation that was the functional matrix of the classic-humanist device. Rather, biological continuism is the very instrument of economic and political biopower.

If we want to look for an alternative route to this main road, we must, in a genealogical way, look for the traces of the discontinuous path of a pluralistic immanent ontology, radically differentialist, wherein animal and human specificities coexist within the multiple organic world without hierarchical levels. An ontological fluidity, open to contamination and hybridization, has today, in the genomic turn of biology, a new legitimacy. Yet the current debate on the Anthropocene, revamping technological activism as a tool to rebalance the nature that man himself has abused, continues to highlight the human biopower that manages the living.

The focus of this chapter is the political meaning of these persistent managerial practices within the modern ontology of immanence, power relations, and devices based on ontological discourses: from humanism, albeit with deviant Renaissance heresies, to biopolitics grafted onto Darwinian evolutionism, up to the emergence of genomic epistemology in the Anthropocene. Discourses and devices operate initially—in full humanism—through the exclusion of the non-human from the specific human species; then they proceed to a series of exclusive inclusions (or inclusive exclusions) within the very concept of human nature (race, gender, abnormality), and finally they lead to the continuism of the biosphere, reaffirming human domination. The problem, from our genealogical point of view, is not which ontology is *true*, but, in Foucault's wake, which powers bear ontological truths (about the separation/continuity of the human being endowed with language with respect to the animal being in general), or, in other words, what biopolitics has been exercised, is exercised, and could be exercised on the basis of ontologies that are qualified for the place that animality occupies in the living.

In modernity, human nature and its distinction from the non-human becomes the primary field of political battle. The ontological order of lives becomes the stakes of political practices (Foucault 1978: 133–159). Power (obviously human) mobilizes the animal imaginary opposed to, or continuous with, the human, in order to govern both. And the *political* question of animality can only be the biopolitical use humans make of it, on the basis of Darwinism and post-Darwinism. Ontologies, narratives, and metaphors in modernity cannot but be rooted in the only legitimate discourse that replaces religion: science. And it becomes important to underline the governmental-technical shaping and the relations of power that are not explicit in its claim of neutrality and objectivity.

But is there, or has there ever been, any alternative thought that can have political effects? Is a nonhierarchical man-animal relationship thinkable? Here the problematization of the humanist dichotomy by Renaissance Italian thought is worth tracing (Esposito 2012: 45–104). Thinkers such as Bruno, Vico, and above all Spinoza radicalized these heretical traces, producing an anti-hierarchical thought of immanence (a losing thought, when compared to the mainstream), unsuitable to *serve* domination. From the viewpoint of Machiavelli, Vico, Bruno, and Spinoza—condemned by the dominant powers and morals—animality is power and resource. Machiavelli uses a metaphorical and mythical language populated by animals, which do not represent disability and lack. Through animality, this language, mixing popular and cultured tones, communicates excesses, frictions, and discontinuities that disturb the humanistic construction of Man and the canonical relationship of governor and governed citizen. Animality is not simply degradation, passivity, or threat but rather a reserve of meaning and power. The positive mixture of man and animal, in Machiavelli and Vico, finds in Bruno and above all in Spinoza an accomplished ontology that challenges subjugation. It is no coincidence that the contemporary deconstruction of anthropocentrism makes explicit reference to these thinkers. This deconstruction moves from Nietzsche and Heidegger toward the Frankfurt School and the French poststructuralism of Derrida, Merleau-Ponty, Deleuze, and Foucault and opens onto—above all with Deleuze, reader of Bergson, Simondon, and Whitehead—a radically monist/horizontal ontology of life in its plural singularity, which cannot be hierarchized.

The epistemic revolution of genomics, as I said, offers scientific support to this radical monism, dissolving forever the centrality of Man. But this is also the geological era of Anthropocene.[1] A biosphere separated from social powers becomes unthinkable: biological existence is relational, integrated with the environment, ceaselessly hybridizing human, non-human, and artificial devices. The very ambiguity of the concept of "Anthropocene"—the end point of the humanistic path, between man's repositioning as only one of the elements of the biosphere and the emphasis on the power humans have exercised and exercise over the biosphere to *save* it—pushes us to examine, with a new urgency, the dark political crux of the animal question, which is today the biosphere, the *continuum* of human-animal-vegetable, and the environment. The crux that the term Anthropocene contains and masks is the domination and exploitation that anthropocentric modernity and postmodern neoliberalism, with its unlimited enhancement of the living, affirm and reaffirm even in the crisis of the biosphere: it thus becomes revolutionary to contest the existing order, as difficult as it is to imagine a different one.

No wonder: what wavers and is questioned in the radical rethinking of the human/animal/nature sequence is the very mainstay of Western political thought, as a representation and as a technique of government, both based on mechanisms of exclusion and hierarchy. For this mechanism not to be reproduced, it is not enough to imagine an ontology where logos and language, which legitimize domination, are conceived *within* the differentiation of coexisting singularities; it is necessary to think and practice politics in a way diametrically reversed with respect to the common sense conveyed by the Western political tradition. Human action (even when benevolent) should not be *on*, but *with* nature and animals: a destabilizing goal, radically alternative to a code that has been in force for too long.

2   Troubling the Humanist Dichotomy: Machiavelli, Vico, and the Ontology of Immanence in Bruno and Spinoza

Genealogy finds discontinuities in the humanist mainstream: the different path inhabited by a-conceptual animal figures that attest to the unease (and fascination) of what cannot be thought or said. The Italian Renaissance has an ambivalent relationship with humanism. On humanism and its dualistic and anthropocentric ontology rests the social, moral, legal, and political pyramid that will be effective in all Western culture and in large part of the global world influenced by it. Foucault identifies in it—in the *invention of Man*—the ordering device of the West (Foucault 1989: 335). This means that humanism, which develops in state theory and capitalist economic practices, has built itself—by the drastic division between the angel and the beast—to establish an order of domination. Not that the "great difference" was not already present in the medieval world: here the male Christian man was endowed with a soul, while sinful and demonic animality was ascribed to infidels, women, and monsters as objects of awe. But this division, as Latour suggests, is always functional to create hybrids (Latour 1991: 47). The same holds for the far from univocal relationship between the two terms in the ancient world. And yet the medieval iconology of animality—the allegories of the lion, the unicorn, the horse, the dragon, or the Christian lamb and fish—is used, against the definition above, to designate noble virtues: that is, they empower the Christian order of the world, representing its moral values. As Foucault's meticulous research on madness shows, in medieval practices the relationship between man and animal is unstable and fluid and develops *within*, but also *in spite* of, the frontier established by the Church: there is a widespread mixture, always exposed to reversal (Foucault 2006: 3–38).

The modern age imposes order on such a restless life: this is humanism as a device of power. Ordering things, putting them where they belong, stopping the disorderly coexistence of the multiverse as shown in the myriad of medieval mosaic tesserae and human-animal monsters carved on the facades of cathedrals. A meaningful universe needs to be organized; it needs a representation in the center of which there is a Man who dominates—platonically—his irascible or concupiscent part, the white or black horse, and directs it by means of language and reason. Compared to the rigid humanist dichotomy of Cartesian rationalism, a certain Italian Renaissance with Machiavelli or Leonardo da Vinci, and then Bruno, Vico, or Caravaggio, stands as internal-external, with an original look at the animal part that cannot be reduced to a subalternity to be tamed. The realism of these philosophers and artists signals a gap in the humanist project. The human is more complex and contradictory than the figure outlined in humanist idealism. The animal holds its own truth, neither domesticated nor subordinate, but rather custodian of man's secret nature, of his possible excess. Animal is disorder, fury, and instinctive strategy in the face of the rational cosmos that is being built. The hybridization between the two kingdoms breaks natural law: it is transgressive but fascinating. Unlike ethics, politics (but also economics) has the privilege of having to deal with the irrational *real* and with so-called animal spirits, which cannot be morally condemned. The great modern political theories and, in a different way, the modern practices of political and economic governance of lives do not arrogantly elude animality in the human living beings: they recognize and practice it, but in order to domesticate it.

Machiavelli's political thought challenges the humanist utopia in a different way, thinking politics through hybridization: the positive and powerful persistence of animality within the human subject (Versiero 2004). When one looks at the political struggles and government practices as they happen, a world of theriomorphic beings emerges, of princes who have the cunning and sagacity of foxes and the courage and strength of lions. And they move like the centaur, the ancient man/animal, the intelligent and wise hybrid that mobilizes passion and animality for life. The Machiavellian use of the figurative eludes the straitjacket of logos and enhances what is nonlinguistic, circumventing the logical-linguistic bottleneck into which Western thought is proceeding. The nonlinguistic allows access to a region beyond the symbolic, access to what is common to man and animal, but, in logocentric culture, forcibly divided. The matrices of politics, law, and force are not in contrast with each other. If it is true that the former is man's own and the latter is "the beast's own," this apparent tribute to humanism is revolutionized: both intellect and strength have an animal figure, the *golpe* (fox) and the *lione* (lion): "because the lion does not defend itself from snares and the fox does not defend itself from wolves. So one needs to be fox to recognize snares and a lion to frighten the wolves" (Machiavelli 1985: 69). The canonical dichotomy fox-lion is revolutionized by the coexistence (and not the alternative, as in Cicero and Dante) of the two animals/virtues: to them Machiavelli ascribes not only cunning but also strategic "rationality." This is the topic of the *bestial men* (*i bestioni*) that can be found in Leonardo's drawings and which work across the human duplicity of divine reason and bestial passions in a transversal way, like in the Centaur Chiron. Machiavellian animalism, by mixing rational strategy and feral instinct, revalues the animal condition: political reasoning is rooted in what is low, in corporeality, moods, flesh, and blood, and in the instinctive dissimulation of the *golpe*. The mixture of intelligent animality and passionate wisdom outlines an affirmative biopolitics *ante litteram*. Current Italian studies of biopolitics see in Machiavelli's and then Vico's lessons a dissident alternative to Cartesian anthropocentrism, as well as to Hobbes's political—albeit materialistic—science (Esposito 1984 and 2012). In the Real of history, Vico, the anti-Cartesian *par excellence*, recovers human animality. The *erramento ferino* (feral wandering) of the bestial man, endowed with poetic faculties, pushed by "very violent passions, which is the thinking of beasts," shares behaviors and knowledge typical of animals; "and human nature, as she is common with the beasts" (Vico 1988: § 340, 370) is a founding and inaugural moment of history and of political coexistence that never gets rid of that magmatic, corporeal origin from which man comes. Human animality is also manifested in the most evolved performances (mental, cultural) such as metaphorical thought (Vico 1948: 106, § 378).

These are alternative traces with respect to the great modern pessimistic anthropology, from Hobbes to Canetti, Freud, and Gehlen, in which naturalness is thought in a functional way to a project of order that precedes and governs the very definition of animality. In Hobbes, animality is only the hypothetical stage that justifies the permanent transition to civil life, and it is shaped according to a new mechanical science (as in Descartes): animality as a predictable and manageable machine. A physiological computational reason instead of the strategic flexibility of the fox. The effects are diametrically opposed: in Machiavelli, the conflict is persistent and positive, and the hybrid man-animal is enhancing and *unstable*, preventing the rigidity of an ontological *status*, whereas in Hobbes, human life is predictable, serial, governable through a totalizing device such as sovereignty. The drastic scientific naturalization of the human being, in fact, is shaped with its government as its goal: the political scene is built on the expulsion of a negative animality (fear, death) that besieges it. The sovereign device "spectralizes" animality (which is also its main pillar); it makes a ghost of it and produces the romantic restlessness and the Freudian discontent of civilization. What is qualified as ferocious or bestial in man—signified by animality—becomes the errant signifier of different metaphorical functions: power or passivity, fear or courage. Thereby it is charged with anguish but also with hope: it has the fascination of the Lacanian *Real*, which mere language and reason leave unsolved (Lacan 2009). The pillar of the modern humanist building is therefore animality, a phantom removed, mute, betrayed by the order of the symbolic and normative, which supports the imaginary construction of the rational man. A ghost that, like all ghosts, returns.

A brief mention is in order, in this genealogy of alternative traces, of *the counter-ontologies of immanence* that, at the philosophical level, reject in early modernity the anthropocentric paradigm. Brief because their effects are limited: these ontologies, unable to structure political organizations, support rather singular, libertarian, and discontinuous counter-conducts. Spinoza, above all, imagines a radical immanence, freeing the whole living being from the burden of normative judgment: life, common to the living as a whole, has no measure that overcomes it (Spinoza 2002). This is a losing ontology, much like the ontology and science of Giordano Bruno, in the face of Galilean and Cartesian science. Also Bruno, like Spinoza, gives the name of Life to the coincidence of God and World. Impersonal Life—like the indistinct *ingens sylva* (massive forest) of Vico, populated by beasts—removes the subjectivity of the *Anthropos*, replacing it in the infinite circle of the living: like earth among the thousands of planets (Bruno: 1950).

Like Hobbes, Spinoza believes in the naturalization of the living, but not marked by a lack to be compensated for: the *conatus sese servandi* (self-preservation drive), an impulse desire that moves every living being. Spinoza's choice is the radical rejection of the great dichotomy: all bodies are equivalent with respect to being—the stone, the madman, the rational man, the animal. The immanent measure is, as Deleuze will comment, the power that each living being has to exist and act according to its own specific nature, that is, according to the power inscribed in its own existing body: certainly self-preservation but also powerful expressiveness (Deleuze 1992: 217). Animality dissolves in the multiple monism of nature/life. This underground current of immanentism is referred to by the most radical post-anthropocentric currents of animal philosophy, which go beyond simply extending to animals the modern and liberal ethical framework of rights, but aim at a global rethinking of ontology in which body and mind, animality and humanity are not only not-split but are rather dissolved into a multiform and plastic immanence of living: an ontology in which animality points in the direction of a different way of life and a different coexistence.[2]

3 PRACTICES AND THEORIES OF HUMAN ANIMALIZATION: ANIMALITY AS MANAGEABILITY

The hard Cartesian separation between *res cogitans* and *res extensa*, rational thought and reified body of animal/machine, supports the liberal attribution of rights, autonomy, and reason to the unencumbered subject, while the scientific objectification of the body as a mechanical *res* sets up the planning of its productive exploitation. The paradox of the anthropological machine which divides the human living being from the subordinate animal "by means of an exclusion (which is also always already a capturing) and an inclusion (which is also always already an exclusion)" (Agamben 2004: 37) emerges right away in the techniques of exercising power. The division indeed leads to hierarchies within humanity itself. The anthropological device excludes from man his *mechanical* part, the animal, corporeal, destined to serve the upper part, the linguistic, logical, rational. Rationality, anyway, is de facto only provided for some who can afford to live a fully human, liberal life (nonanimal, nonmechanical). They are therefore citizens, distinguished by education, wealth, and class.[3] The great division unfolds in the hierarchy *within* humans, between those who know, think, own capital, and buy the work of men and animals and those who do not know, who have only the body/machine for productive and reproductive work.

The capitalist factory places the bodies of the manufacturing workforce—man, woman, child, work-animals—in functional spaces, planning their mechanical gestures, as well as in animal breeding, where there are pre-established paths of foraging, reproduction, exploitation, and slaughter. The dichotomy further separates man from himself, on the basis of economic utility, that is, for the many, the animal lower part, destined to be directed to *serve*, to function, and to work for survival. These many reproduce and produce in an exchange that capitalism claims to be free, so that the responsibility falls on the exploited themselves, who have nothing but their labor-power to sell. A labor-power simply extracted from (their) animality. The discontinuity between human beings in their own right and those who live and are managed almost like animals produces a paradoxical *continuum* between these subjected subjects (humans and animals alike) and enhances in both passive traits typical of the animal. Imagination and indocility (which the Renaissance hybrid extolled) are regarded as deviant; they must be corrected, disciplined, or marginalized and segregated. The disciplinary techniques start from the exceptions, from the unproductive anomalies of those not animalized enough: madness, diversity, degeneration—equated to wild animality—are observed and monitored in order to be standardized for the specific functions that production requires (Foucault 1977: 167ff). In the factory, the human/animal workers are subjected to training; in the fields, houses, and stables, sexual conduct for reproduction is normalized; and brutalization, suppression, and segregation await those who are abnormal, insane, wild, and unrecoverable to the economic project.

This incongruous outcome of the great project of the Man-as-Centre-of-the-Universe, king of the animal kingdom, finds new force in the scientific revolution of paleontology and evolutionary biology in the nineteenth century, which proudly affirms itself against the reactionary remnants of religious ontology, and also troubles traditional rationalist humanism. The very definition of biopolitics as the intersection of power technologies with the biological mechanisms of life is based semantically on the double assertion of the subjugation of animals—which the humanistic separation legitimates—and of the animalization of man (Foucault 2003: 239–260). Legal universalism yields to a trivialized use of evolutionary science, which justifies differential clusters of living beings—the statistical-zoological term is *populations*—based on the manageability, potentiality, or risks of biological endowment and presumed racial and gender trends. Obviously, the epistemic framework of evolutionism is much more complex. But the trivialization of the stages of the evolutionary chain lends itself to differentiations that are naturalized and racialized. A hierarchy of the living emerges, which selects the human from the "less human": the black or the Jewish, less than men, explicitly equated to animals. And among the animals, to those that are infamous, harmful, and filthy: parasites, viruses, ticks… The biopolitical violence of the animalization of the human being that legitimates racism on a biological basis cannot be overlooked.

Western science always objectified the living in view of the Baconian power to act on it, and it would therefore be naive not to call it "true." The problem is—given the link between truth and power—to understand the contradictions and ambivalences involved in the animalization of the human being, with the narcissistic wound that Darwinism inflicts on man descended from apes. The answer to this wound has been to reproduce, *within* the continuum, the hierarchy of humanist discontinuity by means of *animality*, a category that goes from blind, ferocious, and a-rational violence to stupidity, inability to rebel, passivity, and subjugation. In just one double movement, human nature is animalized and hierarchized "inside itself," and the rational exploitation of inferior animals, subhuman and non-human, is justified. Governmental techniques have been settled on the ganglia of human animality: birth and morbidity, health, sexuality, strength—all variables that can be modified. The government works on life by separating, evaluating, and selecting the right lives from the deviant, dissolving, anti-vital ones.

A decentralization of *Anthropos* begins with Darwinian evolutionism, which the new molecular and genomic epistemology radically increases. The transition, in the second half of the twentieth century, from a biology still rooted in the macroscopic and anthropomorphic dimension to the molecular biology of the Genome project, marks an enormous step toward de-anthropologization. Man, the top of the evolutionary process, is dissolved into the mere flow of the living organic genome (Kevles and Hood 1992). This fluidity dissolves states and stages of the hierarchy of the living beings, but—as an effect of the power of science—it also opens up to engineering projects on *bios*, hybridizations, and contaminations: projects and actions that enhance the governmental biopolitical control *over* life. Whereas the organic-genomic *continuum* could provide scientific resources to the neo-Spinozist ontology of Deleuze, which aims to preserve the differences between living beings overthrowing biopower, transgenic practices with animals are linked to exploitation and commercial valorization (Kevles 1985). Even if they are *benevolent*, aimed at empowering the living, they occur under the mark of pastoral biopower, "the great domestication of being" (Sloterdijk 2016).

The Foucauldian biopolitical framework, therefore, despite great changes in the techniques of domination, remains actual and even grows in accuracy. Not only is most of the material of genomic experimentation in genetic engineering practices provided by *inferior* animals for the purpose of human empowerment, where humans are clearly the strongest form of life in the biological continuum; but also the technocratic approach conditioning life persists and even intensifies, as it is the *real* political crux of the animal question even in declining anthropocentrism (Peterson and Somit 2012). The biopolitical turn of the anthropological device strips the living of its surplus, which is not linguistic and rational at all, but rather is all that is not serial, governable, optimizable, and exorcised by the predictive patterns, and is therefore stigmatized as pathological, malignant, anti-vital, and uncivil. Animalization, as biologization of the human, does not free the animal part, but dominates it, reducing the potential of the living and its very reason to the economic-utilitarian matrix. The philosophers of the Frankfurt School remind us about this point (Horkeimer and Adorno 2002), but the utopian horizon of their proposition of an emancipation of nature, of a *good* animality, is destined to fail since it is enclosed in the same dichotomy from which we must instead radically escape.

4   RADICAL CRITICISM OF DOMINATION: CAPITALOCENE AND INTERSECTIONALITY OF OPPRESSION

Against the nefarious features of biological continuism, Derrida provocatively claims discontinuity, "rupture or abyss between those who say 'we men', 'I, a human' and what this man among men who say 'we', calls the animal or animals" (Derrida 2008: 30). The beasts themselves, Derrida claims, know this difference, and they know it as soon as the human being stands as master, exploiter of animal life. It is therefore the practice of man's dominion over the animal, rather than their respective ontological status, which marks the effective discontinuity in the biological continuum, justified by the most sophisticated endowment of man that just emphasizes what other animated beings lack: the high endowment which still regulates the experimentations and hybridizations of the posthuman. The fact that every differentiation is always marked by effects of domination, subalternity, and exploitation—whether of animals, women, foreign immigrants, or non-white races—redraws the question of the living, not in ontological but rather in political and economic terms, within the historical relationships of domination, then reverberating in ontological terminology.

By reversing the order of priority, as Derrida suggests, the signifier "animal" signals the subordinate position, nothing else. If Man is a recent invention, so is Animal, and its signifier signals relations of power. The two terms—animal and man—are therefore *political* signifiers. If we apply a Lacanian approach, we could say that the term animal reveals the gap, the failure of the symbolic: the *Real* that it cannot say about the human. According to Lacan, the Real is the irreducible difference that language (the normative bearer) tries to tame through the generic signifier: the animal in general or *animals*—nonexistent, as such—beyond the concrete singularity-plurality of existing living beings (Derrida 2008: 31). If the pivotal point of the linguistic device is the relation of domination, the unspoken *real* of the signifier *animal* is the violent, painful asymmetry of this relation. Derrida indeed criticizes and deconstructs this representational mode, but does not dig into the historical devices that produced it: he simply questions them through the silent and disturbing gaze of the cat that observes man, his nakedness, and the poverty of his mastery. The Derridean way suggests the recovery of the limit, of the Otherness that Man, ascribing the term Animal to absolutely different living beings, nullifies. By contrast, in *The Open* Agamben, in the wake of Heidegger, counterposes the inoperative power to which the animal attests to the totalizing biopolitical and technical mechanism of the Western anthropological machine and projects it into a forthcoming messianic community (Agamben 2004: 85–92).

Evanescent and totalizing, like its reverse: it is the Anthropocene, without externalities. To the catastrophe of the biosphere he caused, Man—this exorbitant geological element that marked a new era—replies from within its system of domination, by means of a technocratic rather than political approach, which is precisely the problem. Maybe it is the totalization of the domination device—for which "Anthropocene" is a felicitous term—that must be politically and historically disjointed. In fact, the biopolitical management of lives (human and animal alike) is a complex device, which adapts to the changing living forces on which it is installed. It often finds its power on the cooperation of the governed themselves. It is not a form that contrasts with life, but a form of life that is historically planned through specific biological and economic natural differences. Naturalization crystallizes the dissymmetry of the relationships of subjugation. But being subjugated does not erase the complexity of the relationship nor the vital, surprising power of the governed, active all over the chain. These are surely now weaker powers, but they are nevertheless positive powers, able to arrange new relations. This ambivalence specifically marks the current form of life that biopower exercises over nature, the Capitalocene, a political-economic term that I personally favor over the anodyne and technocratic one of Anthropocene (Moore 2016).

However, if it is true that on the living *continuum* capital installs commodification, the matrix of the productive and reproductive surplus (for farm animals), there is an ontological and of course political limit to absolute subjugation. The power relationship is ambivalent, being exposed to the active power of those very living beings who produce value, work, and life and to natural forces, which can force subalternity to change form. Of course, the famous image by Horkheimer—often quoted by antispeciesist literature—of the skyscraper summarizing the connection between human and animal exploitation remains valid: after having listed the different floors, from the top of the capitalist magnates to the bottom of the prole-tariat and, under them, the colonial coolies, Horkheimer places in the slaughterhouse of the cellars "animal hell in human society, sweat, blood, despair of animals" (Horkheimer 1978: 66). Man's dominion over man is rooted in man's dominion over nature (both external and internal to him-self). The negative dialectic from Frankfurt—which certainly has the merit of having highlighted this connection between dominion over humans and dominion over animals/nature—places it, however, in a horizon of catastrophic totality and opposes to it, like all totalizations, its utopian reverse of a redeemed human nature.

But this double subalternity (men over men and men over animals) never exhausts—and I am going beyond Frankfurt here—the potential of the governed, human or non-human, as long as it is alive. True, the two levels of exploitation are parallel and concurrent, and there is no dominion over man without surplus accumulation rising from the exploitation of nature. They mutually influence and reinforce one another and so must be faced and criticized together as one single issue. But they change accord-ing to the context, taking different positions, oppressions, and resistances. An ontology of the Spinozian and Deleuzian kind, which is non-dialectical but multiplying and immanent, emphasizes just this instability of the sta-tuses and their incommensurability, which cannot be reduced to market equivalence. The different and specific struggles are linked crosswise; dif-ferent living beings divest the essential identity they are ascribed and tem-porarily take on the forms of the oppressed minor living being: becoming a woman, becoming an animal, a nigger, a minor... (Deleuze and Guattari 1987: 281–82). The Machiavellian trace of animality, as a secret resource of the human and the unstable flexibility with which it can be taken on, returns on stage. These animal metamorphoses—the hybrids that emerge by resisting oppression—are not afraid to abandon the privileged status of man, male, white, autonomous, and rational.

5   The Libertarian Trace of the Cynical Parresiast

A pluralistic ontology of the living, compared to any catastrophic and par-alyzing totalization, disseminates the political scene in a myriad of contin-gencies and settings where continuous variations of the meaning of the great signifiers, *animal* and *man*, occur. An unusual intersectionality of the struggles against oppression involves animals and environment, dis-puting the moral antispeciesism whose weakness lies in separating the aim of animal liberation from human liberation. In the indistinctive ground where the difference man-animal is lacking, subalternity binds in a common plan-sequence—which is not anthropological or ontological but rather historical and political-economic—humans (as the bodies of workers, of women, new slaves, new exploited of temporary jobs) with the animal bodies from which value is extracted. But no subalternity is sealed or absolute. The exercise of resistance precedes and produces the signifier. Practices and attempts at alternative forms of life can produce (and then be conditioned by) a nonhierarchical, pluralistic ontology such as that announced in Spinoza and in Bruno, which delegitimizes any exploitation.

Foucault, the great genealogist of the human animality governed by a shepherd, recognizes a form of free and alternative hybridization in the cynic philosopher, who proudly takes on the anarchic, provocative, and unpleasant behavior of the street dog in order to contrast the dominant logos (Foucault 2011; Bazzicalupo 2013). A libertarian practice of resistance is explicitly achieved by taking on an animal life form, indocile and resistant to pastoral control. We must remember that Foucault does not engage in a naturalistic ontology. We should not demand from him a naturalism of the absolute immanence of life, such as in neo-Spinozian ontologies. The genealogist starts from historical experiences that attest to an active biopolitics: living bodies that cannot be reduced to the practices of domestication, passing through the Logos (language and reason), man's flag. The cynical parresiast chooses instead the example of the dog without a master, who follows its own instinct, finding its own rule of conduct without discipline (Goulet-Cazé 1993). He escapes the Logos that rules by representation, classification, evaluation, and selection; he forgoes the reason that makes the animal-man a different animal, able of subduing and being submitted; he shows a glimmer of a new way of life. Diogenes mimics the dog, and the freemen mimic Diogenes; they take him as an example. Diogenes, the dog, Diogenes the crazy Socrates: crazy because he does not use logos, but mimics animals. His counter-conduct is inarticulate: he does not speak nor produce arguments; it is *ergo* not *logo*. It is a way of life: "bearing witness to the truth by and in one's body [...] life as the immediate, striking and unrestrained presence of the truth" (Foucault 2011: 173).

The *kynikos bios*, says Foucault, "could not be more opposed to Platonism" (Foucault 2010: 286). And if the latter is an archetype of anthropocentrism, it contrasts with a form of life that is really, materially reversed into natural nudity, devoid of that consciousness and soul that are the matrix of all humanistic governability. Who would rule a stray dog who does not accept the limits imposed by shame, who "feels free to say frankly and violently what he is, what he wants, what he needs" (Foucault 2010: 287), just like a stray dog? Subjection passes through all the social virtues and sense of honor that are the ornaments of humanism. The *kynikos bios* instead "barks" against its enemies and distinguishes good from evil on the basis of *physis*. This fragment of a free animality unaffected by any management is therefore not wild. Rather, it is produced by self-government, *askesis*, physical and moral discipline, aimed at reducing the heteronomous government on one's own body, even if it means simplifying and minimizing its claims, its pleasures. Finding one's own power and freedom to say no to those who pretend to rule you for your own well-being. Animal power and animal freedom then become not the low point of subjugation but the ethical goal of a process and a project: the "running ahead of humanity" (Foucault 2011: 167).

NOTES

1. The term Anthropocene was coined by Eugene Stoermer in the 1980s. See also Will Steffen, Grinevald, Crutzen, and McNeill (2011) and Latour, Stengers, Tsing, and Bubandt (2018).
2. The main philosophical reference is still Deleuze; see also Braidotti (2002).
3. Already the Greeks considered banausic workers noncitizens; Arendt (1958: 79ff) qualifies them as *animal laborans*.

WORKS CITED

Agamben, Giorgio. 2004. *The Open: Man and Animal*. Trans. Kevin Attell. Stanford: Stanford University Press.

Arendt, Hannah. 1958. *The Human Condition*. Chicago: University of Chicago Press.

Bazzicalupo, Laura. 2013. Animalità: il crocevia del pensiero di Foucault. In *Animal studies: Rivista italiana di antispecismo* 2.4, special issue on *Gli animali di Foucault*, 17–30.

Braidotti, Rosi. 2002. *Metamorphosis: Towards a Feminist Theory of Becoming*. Cambridge: Polity Press.

Bruno, Giordano. 1950. On the Infinite Universe and Worlds. In *Giordano Bruno. His Life and Thought, with an Annotated Translation of His Work*, On the Infinite Universe and Worlds, ed. Dorothea Waley Singer. New York: H. Schuman.

Deleuze, Gilles. 1992. *Expressionism in Philosophy: Spinoza.* Trans. Martin Joughin. New York: Zone Books.

Deleuze, Gilles, and Felix Guattari. 1987. *A Thousand Plateaus. Capitalism and Schizophrenia II.* Trans. Brian Massumi. Minneapolis: Minnesota University Press.

Derrida, Jacques. 2008. *The Animal that Therefore I Am.* Trans. Marie-Luise Mallet. New York: Fordham University Press.

Dodds, Eric. 1951. *The Greeks and the Irrational.* Berkeley: California University Press.

Esposito, Roberto. 1984. *La politica e la storia. Machiavelli e Vico.* Naples: Liguori.

———. 2012. *Living Thought: The Origins and Actuality of Italian Philosophy.* Trans. Zakiya Hanafi. Stanford: Stanford University Press.

Foucault, Michel. 1977. *Discipline and Punish: The Birth of the Prison.* Trans. Alan Sheridan. New York: Vintage.

———. 1978. *The History of Sexuality, Volume I: An Introduction.* Trans. Robert Hurley. New York: Pantheon.

———. 1989. *The Order of Things: An Archeology of the Human Sciences.* London: Routledge.

———. 2003. *Society Must Be Defended: Lectures at the Collège de France, 1975–76.* Trans. David Macey. New York: Picador.

———. 2006. *Madness and Civilization.* Trans. Jonathan Murphy and Jean Khalfa. London: Routledge.

———. 2010. *The Government of Self and Others.* Trans. Alan Milchman. New York: Picador/Palgrave Macmillan.

———. 2011. *The Courage of Truth (The Government of Self and Others II).* Trans. Graham Burchell. New York: Palgrave Macmillan. Picador.

Goulet-Cazé, Marie-Odile. 1993. Les premiers cyniques et la religion. In *Le Cynisme Ancien et ses Prolongements: Actes du Colloque International du CNRS,* ed. Marie-Odile Goulet-Cazé and Richard Goulet, 117–158. Paris: Presses Universitaires de France.

Horkeimer, Max, and Theodor W. Adorno. 2002. *Dialectic of Enlightenment.* Trans. Edmund Jephcott. Stanford: Stanford University Press.

Horkheimer, Max. 1978. The Skyscraper. In *Dawn and Decline: Notes 1926–1931 and 1950–1969.* Trans. Michael Shaw, 66. New York: Seabury Press.

Kevles, Daniel. 1985. *In the Name of Eugenics: Genetics and the Uses of Human Heredity.* New York: Knopf.

Kevles, Daniel, and Leroy Hood, eds. 1992. *The Code of Codes: Scientific and Social Issues in the Human Genome Project.* Cambridge, MA: Harvard University Press.

Lacan, Jacques. 2009. "*L'étourdit.*" A Bilingual Presentation of the First Turn. Trans. Cormac Gallagher. *The Letter* 41: 31–80.

Latour, Bruno. 1991. *We Have Never Been Modern*. Trans. Catherine Porter. Cambridge, MA: Harvard University Press.

Latour, Bruno, Isabelle Stengers, Anna Tsing, and Nils Bubandt. 2018. Anthropologists Are Talking – About Capitalism, Ecology, and Apocalypse. *Ethnos: Journal of Anthropology* 83 (3): 1–20.

Machiavelli, Niccolò. 1985. *The Prince*. Trans. Harvey C. Mansfield. Chicago: The University of Chicago Press.

Moore, Jason, ed. 2016. *Anthropocene or Capitalocene?: Nature, History and the Crisis of Capitalism*. Oakland: PM Press.

Peterson, Steven, and Albert Somit, eds. 2012. *Biopolicy: The Life Sciences and Public Policy*. Bingley: Emerald.

Sloterdijk, Peter. 2016. Rules for the Human Park: A Response to Heidegger's 'Letter on Humanism.'. In *Not Saved: Essays After Heidegger*, Trans. Ian Alexander Moore and Christopher Turner, 193–216. Cambridge: Polity Press.

Spinoza, Baruch. 2002. Ethics. In *Complete Works*, Trans. Samuel Shirley, 213–380. Indianapolis: Hackett.

Versiero, Marco. 2004. Metafore zoomorfe e dissimulazione della duplicità: la politica delle immagini in Machiavelli e Leonardo da Vinci. *Studi filosofici* 27: 101–125.

Vico, Giambattista. 1948. *The New Science*. Trans. from the 3rd Edition (1744) by Thomas Goddard Bergin and Max Harold Fisch. Ithaca, NY: Cornell University Press.

———. 1988. *On the Most Ancient Wisdom of the Italians: Unearthed from the Origins of the Latin Language*. Trans. Lucia M. Palmer. Ithaca: Cornell University Press.

Will, Steffen, Jacques Grinevald, Paul Crutzen, and John McNeill. 2011. The Anthropocene: Conceptual and Historical Perspectives. *Philosophical Transactions of Royal Society A: Mathematical, Physical and Engineering Sciences*. https://doi.org/10.1098/rsta.2010.0327. Accessed 26 Sep 2019.

CHAPTER 14

The Animal Is Present: Non-human Animal Bodies in Recent Italian Art

Valentina Sonzogni

1

This chapter aims at exploring the latest developments and ethical implications of the use of living or dead animals in art, with a special focus on Italian art. Through the analysis of some celebrated artworks and relevant case studies, I intend to highlight a latent tendency on the part of art professionals (meaning not only artists but also art historians, curators, and so on) to justify behaviors that are hardly acceptable according to shared ethical standards in other fields. If, from an animal welfare perspective, it seems equally unjust to use an animal for medical experiments as much as for an artwork, the reactions within public opinion are quite different. People who would not permit that a Damien Hirst work featuring a bisected cow is exhibited in the local museum nevertheless support animal experimentation or meat consumption because they perceive them as necessary. This kind of schizophrenic approach—that Melanie Joy has significantly described as "carnism" (Joy 2009)—allows the same person to confer moral status to certain animals to the detriment of others, as well as

V. Sonzogni (✉)
Archivio Piero Dorazio, Milano, Italy

© The Author(s) 2020
F. Cimatti, C. Salzani (eds.), *Animality in Contemporary Italian Philosophy*, The Palgrave Macmillan Animal Ethics Series,
https://doi.org/10.1007/978-3-030-47507-9_14

to accept different treatment standards for different species (e.g., pigs are food, while dogs are pets). In her most well-known example, Joy describes a dinner party in which the guests are tasting delicious meat and ask the host for the recipe. The answer given is that the secret lies in the meat and that one must use "three pounds of well-seasoned... golden retriever!," which provokes general offense and disgust in the guests (Joy 2019).

This way of "knowing without knowing," of having a split perception of phenomena, often affected by perceptive and cultural bias, is at the base of the altered understanding of the animal presence in contemporary art. This altered understanding is exacerbated by a largely shared opinion according to which contemporary art is difficult, inaccessible, a matter only for "specialists." This gap distances the spectator even more from what is looked at, whether it be a painting, an installation, or a performance. The distance acts on the spectator to the point that she/he is no longer able to distinguish whether the animal is real or fake or whether an animal has been killed to produce a film or an artwork, as new technologies help to build images perceived as real.

However, the truth is that animals[1] have always been present in the history of art, obviously as represented subjects from classical art onwards, but also in the earliest works of art that humanity remembers, for example, in the Lascaux Caves paintings. Less obviously, the animal with its body is present in art as raw material, from the egg derivatives used in the tempera paint to the bovine serum contained in Prussian blue and the brushes made of pig bristles; hence animal bodies have always been present in classical to contemporary art (Caffo and Sonzogni 2015a). But it is only recently that antispeciesism,[2] the various animal welfare associations, and animal studies theory have crossed paths, sometimes clashing, more often ignoring each other while trying to position their instances within the field of visual art. Some cases are quite famous and worth recalling here because they have contributed to the advancement of the debate on the use of animals in art.

2

Damien Hirst (born 1965) is one of the golden boys of the Young British Artists, a group that became famous in the early 1990s thanks to the support of the wealthy collector Charles Saatchi. Since the beginning of his career, Hirst has used dead bodies or living animals in his artworks. *The Physical Impossibility of Death in the Mind of Someone Living* (1991), for example, featured a tiger shark immersed in formaldehyde. Hirst's shark can be considered the most evident case of a tendency to use animal bodies in art that has continued increasing in the following decades. Celebrated artists like Pierre Huyghe, Mark Dion, and Marina Abramović, just to cite a few, have used animals—or their remnants—in their artworks. As Giovanni Aloi rightly points out: "The main difference between the traditional representation of animals in art and the use that Damien Hirst makes of the shark lies in the fact that the animal we see in *The Physical Impossibility of Death in the Mind of Someone Living* is an animal body, not the man-made depiction of one […]. Hirst's shark gives us the opportunity to experience an intensity that a painting cannot quite conjure: that is the encounter with animal matter" (Aloi 2012: 1–2, 4). Further, Steve Baker points out one characteristic of what he defines as the postmodern animal: "one characteristic of much post-modern animal art is its refusal of symbolism, its insistence on carving out a space in which the physical body of the animal – living or dead – can be present as itself" (Baker 2013: 8). The material presence of the animal body seems to be a trait in common between the international and the Italian artistic experiences, as we will see later in this chapter.

In 2012, at the Hirst Tate Modern retrospective, for the work *In and Out of Love*, 9000 butterflies were used for the duration of the exhibition from April to September. The work consisted in a replica of Hirst's first solo show in 1991. Then, he installed white canvasses on which he glued butterfly pupae. Emerging butterflies would continue their life cycle in the room, fed with fruits and sugar. The surviving butterflies would mate and produce eggs starting a new cycle of life. On average, 400 butterflies died per week for this show. This symbolic—and real—use of animals in art has prompted a general opposition to host artworks by Hirst in museums, especially in cities that have strong associations with animal welfare.

Another recent and meaningful event relating to art and animals has been the cancelation of Hermann Nitsch's show scheduled at the Jumex Museum in Mexico City in 2015 and the recent polemics that have followed the exhibition "Hermann Nitsch/Katharsis" in Mantua's Palazzo Ducale. As many as 27 animal welfare associations have signed a letter to the mayor of Mantua to halt the exhibition (Anonymous 2019). Nitsch (born 1938) is a celebrated Austrian artist who has worked, since the late 1950s, on his own idea of a performative art in which symbolic and cruel elements from pagan rituals, such as animal blood spilled on the performers, become the center of a *mise-en-scène* that sometimes lasts for days. His *Schütterbilder* (*action paintings*) are in fact painted with a mixture of color and animal blood and are considered offensive of animal welfare, although in his art he uses the same carcasses that are shipped every day from slaughterhouses to butcher shops all over the world.

In its letter-writing campaign, the animal welfare associations stated: "We wish to remind you that Law 189/2004, in Articles 544 bis/ter and quater prohibits, respectively, the killing, mistreatment and exploitation of animals for shows. In this case they are all present" (Anonymous 2019). With the same approach and language, animal welfare associations protest against the organization of Palio di Siena, the traditional horse race that causes injury and often death, the numerous local "sagre" (popular festivals) in which animals are part of pageantry or participate in the staging of the Christmas nativity. By addressing the presence of animal mistreatment in art practices, the major Italian animal protection associations acknowledge that contemporary art is perceived as a form of popular entertainment and, even when the animal is not present with its own body, it is equally unacceptable that the remains of its body are divided up and used as props for an art performance.

The work of these artists and the related exhibitions accompanied by the press and public opinion—who alternatively take sides with artists or with protesters—have led to the deployment of two rather well-defined positions: "one shall never use the body of an animal in any form in art" versus "one shall never set limits to the freedom of art," that is to say that art shall be beyond or above or before ethics. Artists and curators do not openly confront the topic of the use of animals in writing, but they do elaborate literary explorations of the living world or approach it from a visual or an interdisciplinary point of view (Ramos 2016).

One artist who has addressed this issue, publishing an actual list of ideas of how artists should work with what he defines as "the living world," is Mark Dion (born 1961). His conceptual art is characterized by the use of scientific specimen and complex installations reminiscent of a natural history museum or a scientific presentation. Dion's works include stuffed animals as well as other organic materials that are never just symbolic but work in a complex whole that delves into the aesthetics but also the functioning of science and its role in the common imaginary.

In "Some Notes Towards a Manifesto for Artists Working With or About the Living World," Dion states that as humans we "do not stand outside of nature: we, too, are animals, a part of the very thing we have tried to control, whether for exploitation or protection" (2000: 66). He admits the responsibility of artists who decide to work with the natural world and writes in point 4 of the "Manifesto": "Artists working with living organisms must know what they are doing. They must take responsibility for the plants' or animals' welfare. If an organism dies during an exhibition, the viewer should assume the death to be the intention of the artist" (66). Such an assumption of responsibility for the life of sentient beings is a very strong declaration, as it confirms that artists shall be aware of their actions because they belong to the public sphere and the public nature of the artworks (when published, photographed, and exhibited in a museum) bestows upon the artist a social role.

Furthermore, point 10 states: "Nature does not always know what is best" (66). This reveals the self-assumption of the artist as a creative force that can be compared to Nature. It counterbalances, in a way, the previous statement concerning the responsibility of any action that artists operate on the organic, living context. As he develops further in point 11, though, "We reject the notion of the environment as a perpetually stable and self-regulating system, existing in a constant state of balance. The natural world is far more dynamic and intricate and its history, for at least ten thousand years, has been more entwined with human history than notions of natural balance allow for" (66). Are nature and the history of nature far more connected than we believe? With regard to animals, Dion seems to have a precise idea of what they are and why they deserve a dignified treatment: "13. Animals are individuals, and not carbon copy mechanistic entities. They have cognitive abilities, personalities and flexible behavior, which is not to suggest that they exhibit distinctly human characteristics" (66). His understanding makes him state that humans should not project their own narrative onto the lives of animals because they have species-specific traits.

In a recent discussion with Giovanni Aloi (Aloi 2012: 142–51) about his "Manifesto," Dion has further clarified his position toward animals and taxidermy, disclosing his personal view of the matter. The stir of public opinion connected to the killing of animals in art (whereas prior to their use in art exhibits or in actual performances) is, in his view, related to the fact that killing animals for food, for handbags, for their status of pests, or as the consequence of a road accident seems to be inevitable, whereas meaning-making is perceived as a luxury. This statement by the artist is particularly interesting, even because he reinforces this idea by declaring: "Fish die for less" (Aloi 2012: 150). Paradoxically, if a fish is killed in a performance, namely that of Marco Evaristti *Helena & El Pescador*,[3] this can bring attention to the individuality of the animal, whereas the millions of fish killed every day for food consumption or by sea pollution are truly invisible. This paradoxical perspective is not devoid of ethical implications, although Dion does not follow up on this point. If fish die for much less than art, and art is a powerful meaning generator, we should try and avoid blaming artists without considering the overall meaning produced, let alone the intentions of the artists.

But is it possible, and does it make sense, to analyze and/or criticize some artistic interventions by developing a more mature and profound relationship with the world of animals, thus avoiding censorship of the intellectual work of artists but also avoiding an artistic freedom devoid of any reference to ethics? After all, a world in which everything can be done is not necessarily the best of all possible worlds. The variegated scenario of contemporary ethics—and of many disciplines with which the history of art intersects in an attempt to "anthropo-decentralize" itself—can provide us with effective tools to advance the discussion.

3

It must be noted that, in recent years, there has been a visible increase in the use of the actual body of animals, and not only their representations, in artworks. Beyond the aforementioned Hirst and Maurizio Cattelan, whose work will be described later in this chapter, the question of animality traverses the works of numerous artists whose research does not specifically focus on this topic. The use of animals—although, as we will see, also of humans—in contemporary Italian art is not a recent phenomenon but rather dates back to the post-World-War-II avant-garde, namely in the context of Arte Povera. In 1969, Jannis Kounellis (1936–2017), an artist of Greek origins but mainly operating in Italy within the art group known as Arte Povera, "exhibited" the work *Untitled (12 horses)* at the L'Attico galleria in Rome. In order to investigate the reality of the materials used in the production of the artwork and to eliminate the threshold between the real and the representation, he attempted a redefinition of the act of seeing by bringing 12 living horses into the gallery (see Celant 1992: 130; Barbero and Pola 2010: 42–45). The horses were simply standing in the gallery, tied with their ropes to the wall with grass at their disposal, challenging the traditional representations of the horse in art. Whereas horses in art were parts of rampant monuments, together with national heroes and patriots, Kounellis' horses, with their calm presence and the very living matter of their bodies, had the power to change the perception of space. Their natural and casual presence has since then become as iconic as that of painted horses.

The work has been re-enacted recently at the Gavin Brown Gallery in New York, generating a trail of comments and articles completely different from the first time the work was proposed (Smith 2015). Because Kounellis was still alive, the new arrangement was developed in accordance with the artist in a similarly minimal, almost industrial space. But, while almost 40 years ago the work created a stir, its potential now seems to have fallen, like that of a radioactive material that gradually discharges. The horse that becomes a living support for the work of art with its very presence does not suffer and does not rejoice and, in this sense, is not alive in this artwork. Horses are not feeling different than when they are forced to live in a stable. While there is an invisible and ancient act of domestication that makes us believe that horses are pets, the same use of horsess in the artistic context is perceived as an abuse. Both acts, in fact, are neglecting the horse's ethological and species-specific characteristics. No animal was ever born to be domesticated, although most of us believe so.

At the 1972 Venice Biennale, Gino De Dominicis (1947–1998) set up a room that included the "work" (and here the quotation marks seem a moral imperative) *2ª soluzione di immortalità (L'universo è immobile)* (*Second Solution of Immortality. The Universe is Immobile*), composed of several elements, including some of the artist's previous works, as well as a young boy with Down syndrome, Paolo Rosa, trivially known for many years among art historians as "De Dominicis' mongoloid." The work is composed of three objects, already exhibited by De Dominicis before: *Cubo invisibile* (*Invisible Cube*, 1967), that is, a square of tape applied on the floor that describes an invisible cube; *Palla di gomma (caduta da 2 metri) nell'attimo immediatamente precedente il rimbalzo* (*Rubber Ball [Dropped from 2 Meters] the Moment Immediately Before It Bounces*, 1968), a simple plastic ball; and a stone titled *Attesa di un casuale movimento molecolare generale in una sola direzione, tale da generare un movimento spontaneo della pietra* (*Awaiting a Casual Molecular General Movement in Only One Direction, Such That It Generates a Spontaneous Movement of the Stone*). Rosa sat in the same space, peering at the objects, facing the spectators. In the intent of the artist, the person with Down syndrome should have symbolized immortality because of his absent past memories and future projections. He was soon to be replaced by a small child (Charans 2012).

In 1970, at Galleria Toselli in Milan, De Dominicis had already exhibited a living kitten accompanied by a caption relating to the investigation of immortality in the work with the same title as the abovementioned *2ª soluzione di immortalità*. In 1975 he then organized an art exhibition for animals titled *Ingresso riservato agli animali* (*Admission for Animals Only*, 1975), in which the public was composed of, among others, an ox and a donkey. While the animals were left to move freely in the space of the gallery, the "real" human public was standing outside the show at Lucrezia De Domizio gallery in Pescara. For De Dominicis, individuals like Paolo Rosa, the kitten, or the ox invited to the gallery conveyed the idea of unconsciousness and unawareness in the face of death, therefore embodying the concept of immortality.

Following the outrage of public opinion, the parents sued the artist, citing as a reason the fact that they did not understand the real purpose of their child's involvement. The artist was acquitted "because there was no substance to the fact." The resounding words Pier Paolo Pasolini wrote in the newspaper *Il Tempo* on June 25, 1972, denounced the emptiness of Italian (sub)culture and the devaluation of humanity made in this context: a context characterized by the most unbridled indifference and by an ideological confusion which had allowed a person with Down syndrome to be left—by the artist, but also by his own parents—at the mercy of a situation that he could not fathom (Pasolini 1999). This work was also re-enacted at Frieze Art Fair in London in 2006 in the Wrong Gallery booth, curated by Maurizio Cattelan, Massimiliano Gioni, and Ali Subotnick, and performed by an actress with Down syndrome, perfectly aware of the role she played in the performance (Somers Cocks 2006). Of course, in re-enacting it "ethically," the artwork's original intention has changed and therefore the artist's freedom of expression is in question. Since Gino De Dominicis died in 1998, whether he would have endorsed a reenactment of his Biennale room cannot be known.

With regard to paying respect to sentient beings, in life and in death, it is relevant to analyze the complex set of reactions triggered by Maurizio Cattelan's 1997 work *Novecento* (*1900*) (see Caffo and Sonzogni 2015b: 21–26). Cattelan (born 1960) has used stuffed animals in several works, perhaps the most famous being *Turisti* (*Tourists*), an installation composed of 200 stuffed pigeons exhibited at the Venice Biennale in 1997. In the empty space of the pavilion, the animals were placed on the ceiling beams, and their presence was revealed to the visitors by their fake dejections on the floor. *Turisti* was one of the first artworks in Italy against which public opinion took a very definite strong position.

In *Novecento*, the subject of the "artwork" is a taxidermy horse whose legs have been elongated during the process and which is hanging from the ceiling. It is an impressive image of extraordinary power, which makes visible, at once, the force of gravity and the uselessness of energy, while, through its title, it cruelly mocks the past centuries in which horses were used as the central elements of celebrative, heroic monuments. Contrary to what the animal rights advocates claimed in several messages sent to the Castello di Rivoli Museum of Contemporary Art in Rivoli (Turin), where the piece is part of the museum collection, the horse was chosen for the artwork when it was already dead. It is revealing to note that some people, who have only seen a photograph of the work, ask if the horse is alive and, by questioning the very nature of the artwork, make visible a dilemma that will afflict the museum in the future, that is, what can be done with the body of an animal. Today this limit is not set, and visitors or occasional readers cannot make sense of what they are looking at. In this sense, Cattelan's horse is a living emblem of an era at its end and of the beginning of another era in which the question of the animal has to be investigated.

The question whether the very nature of the artifacts changes the meaning of the artwork in the context we are analyzing can be answered if we compare the use of animal bodies in contemporary cinema productions or advertisement. The use of automata, special effects and motion capture has integrated the digital element in the film narrative without transforming a film into something else. Equally, a new frontier of art can be the use of digital technologies with respect to the involvement of real animals in the installation. An early case in Italian art is a video animation by artist Diego Perrone (born 1970), *Vicino a Torino muore un cane vecchio* (*Near Turin an Old Dog Is Dying*, 2003). In this video the artist digitally simulates the death of a dog, an event that—were it filmed in analog reality—might have been greatly criticized by the public, especially since its subject is a dog, an animal considered by most people to deserve a human-like treatment. In the video one can see the slow agony of a hairless dog, very thin and worn out, gradually dying under the relentless technological eye of a camera that, because this is a digital animation, is not there (Caffo and Sonzogni 2015b) (Fig. 14.1).

The origin of Cattelan's horse instead is similar to that of the mummies at the British Museum: it is dead matter, cadaver, corpse. The question

Fig. 14.1 Diego Perrone, *Vicino a Torino muore un cane vecchio (Near Turin an Old Dog Is Dying)*, 2003, still from computer animation (length 5 min and 20 sec). (Courtesy Massimo De Carlo)

then is whether the mummies in the British museum really disturb the spectator in their very essence as corpses, and why then millions of tourists have taken a souvenir photo of those obscene teeth protruding from the skull. Whereas it was previously accepted that a mummy is a displayed exhibit and photographs were allowed, this sensibility has recently changed. The mummies of the Egyptian Museum in Cairo have been placed in a special wing where photography is absolutely forbidden. A museum exhibit has once again become what it has always been: the corpse of an individual who deserves a respectful attitude based on a shared cultural and scientific approach but, also and above all, on ethical considerations. Even for the horse of *Novecento* someday—and perhaps not too far in the future—it will be so. Already today, the room where it is installed is preceded by a plaque that warns the viewers of a possible disturbance of their sensibility.

The use of living animals in Italian art has increased since *Untitled (Donkey)*, by Italian artist Paola Pivi (born 1971). This work, along with performances, a large installation, and sculptures, was part of the Venice Biennale in 2003, where it was installed on the back of a building enlarged to the size of a commercial billboard. Although it should be mentioned that Paola Pivi has recently declared publicly that she is no longer willing to use living animals (Pivi 2017), her work has largely been based on the use and presence of the animal body. Pivi has used animals in her work since the beginning of her career, exploiting their subversive power of displacing the observer who would not know what she/he is looking at, wondering whether it is a dream, an apparition, or a reality. *Untitled (donkey)* is a photographic work that, like a few other works by this artist, was realized by photographing a real donkey on a boat in Alicudi, Sicily, and not by digital retouching or photomontage. In a work from 2015 titled *Yee-haw (horse)*, she has transported purebred living horses up to the Eiffel Tower in Paris and photographed them unleashed while exploring the architectural structure. The result of the performance is a series of photographs with the same title exhibited at Galerie Perrotin the same year. Such works reveal that animals are used as mere components of her installations. They are transported to unusual places in order to stir surprise in the spectator. The horse is not just in a stable or placed in a gallery but brought to the top of a man-made tower, causing a total alteration of what is perceived as "natural."

Another well-known recent case that had significant consequences in Italy is that of the German Pavilion at the 57th Venice Art Biennale in 2017. The celebrated German artist and choreographer Anne Imhof (born 1978) created an installation-performance entitled *Faust*, staging an artistic narration of the capitalized body. Dancers and dogs move in the German Pavilion at Giardini della Biennale di Venezia, temporarily redesigned by installing cold glass walls and floors. *Faust* is a performance lasting over 5 hours, for which the architecture of the pavilion has been heavily modified. The dogs' enclosures and the heavy glass panes that cover even the floor close the classic pronaos of the German Pavilion, where performers are scattered throughout the space, even on the rooftop. They inhabit the space, almost like contemporary zombies moving around and popping up in between the visitors. In this context, the artist stated that the dog can be understood as a domestication of the Pavilion, in the sense that the latter becomes a domestic space. On the other hand, the dog can also express the domestication of itself or can be conceived as a guardian: "The dog in the kennel, the dog and its master, the dog and its companion – these pairings are evidence of how cultural change has altered power relations. They are a symbol of the changing constructions of nature: Where there used to be a dualism between nature and culture, the world now presents itself as a kennel" (Imhof 2017: 5). Promptly removed from the installation following the protests of the Gruppo Vegan Venezia (Venice Vegan Group) and LAV, Lega Anti Vivisezione (Anti-vivisection League) (Anonymous 2017), the dogs, in this case six young Dobermans lent by a breeder, had been locked up/exposed two at a time in a sector of the German Pavilion and, according to the statement presented to the municipal police, subjected to a much too intense stress during the vernissage of the artistic event. In this case, dogs were the unaware co-protagonists of an artistic performance and not subjected to any visible mistreatment. They were left free to express their nature as dogs, even in a confined environment (not very different, in fact, to that of a breeding dog farm). It was perhaps, again, the fact that the animal, when it becomes the material of an artwork, brings up questions and expresses contradictions in the categories of what is moral and what is tolerable.[4]

While some artists explore the living matter of animals as the substance of their artworks, others deal with their presence, counterbalancing their virtual exploitation with an inquiry of their silent role as human companions. The 56th Venice Art Biennale, inaugurated in 2015, presented two pavilions connected to John Berger's essay "Why Look at Animals?" (1980). The encounter with the art of Joan Jonas (born 1936), the artist selected for the US Pavilion, was poetic and exciting. The title *They Come to Us Without a Word* seemed an answer to Berger's question about what we are really seeing when we look at animals. The lyric investigation of Jonas used the many layers of the installation that combined videos, performative props, drawings, lights, music, found objects, and quotes from Berger and from the Icelandic writer Halldór Laxness, conducting a survey of the spirituality of nature, a quest for the animal soul. In each of the four rooms of the American Pavilion, two video projections were installed, one presenting the main theme of the room and the other a hidden secondary narration, a red thread running through the exhibition spaces. Free-standing rippled mirrors were installed alongside Jonas' distinctive drawings and kites, as well as a selection of objects that were used as props in her videos. The installation was conceived as a succession of theatrical stations staging the interaction between man and animal: the beloved artist's dog, who becomes the vehicle for narrating the world, and children interacting with animals, projected onto the screen while they caress them or carefully following the dogs' profile with their fingertips. At times truly moving, Jonas's work marks the importance of the role of artists in changing the perception of things, providing special access beyond habits, in closer contact with emotion. She explains: "Although the idea of my work involves the question of how the world is so rapidly and radically changing, I do not address the subject directly or didactically, rather, the ideas are implied poetically through sound, lighting and the juxtaposition of images of children, animals and landscape" (Jonas 2015).

A few steps away, the Greek Pavilion, titled *Why Look at Animals?*— referring again to Berger's text—presented *Agrimiká*, a project by Maria Papadimitriou (born 1957), whose moods and colors were different from those of Jonas. For the Biennale, the artist transferred a tannery of animal skins, still operating in Volos, a coastal port city in Thessaly, to the Greek Pavilion space with the aim of reconstructing a micro-history in a very difficult time for Greece. The installation contains skins from animals, family photos, magazines, newspapers, and many other objects that were contained in the actual shop when it was dismantled. The Greek word a*grimiká* that the artist used as the name of the reinstalled shop describes a type of liminal animal, living on the city borders but which, despite its long coexistence with humans, has not managed to be domesticated. Symbol of the unknown *par excellence*, this type of animal embodies the "absolute otherness" that becomes a sounding board for our ancestral fears: bats, wolves, and bears are the protagonists of the symbolism of fear—from the fables of Aesop up to Goya's engravings—and beyond. The subtitle of the Greek project refers to the famous question posed in Berger's essay in which the animals, exposed to the gaze of the public in zoos, are actually the "absent referents" (Adams 1990) of these places and disappear behind their real and tangible essence. Reduced to trophies installed on the walls, the animals or their remains populated the space of the pavilion. A plaque at the entrance signaled that all exhibition material came from findings, without any abuse on living animals. And yet the dead bodies of the animals, though not for the exhibition's sake, were there. Like martyred saints, they stood there, beyond time, staring at the spectators from behind their glass pupils. Animals are everywhere, yet highly invisible. The poetic gesture of children who caress the screen on which the faces and bodies of other animals flow in Joan Jonas' work is perhaps their only hope for a different future. Today's children will save the animals of tomorrow. The animals of today, while you are reading, are no longer here.

4

Exhibitions have started to include the sign of animal presence, or to peek at environmental topics, although with a rather moderate approach, often devoid of all the ethical and political stances with which animal studies are imbued. As one can tell from the examples listed in this chapter, Italian artists, museums, and curators are starting to research and make exhibitions revolving around the topic of animals, but it is only abroad that monographic shows have focused on this theme. Exhibitions like *Animal Lovers* at the Neue Gesellschaft für bildende Kunst (nGbK) in Berlin (2016), *Animals/Harmony/Subjugation* at the Museum für Kunst und Gewerbe in Hamburg (2017), and *Becoming Animal* at Den Frie Centre of Contemporary Art in Copenhagen (2018), just to cite a few, prove that an autonomous field of animal-related research in art history has not yet fully developed in Italy. In order to achieve a larger participation of scholars and researchers in this field of study, it would be necessary to open the boundaries of art historical research to other disciplinary areas such as— but not limited to—philosophical ethics, ethology, and sociology. This would lead to in-depth research on the role of animals in contemporary culture and to the production of exhibitions dedicated to this topic. In the last few years, three journals have contributed to launch and animate the animal studies debate in Italy, all of them with a more or less strong accent on animal rights and activism. *Liberazioni. Rivista di critica antispecista* (*Liberations: Journal of Antispeciesist Critique*[5]), published since 2009, has, from its very beginning, kept the promise it makes in the subtitle, with a strong preference for antispeciesist themes and approaches, occasionally focusing on art, especially on cinema and literature. *Animal Studies. Rivista italiana di antispecismo* (*Animal Studies: Italian Antispeciesist Journal*[6]) was founded in 2012 with a similar attitude, focusing on antispeciesism and animal rights but grouping relevant articles under monographic issues, thus providing a resource not only limited to the field of animal studies. Under the direction of Roberto Marchesini, it has shifted to the field of ethology. Finally, *Animot. L'altra filosofia* (*Animot: The Other Philosophy*[7]), published since 2014, has devoted monographic issues to art-related matters such as cinema, architecture, and literature, with a special attention to the visual arts. Each number, in fact, hosts the work of an artist, especially developed for the journal, in collaboration with the curator. Despite their differences, these journals have contributed to the development of the discipline in Italy providing, sometimes for the first time, translations of seminal writings not only related to art but also of philosophical texts that serve as a basis to establish animal studies as an independent field of research and knowledge production.

Besides the aforementioned artistic interventions at the Venice Art Biennale, only recently a few Italian critics and curators have started to shift their attention to the animal presence in the visual arts, mainly in connection with the urgency felt under the pressure of precipitating climate change and the Anthropocene debate. Posthuman studies have also contributed to make the presence of animals in the arts more visible as symbols, but even more as bodies, as former living matter that has become the support for artworks. Posthuman studies in fact, especially through the writings of Donna Haraway, Judith Butler, and Paul B. Preciado, have included the animal body and its relation to the human body in their field of interest. Also queer theory, feminist theory, and postcolonial studies share with animal studies the idea that the subject of the discipline is the oppressed. For instance, pattrice jones (it is mandatory to write the name in lower case) is a feminist and animal activist as well as the founder of an animal sanctuary.[8] In the field of Black Studies, there are intersectional positions such as that of Black Vegan Sistah, combining black identity and veganism issues.[9]

As for the art world engagement with animals, it is worth mentioning the only Italian experimental residency for artists in which this topic is confronted: RAVE, based in Trivignano Udinese (Udine) and initiated by Isabella and Tiziana Pers. Both are artists, and the latter has worked at the intersection of painting and performance. The actions of Tiziana Pers consist in exchanging an animal destined to the slaughterhouse with one of her paintings, in order to rescue the animal while avoiding paying for its life. The absence of money in the transaction allows Pers to operate respecting her antispeciesist beliefs, that is, that animals can only be liberated or rescued and never purchased. The rescued animal, now safe in a stable, is involved in the RAVE artist residency, as the selected artist will be confronted with its presence and its story as a single individual. The project has so far hosted renowned artists, among others, Regina José Galindo and Diego Perrone[10] (Figs. 14.2 and 14.3).

Fig. 14.2 Tiziana Pers, *Art History Vucciria*, 2018. (Photo by Umberto Santoro, color print on Hahnemühle paper. Courtesy aA29 project room)

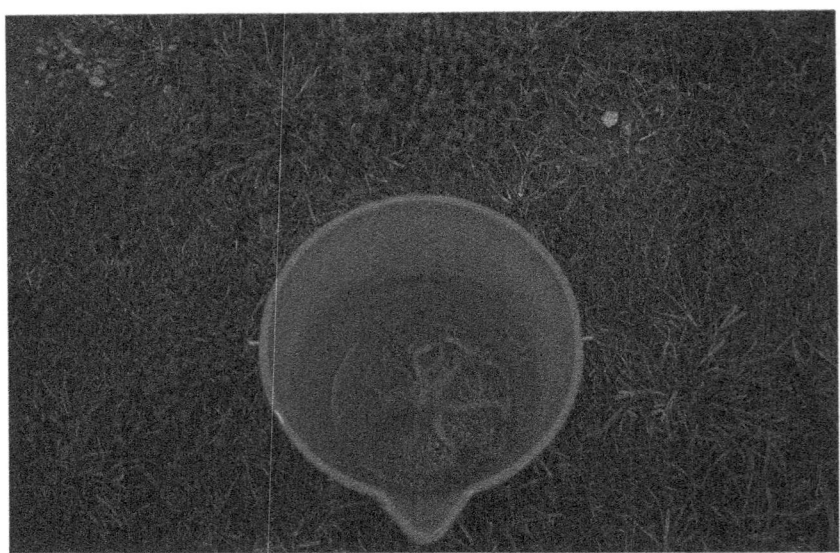

Fig. 14.3 Tiziana Pers, *Art History Vucciria*, 2018. (Photo by Umberto Santoro, color print on Hahnemühle paper. Courtesy aA29 project room)

A separate investigation should be carried out into the international laws that regulate the presence of living animals in museums and galleries, when art works or art performances include living animals. The rationale generally followed is that the animals in art spaces are compared to animals in zoos or used for public events or performances such as films. A veterinarian is consulted to ascertain that the basic welfare conditions of the animal are respected, that is, the temperature of the room, food to eat, and space to move around freely. If this seems acceptable to most people, it should be noted that the guidelines for the welfare of zoos, laboratories, and animal farms never match the real needs of the animals and mostly serve human needs that are satisfied by the use of those animals.

It is more likely that a museum cancels a show because of the pressure of public opinion than because animals in the performance are exploited to the limits of tolerance. In 2017 the Guggenheim Museum had to withdraw three major works from the exhibition *Art and China After 1989: Theater of the World* (Haag 2017). Among these works was *Dogs That Cannot Touch Each Other* (2003) by Sun Yuan and Peng Yu, a seven-minute video where eight American pit bulls on eight treadmills were trying to fight each other but never reached one another because they were strapped to the treadmills. The violence in this video is palpable, and it is certainly not diminished by the fact that it is not a live performance but a video of a past one. This was no less insulting to animal welfare groups that lobbied against the inclusion of this video in the show.

When dealing with the welfare conditions of the animals, professionals have to refer to ICOM (International Council of Museums) Code of Ethics for Museums.[11] Point 2.25 regulates the Welfare of Live Animals as follows: "A museum that maintains living animals should assume full responsibility for their health and well-being. It should prepare and implement a safety code for the protection of its personnel and visitors, as well as of the animals, that has been approved by an expert in the veterinary field. Genetic modification should be clearly identifiable" (ICOM 2006). The Code of Ethics specifically developed for Museums of Natural History (ICOM 2013) mentions the various behaviors to observe in dealing with animal specimen: to obtain them legally, to limit the animal suffering and distress when killing them, and so on. It also states that if possible, alternative methods should be sought. Finally, ICOM's position on animal remains is remarkable: "Animal remains should be displayed with respect and dignity regardless of the species or its origins. It is understood that 'respect' may be interpreted differently depending on the country, institution as well as the lands, cultures or peoples from which the animal material originated. Institutions should develop guidelines appropriate to their own milieu and audience and apply these consistently" (ICOM 2013).

It would be very worthwhile to develop the topic of the rights of the animal bodies in art with a pool of professionals as often done in other fields, seeking veterinarians for definitions, as well as looking into jurisprudence concerning the biological as well as psychological status of the animal (Sobbrio and Pettorali 2018).

5

It is important to remark that art professionals, galleries, fairs, and, above all, museums should develop guidelines for the use of living animals in the context of art pieces or installation. These guidelines should be up to date and consider the dignity and the species-specific characteristic of the animals, in order to break the tradition of the animal as a machine. If museums endorsed this role of "enlightened" legislators, the whole art world would benefit from their expertise.

In conclusion it is worth reporting some interesting positions within the world of art connected to animal studies that may help to shed light on the importance of animal representation and recent presence in art, especially with the advancing of inescapable questions posed by global warming, animal extinction, plastic pollution, and so forth. If the question of what art would be without animals is unimaginable, because animal presence is so entangled in art history to have shaped it in its very essence, it is far more interesting to raise the question of what animals would be without art (Ramos 2016: 12). Could it be, for instance, that the interest of some artists for animal welfare and for ecological consciousness morphs into a form of commitment that supports the animals? Even Peter Singer, whose commitment to animal liberation cannot be denied or minimized, when commenting on the Marco Evaristti's *Helena & El Pescador*, analyzed earlier in this chapter, states: "It's obviously cruel to keep goldfish in a small sterile container like a blender, and it's horrific to think that people might choose to grind them upon a whim. Nevertheless, I can see that the artist could be making a point about our relations with animals" (qtd. in Baker 2013: 13). Is this really so? Would it not be an advancement in the arts if the ten goldfish were replaced by identical automata showing respect for the tiniest form of life? Would that respect not become part of the larger message that the artwork wants to convey, even if the shock produced by the artwork would then be reduced? Whatever the individual position, according to one's ethical set of beliefs, I think that this is an option that should be encouraged by curators and museums within their relationship with the artists.

A 2015 text by Giovanni Aloi has marked a fundamental point in the international debate that, having abandoned the roundtables and few timid groups of animal studies, is entering the art world at large. In "Animal Studies and Art: Elephants in the Room" (Aloi 2015), Aloi addresses the metaphorical "elephants" that prevent animal studies from openly confronting contemporary philosophy and art. According to Aloi, many of the essays on animals in contemporary art do not take into account the ethical and aesthetic complexity of the topic and, I would add, the incompatibility between many important works of art and the rights of animals as stated by modern jurisprudence. Indeed, it would be enough to follow the guidelines of the *Universal Declaration of Animal Rights*, adopted in 1978 in Paris, to reject most of the contemporary visual production involving non-human individuals. The Declaration, in fact, besides not permitting in any form the exploitation of animals for any human recreational activity (Article 10), also prescribes respect for the corpse of the animal, often used in contemporary performances or works as support material as stated in the Article 13: "1. Dead animals shall be treated with respect. 2. Scenes of violence involving animals shall be banned from cinema and television, except for humane education" (1978: 2). The crucial point that Aloi's text brings to our attention, however, is that the thematic complexity raised by the use of the living animal body or its killing requires that the answers from artists, critics, historians, curators, and so on be equally complex: "Over the past few years, I have noticed a tendency from certain animal studies voices to exclusively focus on these negative instances as if animal studies constituted a sort of cultural 'policing body'. I argue that this approach is reductive. Spotting the mistreated animal, or denouncing the alleged misrepresentation of animals in art should be understood as only one of the potentialities of animal studies within the artistic remit" (Aloi 2015). "Don't throw the baby out with the bathwater," in this case, implies that the works of artists like Nitsch or Hirst should not be judged as inappropriate *tout court*, but that the act of "acknowledging" and the act of "praising" should be kept separate, in order to understand the theoretical knowledge that certain artworks can provide on animals.

Notes

1. For the sake of simplicity, I will use here the word "animal" to signify "non-human animal" and the word "human" to signify "human animal," whether man, woman, or transgender.
2. With "antispeciesism" I refer to the philosophical and political attitude—as defined by Peter Singer in his classic *Animal Liberation* (1975)—set against the speciesist ideology, according to which humans are the only holders of moral consideration and, as such, may dispose of non-human animals as they wish. Antispeciesism intends to undermine the arguments that ennoble violence against non-human animals and demands a structural change in the social order in order to recognize that the boundary of species is not a moral border.
3. In *Helena & El Pescador*, ten Moulinex mixers were placed on a table, each filled with water and containing a living goldfish. The mixers were plugged, and by pressing the button, the fish would be blended. Only one person pressed the button and killed the fish, provoking a call to the police that came to unplug the electricity. This artwork caused a trial in which many experts were consulted, leading to the conclusion that the way in which the fish were killed, compared to other standard methods, was one of the more humane methods.
4. For a confrontation one can analyze Pierre Huyghe's dog in *Untilled*, at Documenta 13, Kassel (2006). This dog, with a leg tinted with nontoxic pink paint, does not do anything extraneous to its species-specific tendencies: sometimes it gnaws on a bone or drinks in a pool of water. It does not perform in the classic historical-artistic sense of the term, but performs what the dog has done with humans for about 10,000 years, that is, it performs a relationship of domestication (see Caffo and Sonzogni 2019).
5. www.liberazioni.org
6. www.siua.it/i-nostri-libri/animal-studies-rivista
7. www.animot.it
8. www.vine.bravebirds.org
9. http://sistahvegan.com/sistah-vegan-anthology
10. www.raveresidency.com/it
11. The ICOM Code of Professional Ethics was adopted by the 15th General Assembly of ICOM in Buenos Aires (Argentina), on November 4, 1986. It was amended by the 20th General Assembly in Barcelona (Spain), on July 6, 2001, retitled ICOM Code of Ethics for Museums, and revised by the 21st General Assembly in Seoul (Republic of Korea) on October 8, 2004.

Works Cited

Adams, Carol J. 1990. *The Sexual Politics of Meat: A Feminist-Vegetarian Critical Theory*. New York: Continuum.

Aloi, Giovanni. 2012. *Art and Animals*. London/New York: I.B. Tauris.

———. 2015. Animal Studies and Art: Elephants in the Room, Special editorial, February 2015. http://www.antennae.org.uk/back-issues-2015/4589877799. Accessed on 03 Oct 2019.

Anonymous. 2017. Cani segnalati in uno dei padiglioni della Biennale. Stop all'uso degli animali. *Venezia Today*, May 11. http://www.veneziatoday.it/cronaca/cani-padiglione-biennale-venezia-2017.html. Accessed on 03 Oct 2019.

———. 2019. Mostra di Hermann Nitsch a Mantova, protestano gli animalisti. *Arte Magazine*, January 9 http://artemagazine.it/attualita/item/8390-mostra-di-hermann-nitsch-a-mantova-protestano-gli-animalisti. Accessed on 22 Oct 2019.

Baker, Steve. 2013. *Artist/Animal*. Minneapolis: University of Minnesota Press.

Barbero, Luca Massimo, and Francesca Pola, eds. 2010. *L'attico di Fabio Sargentini, 1966–1978*. Milan: Electa.

Berger, John. 1980. "Why Look at Animals?" In Berger, *About Looking*, 3–28. London: Writers and Readers.

Caffo, Leonardo, and Valentina Sonzogni. 2015a. Food as Social Object: Ontology, Ethics and Art. Paper Presented at the International Conference *Minding Animals 3*, New Delhi, 13–20 January.

———. 2015b. *An Art for the Other. The Animal in Philosophy and Art*. Trans. Sarah De Sanctis. New York: Lantern Books.

———. 2019. Pulcinella. Estetica animale come estetica prima. *Animot. L'altra filosofia* 9: 70–79.

Celant, Germano, ed. 1992. *Kounellis*. Milan: Fabbri.

Charans, Eleonora. 2012. *Gino De Dominicis. 2ª soluzione di immortalità (l'universo è immobile)*. Milan: Scalpendi.

Cimatti, Felice, Stefano Gensini, and Sandra Plastina, eds. 2016. *Bestie, filosofi e altri animali*. Udine: Mimesis.

Dion, Mark. 2000. Some Notes Towards a Manifesto for Artists Working with or About the Living World. In *The Greenhouse Effect (Exhibition Catalog)*, ed. Ralph Rugoff and Lisa G. Corrin, 66. London: Serpentine Gallery.

Haag, Matthew. 2017. Guggenheim, Bowing to Animal-Rights Activists, Pulls Works From Show. *The New York Times*, September 25 https://www.nytimes.com/2017/09/25/arts/design/guggenheim-dog-fighting-exhibit.html. Accessed on 25 Oct 2019.

Haraway, Donna. 2008. *When Species Meet*. Minneapolis: Minnesota University Press.

ICOM Code of Ethics for Museums. 2006. http://archives.icom.museum/ethics.html. Accessed on 07 Nov 2019.

———— Code of Ethics for Natural History Museums. 2013. https://icom.museum/wp-content/uploads/2018/07/nathcode_ethics_en.pdf. Accessed on 07 Nov 2019.

Imhof, Anne. 2017. Press Release for *Faust*, German Pavilion, Biennale di Venezia.

Jonas, Joan. 2015. *United States Pavilion, Venice Biennale*. http://joanjonasvenice2015.com. Accessed on 08 Nov 2019.

Joy, Melanie. 2009. *Why We Love Dogs, Eat Pigs, and Wear Cows: An Introduction to Carnism*. San Francisco: Conari Press.

————. 2019. *The Secret Reason We Eat Meat*. Video Available on https://www.carnism.org. Accessed on 07 Nov 2019.

Pasolini, Pier Paolo. 1999. "Il mongoloide alla Biennale è il prodotto della sottocultura italiana." In Pasolini, *Tutte le opere. Saggi sulla letteratura e sull'arte*, vol. 2, edited by Walter Siti, 2612–2615. Milan: Mondadori.

Pivi, Paola. 2017. Interview with Rossella Farinotti. *Zero Milano*, April 16. https://zero.eu/it/persone/paola-pivi. Accessed on 03 Oct 2019.

Ramos, Filipa. 2016. *Animals*. London: Whitechapel Documents of Contemporary Art/MIT Press.

Singer, Peter. 1975. *Animal Liberation: A New Ethics for our Treatment of Animals*. New York: Harper Collins.

Smith, Roberta. 2015. Review: Art That Snorts, From Jannis Kounellis, at Gavin Brown's Enterprise. *The New York Times*, June 25. https://www.nytimes.com/2015/06/26/arts/design/review-art-that-snorts-from-jannis-kounellis-at-gavin-browns-enterprise.html. Accessed on 03 Oct 2019.

Sobbrio, Paola, and Michela Pettorali. 2018. *Gli animali da produzione alimentare come esseri senzienti. Considerazioni giuridiche e veterinarie*. Vicalvi: Editore Key.

Somers Cocks, Anna. 2006. Wrong Gallery Re-Enacts 1972 Performance which Outraged Italy and the Vatican. A Man with Down's Syndrome Will Contemplate Three Objects During Frieze. *The Art Newspaper*, October 11, 1.

Universal Declaration of Animal Rights. 1978. https://constitutii.files.wordpress.com/2016/06/file-id-607.pdf. Accessed on 07 Nov 2019.

CHAPTER 15

Animality Now

Leonardo Caffo

It is only for the sake of those without hope that hope is given to us
(Benjamin 1996: 356).

The main purpose of this chapter, written in a simple and literary style, is
to discuss a problem of enormous philosophical relevance: the "animal
issue." There are several ways to do this and, while not claiming that this
is the best one, I have chosen to analyze the matter starting from the tradition inaugurated by Jacques Derrida, who focused on animality as both a
theoretical and a material entity. What is animality? What does it mean to
be animals rather than humans? And what have we done to animals to be
able to think of ourselves as different from them?

These are the questions I will try to answer in the pages you are about
to read. But there is more: what you are about to discover is what I have
found out after learning to see things I could not see before. Meeting the
gaze of an animal for the first time—not looking but being looked at—is
the source of a disarming awareness: we are killers. Even now, right here,
right at this moment. What we have done to animals is not only very serious but will have eternal consequences. We create life to destroy it.

L. Caffo (✉)
Polytechnic University of Turin, Torino, Italy

© The Author(s) 2020
F. Cimatti, C. Salzani (eds.), *Animality in Contemporary Italian
Philosophy*, The Palgrave Macmillan Animal Ethics Series,
https://doi.org/10.1007/978-3-030-47507-9_15

Furthermore, what we do to non-human animals is the ultimate horror—the same that occurs when we discriminate against some human animals. When we talk about genocides, saying for example that Jews were killed "like rats" and Tutsis were massacred "like cockroaches," we are forgetting something crucial: we are comparing isolated historical events in the human sphere to the everyday fate of millions who die day after day, forever, murdered without a second thought.

Thinking about animality means thinking about the humanity we want to become: what we have been, what we are, and, of course, what we could be. A humanity that lives by crushing the corpses of history, like Paul Klee's *Angelus Novus*, is a false humanity: it is bestiality that tries to be different from animality—which is precisely what we should go back to in order to save the world and ourselves from the unbearable lightness of the animal slaughterhouse.

> I am leaving you a place in my home, but do not forget that this is my home. (Derrida in Borradori 2003: 127)

1   PERSPECTIVES

Animality is what we are: it is what is left after deconstruction. That is, after the deconstruction of a social concept: that of "human" understood not as an animal body, but as a specialized life, which is the recent invention mentioned by Michel Foucault (2002). It does not matter whether Foucault was right, as I believe, to argue that the human being coincides with self-thinking. Rather, what we should reflect on, resuming the tradition inaugurated by Derrida's *The Animal That Therefore I Am* (2008), is what it means to be an animal.

First of all, being an animal means being mortal, but above all being one of the infinite mortals of this world. What I will claim in these pages is that our essence, firstly of *quodlibet*—that is, of whatever singularity as understood by Giorgio Agamben (1993)—and secondly of human community, makes our gaze on the world *one of the possible perspectives*, not the only one. The question of the point of view, despite seeming trivial, is central to Western philosophy (and, albeit to a lesser extent, Eastern philosophy), running through it as an impassable limit: the world we look at, which we claim to understand and analyze partitioning it through our ontology, is the world as seen by a featherless biped.

I am a realist—or, as we say today, a "new realist"—and I firmly believe that reality resists our hermeneutical processes, transcending them by preceding them, acting as an immovable substrate by which to verify our statements. But it is precisely this position, the conviction that there is a reality that prescinds from my gaze, that forces me to think about what it really means to observe the world from a body made in a way and not another. In a certain sense, paradoxically, it is hermeneutics itself that calls for realism.

2   DICHOTOMIES

The history of the "man/animal" dichotomy is long and unfortunate. Thinking about Descartes or Heidegger, it is hard not to smile at their convictions: animals are automatons, animals do not die, animals do not speak, they do not suffer, they do not think or self-think, and so on. What these beliefs arouse is a sad smile, of course, but a smile nonetheless. For a long time, it was believed that the human being was able to observe reality, of which he is part, as if it were his creation. In this light, in which we often played the role of demiurge, we have understood animals as a single homogeneous category: this was the first of many discriminations against animals, one that is not only linguistic but ontological. For this reason, Derrida (2008) and many other thinkers like Carol Adams (1990) have suggested we stop using the word "animal," which is way too narrow to describe all animal living beings that are not human.

And yet, this would not suffice. The human-animal dichotomy, in addition to being the sword of Damocles for a violated, massacred, and humiliated animality, originated an arrogant and deceitful philosophy: the belief that everything is given to and for the gaze of man. Jakob von Uexküll (2010), investigating the concept of environment and animal gaze, showed how the same space—imagine a room in a university—is completely different in the eyes of a bee, a human, and a dog (which are only a few among the infinite possible examples). If the three individuals were to talk to each other, describing what they were seeing, they would agree on nothing: the room would have different colors for each of them, different shapes, uneven proportions, and opposite architectural forms. Even the question who is right would have no meaning. The fact that the room can be seen in different ways, in fact, means that the room has fixed and real characteristics, but that every conceptual and visual apparatus, belonging to the given subject and species, locates different visual information and in a different way: it is the natural specialization of every life form.

3   ROOMS

Philosophy is also a way to see a room: it is our way, and only ours, to describe the states of things in physical, social, and abstract reality. For this reason, in my opinion, discussing animality means, first and foremost, discussing the limit of philosophy. This limit, by contrast, comes out more precisely in our desire to understand what other animals are, what they are like, and why they are what they are. Animal cognition, which is a wonderfully Cartesian answer to Descartes, has tried to find talking apes (Koko, Kanzi, Washoe, and others); to understand the bees' "dancing" language; to analyze the syntax used by cotton-top tamarins; to understand the mammals' theory of mind; to prove, when possible, that primates have a higher-order one like ours (they think about the past); and so on.

All these attempts implicitly show that we do not want to understand non-human animals: we only want to prove that some of them are like us. We establish the categories that we have and see if the Other has them too. This kind of reasoning, rightly called "identity thinking," helps strengthen the abovementioned philosophical limit: we keep looking at reality as if everything, including animals, were a term of comparison aimed at reinforcing the idea that humans are perfect. Neither Copernicus nor Kant has achieved a true Copernican revolution, by which man would no longer be the center of the universe: things only seemed to change, but they have actually stayed the same. Animal studies have only partly managed to call anthropocentrism into question, by stating that other animal life forms must be studied as such, without trying to force them to enter a competition that they are bound to lose.

4   LIMITS

But let's start from the limit. This investigation of animality is, first of all, an analysis of how we have denied one of our necessary properties (therefore not one we may or may not have): namely, "animal being." This denial has been described, first by Richard Ryder (2010) and then by Peter Singer (2009), as the prejudicial and ideological attitude that falls under the name of "speciesism." Therefore, I will now make a few considerations on the philosophy of speciesism.

As Levinas showed in relation to Nazism, thinking about brutal and violent rejections means thinking about being, or rather about the experience of being at its purest, which manifests itself in a way that is "desert-like [...], obsessive, horrible" (Levinas 1984: 11). Like Hitler's philosophy, speciesism is elementary, but its power has shaped humanity as we know it, reinforcing the bias of the human outlook on reality and awakening the primitive feelings of specialized life. This rather coarse perspective, which appears to be but a violent drive that should be combated with antispeciesism, is actually philosophically interesting. Indeed, speciesist philosophy goes well beyond speciesism itself, because it lays the foundations of society. When Horkheimer describes the social skyscraper in his notes (1978: 66), he identifies the mass slaughter of animals as the ground on which rests the building of human "civilization," based on different kinds of suffering which we tolerate so far as we are not the ones undergoing it. However, to overcome the man/animal dichotomy, we mustn't make the mistake that even Jacques Derrida committed,[1] namely to remain stuck within a completely human perspective. Let me explain.

5   A Cat

The famous encounter between Derrida and his cat,[2] described in *The Animal That Therefore I Am*, marks the moment when philosophy recognized the animal as a unique, unrepeatable individual that not only can be seen, but can *see* you as such. This might seem like a simple pun, but it signals an important passage: the animal is no longer simply the object but also the *subject* of thought. The completely anthropocentric point of view, the one that characterizes philosophy as biased and limiting, is slowly eroded by this unique and unrepeatable cat, who observes the philosopher in his nakedness. The animal, now, can cause shame—namely the principle of every philosophy which, as Socrates says in the Platonic dialogue *Gorgias* (469a–479e), is the true beginning of the journey toward thought. However, as Žižek points out, Derrida remains profoundly human in this way of being watched, thereby preserving the anthropocentric limit. This human barrier in Derrida's discourse on animals provided a starting point for another philosopher, Matthew Calarco (2008), who sought to understand how to change our thinking taking cue from that unique and unrepeatable gaze.

As I have tried to say in *Flatus Vocis. Breve invito all'agire animale* (2012), it is necessary to abandon—even only for a moment—everything that makes us human to really fall into the category of animality. How can this be done? This is the question: the *caesura* that separates us from everything else has deep cultural roots. So, there is a phenomenological problem in animality: to put it with Husserl, in fact, the latter manifests itself not as "thing in itself," but as something that manifests itself to me. My cognitive structures and my experience shape what I see as something other: a monadic otherness with which "I"—the observing self—have nothing to do. Here it is worth mentioning the analogy between animal action and the question, often stigmatized as "exoteric," of Japanese Zen philosophy.

Despite being hard to define, Zen certainly has one essential tenet: *one must become the thing one does, so as no longer do it but be it*. Think of the famous archery example made by Eugen Herrigel (1999): as long as you try to shoot the arrow while thinking about the target outside of you, you will make mistakes. But if the arrow is perceived as an extension of your arm, you will hit the target almost automatically: your hand will do the job. In this sense, I argue, the question of becoming, or as I prefer to say, of animal action, is a matter of Zen; we must stop thinking of the animal as something else, be it *monstrous* (Derrida) or *indistinct* (Calarco), and understand that every singular animality is an extension of us, as we are an extension of it. Man fills the world with distinctions with his natural propensity for ontological taxonomy, but it is only a survival technique that must be restrained.

6   ZEN

The Zen issue calls for some considerations in relation to the animal. The first is about rationality. Animal ethics, an expression of the analytic philosophy of the 1970s, was considered a practice of *rational* philosophy: a long and winding road between arguments, corollaries, and thought experiments. In the English-speaking analytic tradition—which, as often happens, detached itself from the real world to describe, and then prescribe, Neverland—the animal is always thought of as something other than us, only finding space in philosophical thinking as an exception. However, in order to shift toward animality as a more complex matter, reason is not the right tool at all, precisely because it is a typically human practice used to draw lines and borders between humanity and other life forms.

For this reason, in analogy with Zen philosophy—as described in Bodhidharma's four verses (cf. 1989)—animality can only be truly understood through seemingly empty precepts. And yet the power of immedesimation lies precisely in this emptiness:

1. A special transmission outside the scriptures [writing is the human practice *par excellence*, with which we "mark our territory" and create social reality];
2. Not founded upon words and letters [ever since Descartes, "language" has been the pretext to break away from the animal];
3. By pointing directly to [one's] mind;
4. It lets one see into [one's own true] nature and [thus] attain Buddhahood [Buddhahood, in our case, is animality].

Each of these verses clashes with the dogma of Western philosophy: prove without proof, and you will be disproven without proof (Euclid). But in this sense, as will be clear later on, the point is not to prove that animality is Zen. Rather, it is the principle of all things that must be rejected: herein lies the deconstructive essence of antispeciesism in its investigation of Western metaphysics. As the bow and arrow become the archer's limbs, so the animal other becomes the heart of man. Eating hearts in the boat of Siddhartha's ferryman (cf. Hesse 2012: 83–94) is no longer possible.

7   Life

The second consideration called for by the issue of Zen is the more general one of "life." This line of thought, in my view, corresponds to the Western tension between *bios* and *zoé*. In fact, in a cultural species like *Homo sapiens*, there are two different "forms of life." On the one hand, bodily life in the strict sense, related to matter; on the other, the citizen's specialized life within the constellation of social reality. According to the legend, Buddha, only anchored to the world by his body, was able to spend his days fasting in the shade of a cherry tree. These descriptions, certainly metaphorical, depict the elimination of *bios* in favor of *zoé*: an attempt, in my opinion, to come to terms with one's animality, often violently hidden by the universe of social facts.

Reasoning on the primary essence of the human means questioning the very foundations of existence: something that is far from idle and has an effective political impact, as it means to understand that the way we have interpreted our life form on this planet, as Ludwig Wittgenstein would say, is radically incorrect. In all likelihood (regardless of the data provided by animal cognition), we are the only species that thinks about its own life, makes long-term plans, and, if necessary, radically modifies the ways of existence that characterize a certain epoch and a given portion of the world. This "possibility of change," which is intertwined with free will, is what allows for a discussion on ethics and, in our specific case, on the architecture of animality. The feeling that leads to the inadequacy of living derives from this possibility and is profoundly linked to the concept of *metamorphosis.*

Being born in one of the many existing human societies, which are now all rather similar because of the complex effects of globalization, one has the sensation that artifice has crossed the thin borderline that indissolubly binds the natural to the cultural. The imaginative and naive interpretations of Heidegger's discourses against technology (cf. 1977) have often led to a general rejection of something that lies at the basis of our societies: all this, though, is falsified by human nature itself which, deprived of any animal characteristic, has found its complete realization in the ability to "light a fire" (a famous metaphor of the myth of Prometheus). Heidegger's discourse, rather, aims to analyze the consequences of technology: what happens when the cultural artifice becomes a caricature and leads to the structural alteration of what humanity needs to survive. To understand how, and to what extent, this line has been crossed, it is not enough to sit and observe the roots of the problem from the inside: one must find the best possible position so as to begin to understand.

8   FOR EXAMPLE, ONE MIGHT GO STAND BY A FROZEN LAKE …

8.1   *Lakes*

When transcendentalist philosopher Henry David Thoreau stayed from July 4, 1845, to September 6, 1847, alone on the shores of Lake Walden, near Concord, Massachusetts, he had exactly this intention: to reckon with (his) animality. The reason for this is explained by Thoreau himself in his travel diary: "I went to the woods because I wished to live deliberately, to front only the essential facts of life, and see if I could not learn what it had to teach, and not, when I came to die, discover that I had not lived"

(Thoreau 1906a: 100). Therefore, what drove the philosopher to live in his house by the lake was this idea: the essential facts, the traces left once social reality is completely deconstructed, are what allow us to truly see the human—that is, the human being as an animal, as *bare life*.

And this point of no return is what triggers a *primal metamorphosis*, which is far from metaphorical. Now we only see buildings and streets; in the windows of the offices where we daily store our lives, wrapped in a cling film that envelops time and emotions, we rarely see our reflected image: we see ourselves grow old day after day, in a total alienation from the world that has welcomed us. The gesture of rupture that led Thoreau to solitude allowed him to understand the limits of life as we have conceived it: shaped by a paradigm of production that has now taken the form of post-capitalism, life is the pursuit of a job that, at most, will only produce the chimera of a pension for us to look after our mortal remains and postpone the inevitable end.

In rural landscapes—which still made up Walden's background in Thoreau's time—other forms of life peep out, as transition stages of a successful transformation. These other forms are non-human animals, who exist in endless symbiosis with Nature. The caesura that keeps us on this side of the world, isolated from all other forms of life, is evident for Thoreau: "I wish to speak a word for Nature, for absolute freedom and wildness, as contrasted with a freedom and culture merely civil—to regard man as an inhabitant, or a part and parcel of Nature, rather than a member of society" (Thoreau 1906b: 205). One might think that there is a certain naivete in opposing the "citizen" (the human in society) and the "animal" (the human in nature): this naivete, indeed, has partly guided anarcho-primitivism, which hopes for something absurd, and is satisfied with it. However, as should be evident from the philosopher's words, there is an important point that has been completely ignored: while it is necessary, indispensable, and real that man be an inhabitant of Nature, because he is part of it and exists thanks to it, it is contingent and conventional that man be the inhabitant of society, or a citizen. In technical terms, the citizen is the social object rooted in the physical object, namely the human. The ultimate meaning of this, also in light of the shift of contemporary philosophical thought toward realism, is that this artifice should be seen as such.

Thoreau mainly deconstructed social reality in favor of a metaphysical reality: his goal was to highlight the foundations of our existence, unmasking its deceptions. Indeed, there is a thin red line that connects (the later) Derrida's philosophical work to Thoreau's: the cat who saw the naked philosopher (as mentioned) and the animals at the lake who led Thoreau to reconsider our social role are material metaphors of the fact that the citizen can be human again thanks to the animality of the other. But why should one want this to happen? Why claim animality now, right now?

9   POSSIBILITIES

Animality is necessary because of the proportions of institutionalized violence. Nobody can define with certainty what violence is, as the term is as vague as "baldness" and implies many intermediate concepts—such as self-defense, in which it is difficult to decide what is violence and what is not. But there is an extreme, and it is unequivocal: the immense sacrifice of the many, wanted by a few who choose to trample on countless corpses to ensure progress at the expense of the murdered victims left behind by Benjamin's *Angelus Novus* (2003: 392). All this pain, this cowardice that we call "Western welfare," becomes unsustainable in front of its alternative: the symbiosis with the world that animals teach us, the ability to experience the openness that Rilke talks about (2005: 55), the overturning of Heidegger's animal world-poverty (cf. Heidegger 1995: 176) showing how there is actually no gap between animal and world because the animal *is* the world, while we are its ruin as well as our own.

So what if the ability to become other than the world, to break away from it, was anything but an advantage for *Homo sapiens*? What if the complexity of our brain could not cope with the complexity of the world we live in? Well, there is no need to turn a possibility into a catastrophe: there is no point in changing course when the ship is about to sink. This feeling of end of the world—often hoped for and falsely predicted, due to cyclical economic, ecological, and obvious moral crises in a world where the "most civilized country" is one that still implements death penalty in many of its states—is a false sensation. The end of the world, as we conceive it, is the end of one world and the beginning of another.

There have always been various patterns of existence, ignored by the conveniences of contingency. From that lake, so cold and pure, Thoreau found a first space of action in civil disobedience: if this is to be more than abstract talk, civil disobedience is a necessary practice to demonstrate disagreement with a governmental system by which we do not feel represented but also with a state of affairs that traps our lives and from which we want to escape. The metamorphosis lies not so much in wanting to save the animal because it resembles us or because we share the same destiny, but in becoming animals ourselves: in understanding that only by deconstructing the citizen's false needs can we finally prevent violence from perpetuating itself. The various crises will thus no longer appear negative but reveal themselves as opportunities for a necessary change.

10 Animal Attempts

There is an almost magical passage in *Walden* where Thoreau, even just for a moment, succeeds in becoming other than the city by which he was shaped, seeing himself as completely animal:

> This is a delicious evening, when the whole body is one sense, and imbibes delight through every pore. I go and come with a strange liberty in Nature, a part of herself. As I walk along the stony shore of the pond in my shirt-sleeves, though it is cool as well as cloudy and windy, and I see nothing special to attract me, all the elements are unusually congenial to me. [...] Sympathy with the fluttering alder and poplar leaves almost takes away my breath; yet, like the lake, my serenity is rippled but not ruffled. (Thoreau 1906a: 143)

For a moment, in these beautiful words, the desire for animality becomes concrete, transcending itself, becoming a desire for the world: "Thoreau becomes a lake, becomes an alder and poplar ... and walks serenely". Consider the American transcendentalist's attempt to become an animal: what would such an attempt look like today? Could we understand it? Would we endorse it? There once was another Thoreau, living in our time, in a different hut: Theodore John Kaczynski, a solitary mathematician who retired to Lincoln, Montana, where he led a simple life, with little money, no electricity or running water. For everyone, today, Kaczynski is a terrorist: he is known for killing some subjects that he deemed dangerous for the life of our species in this world. In that hut, isolated from a world that was haunting him, in front of his fireplace, Ted wrote *Industrial Society and Its Future* (1995)—a strong invitation to become animals, weak only in the ways that a young man, after his doctorate in mathematics, adopted to implement it.

Ted's lake was not Walden, but it is clear that he shared the same outlook as Thoreau's: that of a desperate Nature bent by the weight of development disguised as progress. In this solitary journey toward his metamorphosis, Ted understood the essence of our time and summed it up in a few lines:

> The moral code of our society is so demanding that no one can think, feel and act in a completely moral way. [...] Some people are so highly socialized that the attempt to think, feel and act morally imposes a severe burden on them. In order to avoid feelings of guilt, they continually have to deceive themselves about their own motives and find moral explanations for feelings and actions that in reality have a nonmoral origin. We use the term "oversocialized" to describe such people. (Kaczynski 1995: 25)

According to Ted, therefore, morality has become so demanding that it has collapsed, turning into moralism: a pure act of self-deception by which even what is immoral is to be understood as ethically correct—for our society, falsification is a point of strength. Ted is now rotting in jail, while US soldiers slaughtering the Fertile Crescent are glorified as heroes. A project of deconstruction such as the one proposed here must take this falsification into account. Reality exists, resists us, and is independent: unreality, or rather social reality, the product of conventions, written acts and functions, must be changed.

Gilles Deleuze's animal gaze has to question this threshold of falsehood, this myth of a one-dimensional man: the animal gaze modifies by observing. There is little use for a construct that idolizes the white, monogamous working male against all the infinite diversity of life. Once again, from that lake in Massachusetts, during a winter sunset, we can see reality regardless of any hermeneutics:

> But possibly the day will come when [the land] will be partitioned off into so-called pleasure-grounds, in which a few will take a narrow and exclusive pleasure only,—when fences shall be multiplied, and man-traps and other engines invented to confine men to the *public* road, and walking over the surface of God's earth shall be construed to mean trespassing on some gentleman's grounds. (Thoreau 1906b: 216)

Because this is the natural consequence of this pain—the privatization of the public, the appropriation of the lives of others, the end of fundamental freedoms. Even from an architectural standpoint, in fact, power shows itself in the continuous gesture of fencing, of discriminating between outside and inside, privileging what pleases us and resembles us. The same goes for the fence we called the Berlin Wall, and all the barriers standing between Israel and the West Bank, those around the European Union or on the border between Mexico and the United States. That day that seemed so far away to Thoreau has now arrived: now only a few have heard of the sun shining through the fence. For this reason the enclosures, which are the material practice of falsification, must be deconstructed: because their forms have nothing in common with ours.

11   (Bio)powers

"Everyone hates the power he is subjected to," said Pier Paolo Pasolini (Bachmann 1975–1976: 41) and, in our time, the power that we must strongly hate is "biopower": the kind that expresses itself in controlling the bodies and forgetting animality and nature. To really hate power, to subvert every form of hierarchy of life and death, today we must be antispeciesists. In *Salò*, through the emblematic act of "eating shit," Pasolini described the vital path of the citizen suspended in the limbo of a power that homogenizes diversity, stimulates false needs, raises consumerism to a new value, and, last but not least, massacres anyone who is foreign to this system: whoever does not bend is broken. We still exist, but only "biologically." Culturally, we are "corpses."

12   Ante-inferno

Or something more than that. It takes luck to be able to see the mortification of existence, from the fancy salons of life. Showing solidarity with animals, sharing their fate of marginalization and humiliation, is certainly not difficult: it is almost obvious that slaves, transsexuals, and women—among others—should be, by nature, more sensitive to the animal issue. On the other hand, the disinterestedness of the rich and healthy white man, gloating in the Freudian logocentrism of well-cooked meat, of the female object and of the world that magically opens up at his moccasins is depressing. Considering animality, the ultimate foundation of human life, means using a cultural "level."

Hence the decisively democratic value of animal philosophy: it is regulated by the obvious and yet questioned principle that every animal life has the same moral value, namely none. Indeed, the attribution of values and ethical judgment is also a product of *homo*'s upright position—a position that has become a throne from which to command all that is other. Animality is circular, however, in an oddly selective way: it establishes a relationship between being and imagining whereby being a beast establishes bestiality and this, in turn, defines the being of the beast. The history of a term, says Agamben, often coincides with the history of its translations or its use in translation (2013: 3), and it is perhaps no coincidence that the term "animal" is used, in all languages, in a way that ontologically compresses the infinite variability of life into a single and mortifying Aristotelian word.[3]

13   INFERNO

But as much as one can believe that the *Ante-inferno*, or our fate, is similar to that of animals, who actually live in hell, reality will always prove one wrong. Looking into the eyes of the other is not the first way to reunite with it. The first way is to look at the other in its entirety, to understand what we have done to it—and to recognize what we have done to ourselves with that human violence. Because violence as such is human, and the syntagmatic expression "inhumane violence" is actually an oxymoron. What is inhumane, namely improper of the human, cannot be violence. Because violence itself, as a brutal concept, develops within the confines of our theories. By exercising violence, by exercising this theoretical concept on the territory of intentional actions, we have transformed the concept itself into a goal. Violence for violence. This intentional action has produced the tragic event of the ontological fracture between us and nature, between us and the different, between us and the animal—between us and the other. The very act of classifying is violent. Because the one who classifies is always someone, and this someone is always human, who sets himself as the best on the list, the archetype of the indestructible subject, the most evolved of the living, the most intelligent of animals, the worthiest of entities.

And why? Because he is the one who classifies: an abyssal tautology by which only those who define themselves as such are worthy. But in these continuous violent acts, some of the "worthy" individuals have stayed on the edges of the classifications—an error of assessment of the capitalist system. Through these edges—these few ontological margins left unclaimed—some humans have remembered that they are, above all and mostly, "such stuff as dreams are made on." And this awareness led them to lean out and look at what we left behind the barricade in which they told us, poetically, that it is "love that moves the sun and the other stars" (Alighieri 2005: 91). Beyond the barricade was hell, because hell is the only way we can describe what we have done through the use of violence. Beyond the polished borders of the human, some individuals of the monstrous species have observed the other, or at least what was left after the classification that we often hear called "evolutionary ladder." That terrible moment—that disarming awareness—was like a violent shock that prompted a caress. The name of that caress is *antispeciesism*.

14   DESCRIPTIONS OF HELL

Hell appeared to be indescribable. The bare suffering of non-human animals was a shapeless mixture of screams, blood, and sweat symbolically compressed in the generic and aberrant word "animal." And so, first, those individuals left on the edges had to break down the word "animal" and, unmasking it as a horrible way to signify an absent referent—beasts and monsters—they found the infinite variety of what is different from us: the *Animot*, "term coined by Derrida to speak about animals without establishing, right in the very act of naming them, a perspective of complete oppression" (Filippi 2011: 87). In this discovery lies the beginning of what we can define the *end of history*. The very moment when man, albeit minimally and through a few representatives, realizes on *what*—at the expense of *whom*—he has built his path as an entity becoming in history, he chooses to stop the flow of events. Nothing can really continue in the midst of the suffering that has made the world cry.

The history of these *conscious* humans ends because they become animals again and, as Kojève notes (cf. 1980), love, art, and games can also become natural and therefore lose the anguish of a story, the anxiety of narrating to improve, building nests like sparrows, and being satisfied with this. However, for those humans—who are now only *individuals* who return to *bare life*—this awareness is the beginning of a battle aimed at bringing everyone closer to the edge where they built their nests. This way by which man gets closer to the other is a paradoxical dance. It forces one to stick within the edges in order to unfold them; it makes one delete the boundaries by closing one's eyes in front of the other, whispering softly and with shame: "wait, I'll come back here for you, and there will be more of us."

But in that relentless homecoming, the face of the one awaiting has already changed, leaving the place to a *nameless* other, in that timeless hell that is the slaughterhouse. And the question becomes inevitable: what is a homecoming? What if there is no place for me to go back to? "What remains in suspense, what dangles in thought?" (Agamben 2006: 107). Yet we must return, if only to make history really stop, beyond the edges where those individuals live. And in this continuous *coming and going* the anguish unfolds inexorably inside our mortal bodies. The system we are fighting against is not part of history: it is history itself. We are in it, like everyone. We are in it even if it has stopped at the edges where we live. And so we do not try to understand this choice, but we challenge it to safeguard the fictitious spirit of this world.

Why animals? Why animals and not plants? Plants are also living beings, aren't they? And the anguish becomes even heavier, like a burden that conditions every breath of this windowless existence. Of course plants, as Aristotle observed, hold in themselves such a power that they appear to us as living. But certainly their living is different from animal living. Aristotle himself, in fact, highlighted that they have no *power of the soul* other than the nourishing one (*De anima* II, 4, 415a23–25). In their being lacking and repeatable lies the inexorable fortune of immortality, of not being able to know what dying is. And if it is therefore true that we share death with plants, it is equally true that we do not share *dying* with them.

But the same unchanging answer to these provocations of history pushes us further from the edges. So we look for different strategies, for different communications. And in the meantime the edge shrinks, increasing our anguish. Now that the empathy with the *other* has become complete, we die with them while remaining alive, we die *for them*. Because dying, for us who are on the edges, has already begun, even if death is still far away. And so our *coming and going* has become spasmodic, continuous. We must act in compromise while aiming for the lucid clarity of the goal. We must do it because otherwise history will continue, and suffering will accompany every step of its path.

15   Malaise

This malaise, which is called anguish, oppresses thought, but also makes it *powerful*—capable of going beyond the boundaries of this reality: exploring, one by one, the territories of the possible worlds in which history has not taken this terrible shape. Some individuals, among what remains beyond the edge, accompany us on these journeys with their mute gaze, which answers every question in silence. These individuals—subject to the relative and absolute evil that is *domestication*—make possible the otherwise impossible encounter with the creatures that sleep latent and nameless in the lagers of suspended time.

Through the help of those I gave a name to, I was able to get close to the border between the many worlds that make up the universe of thought.

This was possible thanks to their names, which will survive even when they leave this world and which will be my Charon's ship. Thanks to the malaise of this condition of life I was able to investigate the possible worlds through universal accessibility. I went, and they appeared. In one of these worlds, light years away and lost in the labyrinths of the mind, the end of history lies inert. As in Ezekiel's vision (Ezekiel 1, 4–28)—for example, in the miniature contained in a thirteenth-century Hebrew Bible, now in the Ambrosian Library at Milan (cf. Agamben 2004: 1–2)—the metamorphosis that makes us all identical in deferral will be accomplished. Sitting all at the same table, we will no longer need to speak, the speed of darkness will have become inexorable—and in silence we will be able to hear the nature of this world. Unlike what Heidegger posited, what lives, *animality*, will become the most intrinsic essence of *Being*: "the easiest to think." Dasein will be animal, and the animal will be Dasein.

16   THE END

We who live on the edges will never see this end, because even in the face of history we are unnamed. And then in those homecomings we will no longer find those for whom we have gone, and they will no longer find those for whom they have remained. But this alternation brings out an authentic feeling called *education*. If our living on the edges is not to be in vain, we have to leave traces. And those who read these traces will get closer to the end of history. They will experience the end and climb over these miserable edges. They will walk together, with the other, on the infinite paths of existence: in a word, they will *live* again. This is not a utopian path, nor is it the highway to goodness. In this hell we cannot know what the good is, because evil has covered every manifestation of it.

Trying to describe goodness with logic is a vain and useless attempt. Goodness is connected to history, because history has made it *other than itself*. Then it is against evil that we must fight, against the history that has brought this evil to light. You ask me what goodness is and I tell you that you will have to eliminate evil to find an answer. Antispeciesism is a movement that stands suspended on the flow of events. By delivering these events—these already done deeds—from evil we will be able to look the other in the eye, and only in their eyes will we be able to see goodness again. And this goodness is called *end*. Rationality, as we have already said, is the border of this discourse. Without it we would never have lived, killed by *the brutal horror* of the manifestation of nature. And yet, having survived this horror, we have decided, by the hand of rationality, to smash everything that might have frightened us, but which had not yet done so.

This precautionary mechanism called reason limits us while making us magnificent. We can certainly describe, *with reason*, the way to reach the other. Peter Singer and Tom Regan have inhabited this edge with us and tried to describe how to change through reason. But this description seeks to integrate what is other only because perhaps it is not completely such. However, in this mechanism for safeguarding animal rights, some gear jammed. Welcoming the other because the other is like you means once again perpetrating the fracture.

But a new generation can be seen on the horizon. A generation whose voice will now be animal, and with which the place of negativity will finally become a memory of a distant, lost philosophy. And with the definitive death of the human voice, even philosophy, which is Oedipus's soliloquy, must come to an end.

> The animal world is made up of silences and leaps. I like to see them lying down to rest, when they regain contact with Nature, receiving nourishing lifeblood in return for their abandonment. Their rest is as accurate as our work. Their sleep is as trusting as our first love. (Grenier 2004: 224)

Notes

1. As shown by Slavoj Žižek (2012: 408–416)
2. Derrida (2008: 4ff.) talks about an encounter with his cat when the latter saw him naked, making him feel ashamed (which is no coincidence: shame, ever since Plato, has been the principal sentiment of early philosophy). In that encounter, Derrida looked at the cat and the cat looked back at Derrida, in an emblematic passage from object of the gaze to subject of the gaze. Moreover, in this encounter, we shift from the idea of "a cat" to the disarming awareness that the animal is a singular unrepeatable uniqueness, which cannot be reduced to the aberrant "nameless" term "a cat." It is in front of the animal, of *that single* animal—in an apparently Sartrean but actually Heideggerian anecdote—that Derrida discovered the discomfort, the shame, the embarrassment of feeling naked and being mirrored and observed by the nakedness of the other, in a nakedness that we mistakenly believe to only belong to animals.
3. In this regard, see Derrida's analysis (2008).

Works Cited

Adams, Carol. 1990. *The Sexual Politics of Meat: A Feminist-Vegetarian Critical Theory*. London: Bloomsbury.

Agamben, Giorgio. 1993. *The Coming Community*. Trans. Michael Hardt. Minneapolis:University of Minnesota Press.

———. 2004. *The Open: Man and Animal*. Trans. Kevin Attell. Stanford: Stanford University Press.

———. 2006. *Language and Death: The Place of Negativity*. Trans. Karen E. Pinkus with Michael Hardt. Minneapolis: University of Minnesota Press.

———. 2013. *Opus Dei: An Archaeology of Duty*. Trans.Adam Kotsko. Stanford: Stanford University Press.

Alighieri, Dante. 2005. *Dante's Paradiso*. Trans. Charles Eliot Norton. Stilwell: Digireads.

Aristotle. 1987. *De Anima (On the Soul)*. Trans. Hugh Lawson-Tancred. London: Penguin Classics.

Bachmann, Gideon. 1975–1976. Pasolini on de Sade: An Interview During the Filming of *The 120 Days of Sodom*. *Film Quarterly* 29(2): 39–45.

Benjamin, Walter. 1996. Goethe's Elective Affinities. Trans. Stanley Corngold. In *Selected Writings, Volume 1 (1913–1926)*, ed. Michael W. Jennings et al., 297–360. Cambridge, MA: The Belknap Press of Harvard University Press.

———. 2003. On the Concept of History." Trans. Harry Zohn. In *Selected Writings, Volume 4 (1938–1940)*, ed. Michael W. Jennings et al., 389–400. Cambridge, MA: The Belknap Press of Harvard University Press.

Bodhidharma. 1989. The Zen Teaching of Bodhidharma. Trans. Red Pine. New York: North Point Press,

Borradori, Giovanna. 2003. *Philosophy in a Time of Terror: Dialogues with Jurgen Habermas and Jacques Derrida*. Chicago: The University of Chicago Press.

Caffo, Leonardo. 2012. *Flatus Vocis. Breve invito all'agire animale*. Anzio-Lavinio: Novalogos.

Calarco, Mathew. 2008. *Zoographies: The Question of the Animal from Heidegger to Derrida*. New York: Columbia University Press.

Derrida, Jacques. 2008. *The Animal that Therefore I Am*. Trans. David Wills. New York: Fordham University Press.

Filippi, Massimo. 2011. In morte degli animali. In un tempo sospeso. *Liberazioni. Rivista di critica antispecista* 5: 83–90.

Foucault, Michel. 2002. *The Order of Things: An Archaeology of the Human Sciences*. London: Routledge.

Grenier, Jean. 2004. Mouloud the Cat. Trans. Toby Garfitt. *Comparative Critical Studies* 1.1–2: 224–33.

Heidegger, Martin. 1977. *The Question Concerning Technology and Other Essays*. Trans. William Lovitt. New York: Garland.

———. 1995. *The Fundamental Concepts of Metaphysics: World, Finitude, Solitude*. Trans. William McNeill and Nicholas Walker. Bloomington: Indiana University Press.

Herrigel, Eugen. 1999. *Zen in the Art of Archery*. Trans. R.F.C. Hull. New York: Vintage.

Hesse, Hermann. 2012. *Siddhartha: An Indian Tale*. Cranston: Angelnook.

Horkheimer, Max. 1978. *Dawn and Decline: Notes 1926–1931 and 1950–1969*. Trans. Michael Shaw. New York: Seabury Press.

Kaczynski, Ted. 1995. *The Unabomber Manifesto: Industrial Society and Its Future*. Berkeley: Jolly Roger Press.

Kojève, Alexandre. 1980. *Introduction to the Reading of Hegel: Lectures on the Phenomenology of Spirit*. Trans. James H. Nichols, Jr. Ithaca: Cornell University Press.

Levinas, Emmanuel. 1984. *De l'existence à l'existant*. 2nd ed. Paris: Vrin.

Plato. 2008. Gorgias. Trans. Robin Waterfield. Oxford: Oxford University Press.

Rilke, Rainer Maria. 2005. *Duino Elegies and the Sonnets to Orpheus*. Trans. A. Poulin, Jr. Boston/New York: Mariner Books.

Ryder, Richard D. 2010. Speciesism Again: The Original Leaflet. In *Critical Society* 2, https://web.archive.org/web/20121114004403/http://www.criticalsocietyjournal.org.uk/Archives_files/1.%20Speciesism%20Again.pdf. Accessed 16 Jan 2019.

Singer, Peter. 2009. *Animal Liberation*. New York: HarperCollins.

Thoreau, Henry David. 1906a. *Walden. Volume 2 of The Writings of Henry David Thoreau*. Boston/New York: Houghton Mifflin.

———. 1906b. Walking. In *The Writings of Henry David Thoreau, Volume 5 – Excursions and Poems*, 205–248. Boston/New York: Houghton Mifflin.

Uexküll, Jakob von. 2010. *A Foray Into the Worlds of Animals and Humans: With a Theory of Meaning*. Trans. Joseph D. O'Neil. Minneapolis: University of Minnesota Press.

Žižek, Slavoj. 2012. *Less than Nothing: Hegel and the Shadow of Dialectical Materialism*. London: Verso.

Index[1]

---

[1] Note: Page numbers followed by 'n' refer to notes.